5th International Conference on Science and Engineering in Mathematics, Chemistry and Physics (ScieTech 2017)

Journal of Physics: Conference Series Volume 820

Bali, Indonesia
21 – 22 January 2017

ISBN: 978-1-5108-3905-2
ISSN: 1742-6588

Printed from e-media with permission by:

Curran Associates, Inc.
57 Morehouse Lane
Red Hook, NY 12571

Some format issues inherent in the e-media version may also appear in this print version.

Printed by Curran Associates, Inc. (2017)

For permission requests, please contact the Institute of Physics
at the address below.

Institute of Physics
Dirac House, Temple Back
Bristol BS1 6BE UK

Phone: 44 1 17 929 7481
Fax: 44 1 17 920 0979

techtracking@iop.org

Additional copies of this publication are available from:

Curran Associates, Inc.
57 Morehouse Lane
Red Hook, NY 12571 USA
Phone: 845-758-0400
Fax: 845-758-2633
Email: curran@proceedings.com
Web: www.proceedings.com

Table of contents

Volume 820

5th International Conference on Science & Engineering in Mathematics, Chemistry and Physics 2017 (ScieTech 2017)

21–22 January 2017, Bali, Indonesia

Accepted papers received: 3 March 2017
Published online: 5 April 2017

Preface

Papers

IOP Conf. Series: Journal of Physics: Conf. Series **820** (2017) 011001 doi:10.1088/1742-6596/820/1/011001

The 5th International Conference on Science & Engineering in Mathematics, Chemistry and Physics 2017 (ScieTech 2017),

The Ramada Bintang Hotel, Kuta, Bali, Indonesia

21 - 22 January 2017

Preface

The 5th International Conference on Science & Engineering in Mathematics, Chemistry and Physics 2017 (ScieTech 2017) already held at the The Ramada Bintang Hotel, Kuta, Bali, Indonesia, during 21 - 22 January 2017.

The ScieTech 2017 conference is aimed to bring together researchers, engineers and scientists in the domain of interest from around the world.
The ScieTech 2017 is placed on promoting interaction between the theoretical, experimental, and applied communities, so that a high level exchange is achieved in new and emerging areas within Mathematics, Chemistry and Physics, all areas of sciences and applied mathematics is achieved.

The theme of ScieTech 2017 "The Frontiers of Science and Technology in the Digital World"

This year, we received 205 papers and after rigorous review, 34 papers were accepted.

The participants come from 11 countries. There are 6 (Six) Paralell Sessions and 4 Keynote Speakers.

It is an honour to present this volume of Journal of Physics: Conference Series (JPCS) and we deeply thank the authors for their enthusiastic and high-grade contribution.

We would like to express our sincere gratitude to all in the Technical Program Committee who have reviewed the papers and developed a very interesting Conference Program as well as the invited and plenary speakers.

Finally, we would like to thank the conference chairmen, the members of the steering committee, the organizing committee, the organizing secretariat and the financial support from the conference sponsors that allowed the success of ScieTech 2017.

.

The Editor of the Scietech 2017 Proceedings Editor

Dr. Ford Lumban Gaol
Bina Nusantara University

IOP Conf. Series: Journal of Physics: Conf. Series **820** (2017) 011002 doi:10.1088/1742-6596/820/1/011002

The 5th International Conference on Science & Engineering in Mathematics, Chemistry and Physics 2017 (ScieTech 2017),

The Ramada Bintang Hotel, Kuta, Bali, Indonesia
21 - 22 January 2017

Keynote Speakers

1.

Prof. Dr Igor Kotenko, St. Petersburg Institute for Informatics and Automation of Russian Academy of Sciences (SPIIRAS), Russian Federation

Title: Visual Analytics for Cyber Security: Mathematical Models, Techniques and Practical Solutions

In this talk visual analytics models, techniques and practical solutions for visualization and subsequent analysis in critical operational domains on the example of cyber security situation assessment will be considered. Methods of visual analytics can significantly improve efficiency when working with big data, as they combine the power of intelligent data processing methods, and particularities of the human visual perception of the information. In this talk we analyze existing methods of visual data analysis to solve various problems of protection against cyber attacks. The effectiveness of visual analysis is demonstrated by the example of existing and developed by the authors techniques and tools of visual analytics, including areas of network perimeter monitoring, policy assessment, vulnerability assessment, traffic analysis, attack detection, security metrics representation, fraud investigation, etc. We consider different visual models such as matrices, graphs, histograms, treemaps, spiral view, starburst, clock view, 3D scatter plots, 3D visualization, metaphoric host behavior, etc. The talk discusses a visual analytics technique implemented for displaying a set of security metrics used to assess overall computer network security status and evaluate the efficiency of protection mechanisms. The technique can assist in solving such security tasks which are important for security information and event management systems. The approach suggested is suitable for displaying security metrics of large networks and support historical analysis of the data. To demonstrate and evaluate the usefulness of the proposed technique we implemented a use case corresponding to the Olympic Games scenario and different ways of cyber attack and defense realization. The techniques are based on attack graphs and service dependencies, apply a set of different assessment levels (topological, attack graph, attacker, events, countermeasures and system) and analyze several important aspects (basic, zero day attacks, cost-efficiency characteristics). Such approach allows understanding the current security situation, including defining the vulnerable characteristics and weaknesses of the system under protection, dangerous events, current and possible cyber attack parameters, attacker intentions, integral cyber situation metrics and necessary countermeasures. This research is based on the latest results of the research fulfilled in St. Petersburg Institute for Informatics and Automation of Russian Academy of Sciences (SPIIRAS) by support of the grant of Russian Science Foundation #15-11-30029.

Keywords: applied mathematics and statistics, visual analytics, cyber situational awareness

Short Bio:

Igor Kotenko is a professor of computer science and Head of Research Laboratory of Computer Security Problems of the St. Petersburg Institute for Informatics and Automation of the Russian Academy of Science. He graduated with honors from St.Petersburg Academy of Space Engineering and St.Petersburg Signal Academy, obtained the Ph.D. degree and the National degree of Doctor of Engineering Science. He is the author of more than 350 refereed publications, including several study books and monographs. Igor Kotenko has a high experience in the research on computer network security and participated in several projects on developing new security technologies. For example, he was a project leader in the research projects from the US Air Force research department, via its EOARD (European Office of Aerospace Research and Development) branch, EU FP7 and FP6 Projects, HP, Intel, F-Secure, etc. The research results of Igor Kotenko were tested and implemented in multitude of Russian research and development projects. The research performed under these contracts was concerned with innovative methods for network intrusion detection, simulation of network attacks, vulnerability assessment, security protocols design, verification and validation of security policy, etc. Igor V. Kotenko is a laureate of the St. Petersburg Government award for outstanding scientific achievements in the field of science and technology, a laureate of the program "Outstanding Scientists. Doctors of Sciences of the Russian Academy of Sciences", and a winner of many grants of the Russian Foundation of Basic Research, Russian Science Foundation and several State contracts. He has chaired several conferences and workshops, and serves as editor on several editorial boards.

2. Prof. Drs. Koentjoro Soeparno, MBSc. PhD. Psikolog
Fakultas Psikologi - Universitas Gadjah Mada
Yogyakarta – Indonesia

Title : Social Change and its Impact on Education and Culture

This article will explain how social change effects education and culture. The method used in this paper is a literature review, observation and contemplation. Basically, humans beings seek comfort in their life, therefore any change will be followed by efforts of adjustment. Change is painful As such, change will always produce other changes. There are two (2) sources of social change, namely the industrial revolution in Britain and ideological consciousness of independence. The Industrial Revolution in England led to various innovations and findings in the fields of engineering, computer and information technology.

Discoveries in the field of engineering made industrialization grow which consequently divided society into groups of proletarian class and the bourgeoisie. Industrialization also resulted in the exploitation of natural resources which ultimately caused global climate change. Innovation in the field of computer technology and informatics help generate global acceleration. Globalization is considered a revolution of 4T which is a revolution in telecommunications, trasnportation, tourism and transparency. Computer innovation and informatics create various types of social media that influence education and culture.

The second source of change is the rise of ideological consciousness of independence. History shows that many countries in the world gained its is independence between 1900-1999. Gender struggles in the USA in 1907 and 1960, in Indonesia from 1870 to 1928. Anti-slavery struggle that ended in 1974. These events raise awareness of human rights and gender awareness as a counter culture. When both the source of this social change, the industrial revolution in Britain and ideological consciousness of independence interact, making the world a growing and increasingly rapid and global.

Conclusion Social change is rapidly increasing. Globalization mediated by social media leads to change in lifestyle, culture and finally the educational system. Industrialization interacts with other lifestyles, raises the issue of the exploitation of Natural Resources. The impact is the rise of new style colonization and climate change.

Keywords: social change, the industrial revolution, ideological consciousness of independence, globalization and social media.

Short Bio

Prof. Drs. Koentjoro, MBSc., Ph. D, Psikolog is The Head of Social Psychology Department Faculty of Psychology, Gadjah Mada University.
He is also The Head of Drug Abuse Prevention. Gadjah Mada UniversityProfessor in Social Psychology at the Faculty of Psychology, Gadjah Mada University.
His research Interest
Sexuality, Social and Family Psychology, Social Change and Social Policy, Qualitative Research, Social Change and Problems; Grassroots & Prostitution, Poverty, Community Planning and Interventions.

3. AP Dr Norma Alias
Center for Sustainable Nanomaterials,
Ibnu Sina Institute for Scientific and Industrial Research,
Universiti Teknologi Malaysia

Title : Conceptual Framework for Big Data Emerging on Nanotechnology Theory, Model, Simulation using HPC System

This paper presents six variations to meet the contexts of a conceptual framework for modeling the complex systems involve nanotechnology theory, modeling, large scale numerical simulation in the real world problem. Integrated mathematical modeling and large scale numerical simulations are the tools to solve the complex systems. The conceptual framework is a comprehensive concept in theory, modeling and simulation based on high performance computing (HPC). The main objective is to improve the process of huge computation of the big data modeling and to increase the performance evaluation of parallel programming on HPC platform. The set of concepts contains the revise standard for guiding the scientific algorithm provider to model, discrete and simulate the integration of big data emerging based on parallel computing strategies. The six variations are identifying the parameter and governing the mathematical modeling for large scale complex system, transferring the continuous model, function and equation into discrete counterparts, developing the parallel algorithm and communication model, implementing the parallel program on HPC platforms, converting the discrete solution into continuous feature's existence of exact solutions in any natural phenomenon of complex systems. Last but not least is verifying and validating the numerical approximation and parallel performance evaluations. The framework organizes the idea and step to be considered for solving the integrated theory, mathematical modeling with fast numerical simulation , specific parallel computing strategy , communication software and HPC hardware system which are applicable for solving large scale nanotechnology applications. The matching skills between the numerical simulation of complex system and the computational tool such as communication model, message passing protocol, parallel programming language, memory and processor performance on HPC platform were framed for data scientist and scientific algorithm provider. This framework employs some case study approaches to provide an adaptation of the conceptual framework and satisfy the variation requirements. The examples for variations are presented, the performance evaluations are analyzed and the predictions of the large scale complex system are provided. As a conclusion, the conceptual framework performs more comprehensive perspective based on future software system design dealing with the next-generation of HPC architecture system.

Short Biography

A.P Dr. Norma Alias is Research Fellow at Center for Sustainable Nanomaterials, Ibnu Sina Institute for Scientific and Industrial Research, Technology University of Malaysia. Currently, she is also Associate Professor at Mathematics Science Department, Faculty of Science, Technology University of Malaysia. She received her M.S. degree and PhD from National University of Malaysia, in Computer Science and Industrial Computing. Area of research interest includes mathematical modeling and simulation, big data computing, large scale image processing, and computational nanotechnology on high performance computing system. She has published more than 160 research papers in ISI, SCOPUS, index and non-index journals. She won 9 international and local innovation medals. She has awarded as 40 active researchers in Malaysia and completed 30 research grants.

4. Pramila. Goyal
Distinguished Professor Emeritus
The NorthCap University, Gurgaon-122017, India, E-mail*- pramila@cas.iitd.ernet.in
Formerly Professor at Centre for Atmospheric Sciences, Indian Institute of Technology Delhi, Hauz Khas, New Delhi, India – 110016,

Title : Assessment of Air Quality of Delhi Through Mathematical

This article presents a review of the different Mathematical models based on analytical solution of advection diffusion equation for dispersion of Air pollutants to determine their concentration at different locations in Delhi, which is always used for studying the air quality of Delhi.

Different mathematical techniques have been developed and used for an emission inventory of air pollutants, used as one of the input parameter in dispersion models. The major air pollutants emitted from various sources in Delhi are sulpher dioxide (SO_2), particulate matter of size 10 micron (PM_{10}) and oxides of nitrogen (NOx). The recent study of Air Quality Index (AQI) of Delhi, based on combined effect of SO_2, SPM and NOx is obtained by two different formulations, which defines the status of air Quality and has limits in numbers to represent different classes between good and severe of air quality.

Finally, Air quality, determined by air quality dispersion models in terms of concentration of air pollutants and AQI based on the same are compared and assessed for Delhi atmosphere. It has been observed that most of the Delhi's AQI was bed and severe.

Key Words: Air Quality, Models, Source Inventory, Air Quality Index. Delhi

Short Bio
Dr. Pramila Goyal joined NCU, Gurugram on 20th August, 2015 as Distinguished Professor Emeritus in the Department of Applied Sciences & Department of Civil & Environmental Engineering. Her broad areas of research interests include: Air Pollution Modeling, Environmental Impact Assessment, Emission Inventories, Carrying / Assimilative Capacity of Urban Cities, and Environmental Management Studies.

With a total professional experience of 38 years in teaching and research, she has 55 Publications to her credit – 44 Papers in International and 11 Papers in National Journals, besides, 65 International / National Conference Papers published in the peer reviewed Conference Proceedings. She has mentored 14 Ph.D. and 17 M.Tech students, developed / taught many Courses to B.Tech, M.Tech, M.B.A. and Pre Ph.D students during her stint at IIT Delhi. She has visited numerous countries as Invited Scientist/Speaker / under Collaborative Scientific Research Projects.

Area of Expertise: Air Pollution Mathematical Modelling and Air Quality Monitoring. Air Pollution & Health.

IOP Conf. Series: Journal of Physics: Conf. Series **820** (2017) 011003 doi:10.1088/1742-6596/820/1/011003

Peer review statement

All papers published in this volume of *Journal of Physics: Conference Series* have been peer reviewed through processes administered by the proceedings Editors. Reviews were conducted by expert referees to the professional and scientific standards expected of a proceedings journal published by IOP Publishing.

Projecting the Water and Electric Consumption of Polytechnic University of the Philippines

J D Urrutia[1], J Mercado[2], L A Bautista[3], and E B Baccay[4]

[1]Director, Intellectual Property Management Office
Chief, Center for Statistical Studies, Institute for Data and Statistical Analysis
Office of the Vice President for Research, Extension, Planning and Development
Polytechnic University of the Philippines – Sta. Mesa, Manila, Philippines
[2]Vice President for Research, Extension, Planning, and Development
Polytechnic University of the Philippines – Philippines
[3]Dean, College of Science, Polytechnic University of the Philippines – Philippines
[4]Head of Academic Programs, Office of the Academic Head
Polytechnic University of the Philippines-Parañaque, Metro Manila, Philippines

E-mail: jackieurrutia20@gmail.com, jdurrutia@pup.edu.ph, math_urrutia@yahoo.com.ph, jmercado@pup.edu.ph, merjos2002@yahoo.com, math_baccay@yahoo.com, ebbaccay@pup.edu.ph, labautista@pup.edu.ph

Abstract. This study investigates water and electric consumption in Polytechnic University of the Philippines – Sta. Mesa using a time series analysis. The researchers analyzed the water and electric usage separately. Electric consumption was examined in terms of pesos and kilowatt-hour, while water consumption was analyzed in pesos and cubic meter. The data are gathered from the university limited only from January 2009 to July 2015 in a monthly based record. The aim is to forecast the water and electric usage of the university for the years 2016 and 2017. There are two main statistical treatments that the researchers conducted to be able to formulate mathematical models that can estimate the water and electric consumption of the said school. Using Seasonal Autoregressive Integrated Moving Average (SARIMA), electric usage was forecasted in peso and kilowatt-hour, and water usage in peso and cubic meter. Moreover, the predicted values of the consumptions are compared to the actual values using Paired T-test to examine whether there is a significant difference. Forecasting accurately the water and electric consumption would be helpful to manage the budget allotted for the water and electric consumption of PUP – Sta. Mesa for the next two years.

1. Introduction

For the past few years, the Philippine economy continues to boom in terms of its Gross Domestic Product (GDP), with an average growth rate of 6.3 percent from 2010 to 2014. Together with this growth is the increase in supply and demand across sectors [1]. The growth of the Philippine economy drives the demand for electricity [2]. Electric power consumption in Philippines was last measured at 646.96 kilowatt-hour per capita in 2011, according to the World Bank. Electric power consumption measures the production of power plants and combined heat and power plants less transmission, distribution, and transformation losses and own use by heat and power plants [3].

The aim of this paper is to project the water and electric consumption of Polytechnic University of the Philippines – Sta. Mesa for the next two years to be able to manage the budget of the university for

IOP Conf. Series: Journal of Physics: Conf. Series **820** (2017) 012001 doi:10.1088/1742-6596/820/1/012001

the these expenses. This paper will be helpful to the school administration in allotting budget for the projects and other expenses of the university. Analyzing the water and electricity consumption is not only essential for managing expenses but also for the allocation and proper distribution of water and electricity, as well as conserving them.

1.1. Objective of the Study
The purpose of this study is to formulate mathematical models for estimating the water and electric consumption of Polytechnic University of the Philippines – Sta. Mesa. The researchers are to forecast the water and electric consumption of the university for the target year 2016 to 2017. Moreover, forecasting an accurate water and electric consumption would be helpful to manage the budget allotted for the water and electric consumption of PUP – Sta. Mesa for the next two years.

1.2. Research Paradigm
The researchers make use of the monthly data of electric consumption in terms of pesos and kilowatt-hour of PUP – Sta. Mesa. Also, the researchers input the monthly data of water consumption of the university in pesos and cubic meter.

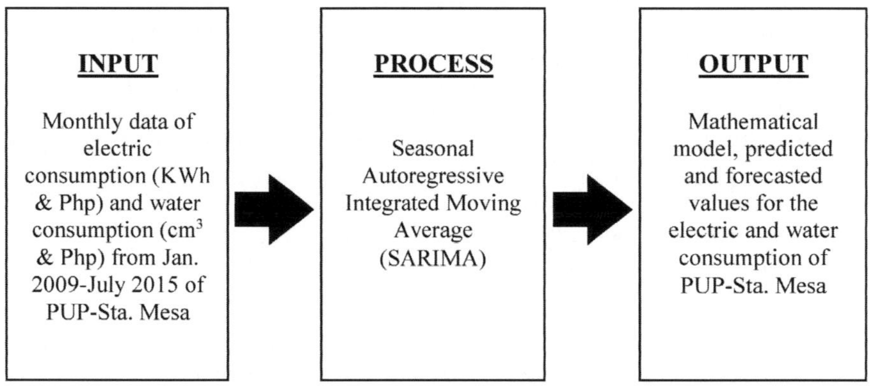

Figure 1. Research paradigm.

They applied Seasonal Autoregressive Integrated Moving Average (SARIMA) in estimating water and electric consumption. The anticipated outputs are the mathematical models, the predicted values, and the forecasted values of electricity and water consumption in pesos, kilowatt per hour, and cubic meter.

1.3. Statement of the Problem
This paper aims to provide the models to forecast the water and electric consumption of PUP – Sta. Mesa. In particular, the researchers want to answer the following questions:

1. What is the behavior of the graph of variables?
    a. Electric Consumption in peso and kilowatt-hour
    b. Water Consumption peso and cubic meter
2. What are the mathematical models formulated using SARIMA in estimating the water and electric consumption of PUP-Sta. Mesa?
3. What are the forecasted values of the university's water and electric consumption for 2016 to 2017?
    a. Electric Consumption in peso and kilowatt-hour
    b. Water Consumption peso and cubic meter
4. Is there a significant difference between the predicted values and the actual values of the university's water and electric consumptions?

1.4. Scope and Limitations
The researchers analysed the monthly data of water and electric consumption of PUP – Sta. Mesa from January 2009 up to July 2015. This research consists of the years including of the actual data retrieved from the university and the forecasted series from the year 2016-2017.

2. Review Related Literature and Studies
The following researches and studies discuss about the electricity and water consumption in different countries like Taiwan, Japan, and Malaysia. These studies aim to forecast or analyze either the electricity or water consumption in a certain area in their country.

In a research entitled "The Determinants of Household Electricity Consumption in Taiwan: Evidence from Quantile Regression," showed that the effects of demographic, socioeconomic, and household dwelling characteristics on household electricity consumption may differ across quantiles and may change over time. This study found that household income and household size were significant in all quantiles for each year. The characteristics of high-electricity-consuming households were also identified. Households with higher income, larger household size, and more elderly members consumed more electricity. In terms of dwelling attributes, larger housing areas, homes with more appliances, and owner-occupied, business-used, and multi-floor houses contributed to higher household electricity consumption [4]. These study infers that the consumers can be factor on the electricity consumption of a household. Being a university, most of the electricity consumers in the campus are students and professors which affects the total power usage since they use electricity most of the time for school activities.

In a study entitled "Electricity Consumption Forecasting in Thailand Using an Artificial Neural Network and Multiple Linear Regression," an artificial neural network (ANN) and a regression model are applied to forecast long term electricity consumption in Thailand. Maximum ambient temperature and electricity power demand are used as inputs to predict electricity consumption. According to the forecasting results by the regression and ANN models of the study, the electricity consumption in Thailand in 2015 is almost 188,552 to 174,394 gigawatt-hour (GWh), and it can reach to 216,986 to 188, 137 GWh by 2020 [5].

A study titled as "A Comparative Analysis of Techniques for Forecasting Electricity Consumption," showed that there is a pattern for electricity consumption over the years in the Universiti Malaysia Sarawak (UMS). Using the actual monthly electricity consumption data over the period between 2009 and 2012, a modified Newton's model was proposed in forecasting electricity consumption. It is found that consumption is high at certain periods and low at other periods and shows an upward trend for electricity consumption. Increase in electricity consumption could be due to the fact that more heating, ventilation and air-conditioning systems are put into use because of increase in student's population [6]. Analyzing the electricity consumption in UMS in Malaysia is quite the same with this study on forecasting the water and electricity consumption of PUP – Sta. Mesa

In a study entitled "Modelling Water Consumption in Kalmunai using Time Series Models," the researcher aims to analysis and identify the best fit time series model to the water consumption in the Kalmunai using time series techniques. Several time series models including autoregressive (AR), moving average (MA) and ARMA were developed to the water consumption data, and it emerged that the most adequate models for the data was ARMA (1, 2) [7]. In this paper, the researchers also used a time series model in forecasting the water consumption in Kalmunai through ARMA models. On the other side, this paper applied seasonal autoregressive integrated moving average (SARIMA) in estimating the water consumption of PUP.

A research in Japan titled as "The Measurement of Electric Consumption of Devices and the Estimation of CO_2 Emission Related to Food Consumption at Households," to clarify the amount of electric energy consumption related to food consumed in households. The first phase of the study was to measure the amount of electric energy used while cooking with the use of rice cooker, microwave and electric pot. They are considered to be the main source of electricity while cooking. The electricity consumption of the three major household electric cooking devises; electric rice cooker, microwave and electric pot were estimated at 276 kWh per household per year, which was 14.4 KWh per year and accounted for 5.0% of Japanese household electricity consumption [8].

3. Methodology

3.1. Statistical Tool

EViews is a statistical software for quickly and efficiently managing data, performing econometric and statistical analysis, generating forecasts or model simulations, and producing graphs and tables [9]. The results of this study were generated through this software, as well as the graphs and tables in interpreting the output of the study.

3.2. Statistical Treatment

3.2.1. Box Jenkins Methodology. Box - Jenkins Analysis refers to a systematic method of identifying, fitting, checking, and using integrated autoregressive, moving average (ARIMA) time series models [10]. In formulating the model, the researchers followed the three-step process of this method which were identifying, selecting, and assessing the mathematical model for forecasting both the water and electric consumption in PUP-Sta. Mesa.

3.2.2. Augmented Dickey-Fuller Test. The augmented Dickey-Fuller test is one that tests for a unit root in a time series sample. The primary differentiator between the two tests is that the ADF is utilized for a larger and more complicated set of time series models. The augmented Dickey-Fuller statistic used in the ADF test is a negative number, and the more negative it is, the stronger the rejection of the hypothesis that there is a unit root [12]. It is expressed as: [13]

$$X_t = c + \alpha X_{t-1} + \beta_t + \varepsilon_t$$

where X_t is the time series, c is constant, t is the time trend and ε is the residual term. This test is used in this study to determine whether or not the time series of water and electric consumption exhibits stationarity. The data must exhibit stationarity to be able to formulate a forecasting model, if ever that it is non-stationarity, it must undergone a regular differencing until it satisfy the said condition.

3.2.3. Correlogram. A correlogram is simply a plot of the autocorrelation function for sequential values of lag allowing the researcher to see the correlation structure in each lag. The main usage of correlogram is to detect any autocorrelation subsequent to the removal of any deterministic trends or seasonality effects. If it has fitted a time series model then the correlogram helps to justify that the model is well fitted or whether it needs further refinement to remove any additional autocorrelation [14]. Autocorrelation function is written as:

$$r_k = \left[\sum_{i=1}^{N-k} (Y_i - \bar{Y})(Y_{i+k} - \bar{Y})\right]\left[\sum_{i=1}^{N} (Y_i - \bar{Y})^2\right]^{-1}$$

where Y bar is the overall mean and n is the sample observation. The correlogram analysis examines the time-spatial dependency within the sample data, and focuses on the empirical auto-covariance, auto-correlation, and related statistical tests. The correlogram is a cornerstone for identifying the model and model order [15]. With the help of the correlograms of each variable that is being analyzed, seasonality patterns are detected. Water and electricity consumption in PUP – Sta. Mesa showed to fluctuate seasonally.

3.2.4. Paired T-test. Paired sample t-test is a statistical technique that is used to compare two population means in the case of two samples that are correlated. Paired sample t-test is used in 'before-after' studies, or when the samples are the matched pairs, or when it is a case-control study. In computing the Paired T-test, the formula used is written as:

$$t = \overline{X}\left(s^2 n^{-1}\right)^{-1/2}$$

where X bar is the mean difference between two samples, s^2 is the sample variance, n is the sample size and t is a paired sample t-test with $n-1$ degrees of freedom [16]. The researchers make use of this test in comparing the actual values of water and electricity consumption to its predicted values which will be obtained using the formulated model.

3.2.5. Seasonal Autoregressive Integrated Moving Average (SARIMA). Seasonal ARIMA is used in the process of formulating the mathematical model for forecasting water and electricity consumption of PUP – Sta. Mesa. It follows some conditions wherein the time series needs to be stationarity. This assumption can be resolved by differencing the data. SARIMA is used in this study, instead of ARIMA because the time series for both variables showed to have seasonal patterns. The model to be formulated follows an equation written in the form of SARIMA (p,d,q) x (P,D,Q)s where (p,d,q) is the non-seasonal part of the model that contain the autoregressive (AR) and moving average (MA) terms, (P,D,Q) contains the seasonal AR and MA terms; and s accounts to the seasonal period.

4. Results and Discussions

4.1. Water Consumption

The water consumption displays an upward trend but has its lowest consumptions on every last month of each year because of holiday vacations wherein students are out of school so water has been consumed less.

Table 1. Test for Seasonality for Water Consumption

F-stat Value	0.856104
@trend F-stat Value	0.000000

The seasonality test of the data resulted at 0.856104 that is higher than the hypothesized F-statistic probability value at 0.001, which the researchers concede it to be indicating the failure of rejection of the null hypothesis that there is no seasonality within the data series, and it implies no presence of seasonality. But the data shows an obvious trend in which the pattern starts to exhibit direct upward at the start of every summer seasons. When at trend, the results shows 0.0000 that is obviously lower than the F-stat probability hypothesized value which indicates the rejection of the null hypothesis, that means that there is a necessity for using seasonal differencing for the data series. SARIMA model, where autoregressive, moving average and the seasonal AR and MA terms are produced using the actual data that has undergone non-seasonal and seasonal differencing, are to be regressed and utilized in forecasting the Water consumption of PUP Main campus in the next two years and five months.

Table 2. Unit root testing for the Water Consumption

Augmented Dickey-Fuller Test		
	Intercept	0.433235832
Level	Trend and Intercept	0.003014275
	None	0.047683014
	Intercept	0
1st difference	Trend and Intercept	0
	None	0

When the data has been differenced, the trend has been minimized making the data stationary and deduced it from displaying seasonal recurrence of the data. This indicates that the data now is stationary. Its variance also does not waver vigorously enough to be considered non-stationary. In the correlogram, the differenced data now, exhibits no seasonality, and it indicates that the processed data now can be used to produce combinations of Autoregressive and Moving average terms to choose the fitted model for forecasting the electric consumption of the PUP Main Campus. Also, the correlogram exhibits a positive Autocorrelation indicating the necessity of having seasonal moving average terms. Autoregressive terms may be also included for balancing the equation to satisfy the model fitting assumptions.

Table 3. Model Combinations, r-Square, Durbin-Watson, and MA&AR Processes for Water Consumption

AR and MA terms	R-squared	Durbin-Watson	MA Processes	AR Processes
ar(1) ar(3) ar(11) ma(1) ma(4) ma(11) ma(12) ma(13) sar(24)	0.85447	2.51686	Non-invertible	Stationary
ar(1) ar(3) ar(11) ma(1) ma(4) ma(11) ma(12) sar(24)	0.72702	2.11599	Non-invertible	Stationary
ar(1) ar(3) ar(11) ma(1) sma(12) sar(12)	0.71217	2.02298	Invertible	Stationary
ar(1) ar(3) ar(11) ma(1) ma(4) sma(12) sar(12)	0.71187	2.00974	Invertible	Stationary
ar(1) ar(11) ma(1) sma(12) sar(12)	0.70830	1.99056	Invertible	Stationary

Among the combinations listed, the model has passed the utmost needed assumptions for choosing model in forecasting, where its R-square probability value was approximately 0.7083 which is quite enough for it to be considered fitted, and the model was also based with the Durbin-Watson, and, the Moving Average invertibility and Auto-regressive stationary Processes. The plotted graph of the actual and forecasted data follows a trend which has gone down quite favorably indicating lower expenses in the water consumption in the PUP Main Campus. Therefore, excess budget for the water consumption can be allocated to other expenses of the university. The graph shows that it has not gone beyond 3 million but slightly fall off lower than 2 million pesos at the end of the forecasted series.

Table 4. Paired T-test for Water Consumption

Value	0.46009248056
Probability	0.64700917653

Paired T-test that has failed to reject the null hypothesis, indicates that the mean difference between the paired observations is in close proximity to zero. It means the observations does not differ that much from each other, or these observations be somewhat have values that are close enough for so to compromise that the model is in the vicinity of being fit for the data.

4.2. Electric Consumption

Electric consumption of PUP Main Campus, has its sudden drop of its value seen in the graph line when ever the last month of the year ends. This might probably be, because of Christmas season which the students are having few weeks vacation. The graph shows the bottommost of the line was on the last part of 2009 and 2010, then the graph continued to waver following a slight trend going upward and emanated to its downward trend started from the end of 2013 until the last of the available data collected from the campus.

Table 5. Test for Seasonality for Electric Consumption

F-stat Value	0.856104
@trend F-stat Value	0.000050

IOP Conf. Series: Journal of Physics: Conf. Series **820** (2017) 012001 doi:10.1088/1742-6596/820/1/012001

The seasonality test shows a probability of F-statiscal Probability Value where it has been conceded that the actual data has a seasonality. At 0.1% alpha considered by the investigators, the results shows 0.00005 which has a lower value than the considered alpha for Probability F-stat Value, it means that the null hypothesis is ponderedly rejected indicating the acceptance of the alternative hypothesis that there the data series exhibits seasonal patterns. The correlogram shows also that the actual data exhibits the seasonality visibly on the 12th lag. Autocorrelation (ACF) has its seasonality on the 12th lag, which is positive, directs that the data needs seasonal moving average term which most of the SARIMA models are. Seasonal autoregressive terms may be necessary for balancing out the equation, for MA terms my not able to be inverted.

Table 6. Unit root testing for the Electric Consumption

Level	Intercept	0.005345186
	Trend and Intercept	0.697542501
	None	0.971813173
1st difference	Intercept	0
	Trend and Intercept	0
	None	0

When the data has been differenced, its graph shows that the trend has been minimized making the data stationary and deduced it from displaying seasonal recurrence of the data. This indicates that the data now is stationary by just looking onto the graph. Its variance also does not waver vigorously enough to be considered non-stationary. In the correlogram, the differenced data now, exhibits no seasonality, and it indicates that the processed data now can be used to produce combinations of Autoregressive and Moving average terms to choose the fitted model for forecasting the electric consumption of the PUP Main Campus. Also, the correlogram exhibits a positive Autocorrelation indicating the necessity of having seasonal moving average terms. Auto regressive terms may be also included for balancing the equation to satisfy the model fitting assumptions.

Table 7. Model Combinations, R-Square, Durbin-Watson, and MA&AR Processes for Electric Consumption

AR and MA terms	R-squared	Durbin-Watson	MA Processes	AR Processes
ar(1) ar(3) ar(11) ma(1) ma(4) ma(11) ma(12) ma(13) sar(24)	0.85447	2.51686	Non-invertible	Stationary
ar(1) ar(3) ar(11) ma(1) ma(4) ma(11) ma(12) sar(24)	0.72702	2.11599	Non-invertible	Stationary
ar(1) ar(3) ar(11) ma(1) sma(12) sar(12)	**0.71217**	**2.02298**	**Invertible**	**Stationary**
ar(1) ar(3) ar(11) ma(1) ma(4) sma(12) sar(12)	0.71187	2.00974	Invertible	Stationary
ar(1) ar(11) ma(1) sma(12) sar(12)	0.70330	1.99056	Invertible	Stationary

The table of combinations listed where the investigators have chosen a model with 71% R-squared Probability assumption for model fitting, where this model is the third highest R-square probability. All three highest probability has passed Durbin-Watson normality test for data distribution and are stationary with their Auto-regressive Processes but only the third model, SARIMA $(2, 0, 4) \times (1, 1, 0)_{12}$, has been conceded for it passed the Moving Average invertibility process. The model has 3 AR terms, 1 MA term, 1 Seasonal AR term and a Seasonal MA term, which will be used to forecast the actual data. The researchers utilized the model to provide values forecasted of electricity consumption, ranges from the following month of the last of the available data gathered until the last month of 2017. Using dynamic forecasting, further two years and five months of forecasted values has been made. The forecasted data has no observable trend but it wavers less vigorous. That indicates the forecasted values, does not increase or decrease in a huge variations. The values also, does not fall lower than 2 million nor increased higher than three million. Predicted values of electricity consumption produced using the model, were almost having the nearest values with the actual. The predicted also establish a sudden drop as how the actual did at every last month of the year and having the same slight trend that the actual data has time-honoured.

Table 8. Paired T-test for Electric Consumption

probability	p-value
0.461022685	0.742368909

Also, the paired T-test shows that it has failed to reject the null hypothesis, indicates that the mean difference between the paired observations is zero. That means the paired observations does not differ from each other or these observations are rather having values that are close enough for so to concede that the model is nearly fitted.

5. Conclusion and Recommendations

Results revealed perceptible model each for the water and electric consumption of PUP-Main Campus and the forecasted series of the actual data for the next 29 months which were considerably favorable. The models were SARIMA $(2,1,2) \times (1,1,1)_{12}$ for water consumption data series while SARIMA $(2,0,4) \times (1,1,0)_{12}$ for the electric consumption data series. These models are used in predicting the future values of the data series while satisfying the supposition in procreating Seasonal ARIMA Models in which these two have abide by.

Researchers recommend collection of new time series data for further studies. The forecasted series may serve as a basis in planning the water and electricity distribution in the campus. It can also help in reducing and minimizing the expenses of Main Campus in terms of water and electricity consumptions to help with quicker and more accurate decision-making regarding the financial statements. Also to give more attention to other priorities school enhances for students' necessities and for the own new developments for the campus itself.

References

[1] Energy Consumption of the Philippines. Retrieved from http://www.canadianinquirer.net/2015/02/06/energy-consumption-in-the-philippines/

[2] Patalinghug, E. (2003). An Analysis of the Philippine Electric Power Industry. Retrieved from http://www.ombudsman.gov.ph/UNDP4/wp-content/uploads/2013/01/An-Analysis-of-the-Philippine-Electric_Patilinhug.pdf

[3] Electric Power Consumption (per capita) in the Philippines. Retrieved from http://www.tradingeconomics.com/philippines/electric-power-consumption-kwh-per-capita-wb-data.html

[4] Wen Hsiu-Hang. (April 2014). The Determinants of Household Electricity Consumption in Taiwan: Evidence from Quantile Regression. Retrieved from http://www.sciencedirect.com/science/article/pii/S0360544215005873

[5] Panklib, T. (January 2015). Electricity Consumption Forecasting in Thailand using and Artificial Neural Network and Multiple Linear Regression. Retrieved from http://ww9w.researchgate.net/publication/271364922

[6] Ozoh, P. (February 2014). A Comparative Analysis of Techniques in Forecasting Electric Consumption. Retrieved from research.ijcaonline.org/volume88/number15/pxc3893841.pdf

[7] Irfana, M. (November 2015). Modelling Water Consumption in Kalmunai using Time Series Models. Retrieved from http://www.researchgate.net/publication/284586231

[8] Sato, K. & Tsuda, T. (January 2012). The Measurement of Electric Consumption of Devices and the Estimation of CO2 Emission Related to Food Consumption at Households. Retrieved from: https://www.researchgate.net/publication/276444742

[9] EViews. Retrieved from: www.msmiami.com/EViews/Brochure/2015-06_FINAL.pdf
[10] The Box – Jenkins Method. Retrieved from www.researchgate.net/file. 401442285675602
[11] Box–Jenkins Methodology. Retrieved from www.mathworks.com/box-jenkins-method.html
[12] Augmented Dickey Fuller Test. Retrieved from economics.about.com/cs/economicsglossary/g/augmented.htm
[13] Unit Root Test. Retrieved from fedc.wiwi.hu-berlin.de/xplore/tutorials/sfehtmlnode53.html
[14] Quant Start. Serial Correlation in Time Series Analysis. Retrieved from www.quantstart.com/articles/Serial-Correlation-in-Time-Series-Analysis
[15] A Correlogram Tale. Retrieved from www.spiderfinancial.com/documentation/numxl/tips-and-tricks/a-correlogram-tale
[16] Statistics Solutions. Paired Sample T-Test. Retrieved from www.statisticssolutions.com/manova-analysis-paired-sample-t-test/
[17] Ables, L., Bartolata, J., De Guzman, M., See, E. & See, M. (September 2010). Electric Energy Utilization in the Households of Albay Province, Philippines: Contexts, Conservation Practices, and Future Efficiency Strategies. Retrieved from www.researchgate.net/publication/259219501
[18] Chujai, P. (March 2013). Time Series Analysis of Household Electric Consumption with ARIMA and ARMA Models. Retrieved from www.iaeng.org/publication/IMECS2013/IMECS2013_pp295-300.pdf
[19] Memon, J. (November 2015). Forecasting the Electricity Consumption of Pakistan: The Way Forward. Retrieved from www.sciencedirect.com/science/article/pii/S0301421515302068
[20] Masuduzzaman, M. (June 2012). Electricity Consumption and Economic Growth in Bangladesh: Co-integration and Causality Analysis. Retrieved from www.mof.gov.bd/en/budget/pathprosperity/chapter2.pdf

Minimization of Food Cost on 2000-Calorie Diabetic Diet

J D Urrutia[1] J Mercado[2] and R L Tampis[3]

[1]Director, Intellectual Property Management Office
Chief, Center for Statistical Studies, Institute for Data and Statistical Analysis
Office of the Vice President for Research, Extension, Planning and Development
Polytechnic University of the Philippines – Sta. Mesa, Manila, Philippines
[2]Vice President for Research, Extension, Planning and Development
Polytechnic University of the Philippines – Philippines
[3]Polytechnic University of the Philippines – Parañaque Campus, Philippines

E-mail: jackieurrutia20@gmail.com, math_urrutia@yahoo.com.ph, jdurrutia@pup.edu.ph, jmercado@pup.edu.ph, razzcelletampis@gmail.com

Abstract. This study focuses on minimization of food cost that satisfies the daily nutrients required based on 2000-calorie diet for a diabetic person. This paper attempts to provide a food combination that satisfies the daily nutrient requirements of a diabetic person and its lowest possible dietary food cost. A linear programming diet model is used to determine the cheapest combination of food items that satisfy the recommended daily nutritional requirements of the diabetic persons. According to the findings, a 50 year old and above diabetic male need to spend a minimum of 72.22 pesos for foods that satisfy the daily nutrients they need. In order to attain the minimum spending, the foods must consist of 60.49 grams of anchovy, 91.24 grams of carrot, 121.92 grams of durian, 121.41 grams of chicken egg, 70.82 grams of pork (lean), and 369.70 grams of rice (well-milled). For a 50 year old and above diabetic female, the minimum spending is 64.65 pesos per day and the food must consist of 75.87 grams of anchovy, 43.38 grams of carrot, 160.46 grams of durian, 69.66 grams of chicken egg, 23.16 grams of pork (lean) and 416.19 grams of rice (well-milled).

1. Introduction

During the last twenty years the prevalence of diabetes has increased dramatically in many parts of the world and the disease is now a worldwide public health problem. Diabetes is a metabolic disease that is diagnosed on the basis of sustained high concentration of glucose in the blood. This high blood sugar produces the symptoms of frequent urination, increased thirst, and increased hunger. Untreated diabetes can cause many complications. Serious long-term complication includes heart disease, stroke, kidney failure, foot ulcers and damage to the eyes [1].

Diabetes in the Philippines affects the lives of 9.7% of the adult population; one out of every five Filipino adults has either diabetes or is at risk with impaired glucose tolerance [2]. Worldwide, there are 371 million people living with diabetes and another 280 million are at high risk, posing a global problem. The total number of people with diabetes is projected to rise from 171 million (11% increase from the previous 154 million) from the year 2000 to 366 million in the year 2030 [3]. Despite all the publicity surrounding new research and new nutrition guidelines, some people with diabetes still believe that there is something called a "diabetic diet." For some, this so-called diet consists of avoiding sugar, while others believe it to be a strict way of eating that controls glucose. Unfortunately, based on dieticians, neither is quite right. This paper mainly focuses on solving the diet problem for persons with diabetes which minimized cost but still satisfies daily nutrients required [4].

1.1. Statement of the Problem

Diabetes is a chronic disease for which control of the condition demands patient self- management. Self- management behaviours include monitoring blood glucose levels, taking medications, maintaining a healthy diet and regular exercise. This paper attempts to look at a research that addresses the diabetic diet at lowest cost with the given food combination which satisfies the daily nutrient requirements. These nutrients are also based on the different age of a diabetic person, both male and female. This paper will attempt to answer the question: what is the food combination that satisfies the daily nutrient requirements of a diabetic person and its lowest possible dietary food cost?

2. Methodology

A linear programming model is used to find the cheapest combination of food items that satisfies the most important daily nutritional requirements of a diabetic person. Whereas linear programming is a technique for the optimization of a linear objective function, subject to linear equality and linear inequality constraints. Given a polytope and a real-valued affine function defined on this polytope, a linear programming method will find a point on the polytope where this function has the smallest (or largest) value if such point exists, by searching through the polytope vertices [5].

2.1. Excel Solver Function

Solvers, or optimizers, are software tools that help users find the best way to allocate scarce resources. The resources may be raw materials, machine time or people time, money, or anything else in limited supply. The best or optimal solution may mean maximizing profits, minimizing costs, or achieving the best possible quality. To use a solver, you must build a model that specifies: the resources to be used, using decision variables; the limits on resource usage, called constraints, and the measure to optimize, called the objective. The solver will find values for the decision variables that satisfy the constraints while optimizing (maximizing or minimizing) the objective [6].

2.2. Simplex Method

A method is described for the minimization of a function of n variables, which depends on the comparison of function values at the (n + 1) vertices of a general simplex, followed by the replacement of the vertex with the highest value by another point. The simplex adapts itself to the local landscape, and contracts on to the final minimum. The method is shown to be effective and computationally compact. A procedure is given for the estimation of the Hessian matrix in the neighborhood of the minimum, needed in statistical estimation problems. As the new data sets were entered into spreadsheets, it was convenient to use the linear-programming spreadsheet solver add-in to solve our modified diet problems [7]. The basic mathematical structure is given by the following linear programming model. The objective function is written in minimize z and is subjected to the following nutrients.

$$\begin{aligned} \textit{Minimize } Z = {} & 11.313x_1 + 6.9x_2 + 4.2x_3 + 26.21x_4 + 4.82x_5 + 2.66x_6 + 5.44x_7 + 5.08x_8 + 14.07x_9 + 2.44x_{10} + \\ & 13.98x_{11} + 5.24x_{12} + 4.76x_{13} + 15x_{14} + 10.14x_{15} + 29.63x_{16} + 4.54x_{17} + 17.09x_{18} + 10.91x_{19} + 5.28x_{20} + \\ & 11x_{21} + 18.64x_{22} + 19.61x_{23} + 3.98x_{24} + 19.8x_{25} + 7.64x_{26} + 4.36x_{27} + 5.28x_{28} + 7.67x_{29} + 13.3x_{30} + \\ & 36.46x_{31} + 24.6x_{32} + 3.97x_{33} + 10.09x_{34} \end{aligned} \tag{1}$$

$$\begin{aligned} \textit{Calories}: 77x_1 & + 100x_2 + 105x_3 + 137x_4 + 25x_5 + 18x_6 + 28x_7 + 52x_8 + 80\,x_9 + 21x_{10} + 110x_{11} + 21x_{12} \\ & + 16x_{13} + 163x_{14} + 160x_{15} + 63x_{16} + 93x_{17} + 108x_{18} + 70x_{19} + 22x_{20} + 136x_{21} + 99x_{22} \\ & + 90x_{23} + 53x_{24} + 278x_{25} + 87x_{26} + 356x_{27} + 67x_{28} + 121x_{29} + 100x_{30} + 91x_{31} + 91x_{32} \\ & + 23x_{33} + 107x_{34} = 2000 \end{aligned}$$

IOP Conf. Series: Journal of Physics: Conf. Series **820** (2017) 012002 doi:10.1088/1742-6596/820/1/012002

$$\begin{aligned} \textit{Calories}: 77x_1 &+ 100x_2 + 105x_3 + 137x_4 + 25x_5 + 18x_6 + 28x_7 + 52x_8 + 80\,x_9 + 21x_{10} + 110x_{11} + 21x_{12} \\ &+ 16x_{13} + 163x_{14} + 160x_{15} + 63x_{16} + 93x_{17} + 108x_{18} + 70x_{19} + 22x_{20} + 136x_{21} + 99x_{22} \\ &+ 90x_{23} + 53x_{24} + 278x_{25} + 87x_{26} + 356x_{27} + 67x_{28} + 121x_{29} + 100x_{30} + 91x_{31} + 91x_{32} \\ &+ 23x_{33} + 107x_{34} = 2000 \end{aligned} \tag{2}$$

$$\begin{aligned} \textit{Protein}: 14.1x_1 &+ 0.9x_2 + 1.2x_3 + 23.1x_4 + 0.9x_5 + 0.5x_6 + 1x_7 + .4\,1.5x_8 + 18.7x_9 + 0.4x_{10} + 20.6x_{11} + 1.7x_{12} \\ &+ 0.6x_{13} + 2x_{14} + 12.4x_{15} + 14.9x_{16} + 2.5x_{17} + 21.6x_{18} + 0.6x_{19} + 0.3x_{20} + 19.8x_{21} \\ &+ 21.2x_{22} + 19x_{23} + 0.5x_{24} + 17.3x_{25} + 1.2x_{26} + 7.4x_{27} + 0.7x_{28} + 19.5x_{29} + 20.4x_{30} \\ &+ 19x_{31} + 18.2x_{32} + 0.6x_{33} + 18.1x_{34} \geq 67 \end{aligned} \tag{3}$$

$$\begin{aligned} \textit{Protein}: 14.1x_1 &+ 0.9x_2 + 1.2x_3 + 23.1x_4 + 0.9x_5 + 0.5x_6 + 1x_7 + .4\,1.5x_8 + 18.7x_9 + 0.4x_{10} + 20.6x_{11} + 1.7x_{12} \\ &+ 0.6x_{13} + 2x_{14} + 12.4x_{15} + 14.9x_{16} + 2.5x_{17} + 21.6x_{18} + 0.6x_{19} + 0.3x_{20} + 19.8x_{21} \\ &+ 21.2x_{22} + 19x_{23} + 0.5x_{24} + 17.3x_{25} + 1.2x_{26} + 7.4x_{27} + 0.7x_{28} + 19.5x_{29} + 20.4x_{30} \\ &+ 19x_{31} + 18.2x_{32} + 0.6x_{33} + 18.1x_{34} \geq 58 \end{aligned} \tag{4}$$

$$\begin{aligned} \textit{Fats}: 4.7x_1 &+ 6.2x_2 + 7.3x_3 + 4.6x_4 + 0.4x_5 + 0.1x_6 + 0.3x_7 + 0.4x_8 + 0.6x_9 + 0.1x_{10} + 3.1x_{11} + 0.3x_{12} \\ &+ 0.2x_{13} + 1.2x_{14} + 11x_{15} + 0.4x_{16} + 0.4x_{17} + 2.4x_{18} + 0.2x_{19} + 0.1x_{20} + 6.4x_{21} + 16x_{22} \\ &+ 1.5x_{23} + 1.5x_{24} + 23.2x_{25} + 0.9x_{26} + 0.5x_{27} + 1x_{28} + 4.7x_{29} + 2.1x_{30} + 2x_{31} + 4x_{32} \\ &+ 0.2x_{33} + 3.8x_{34} \leq 65 \end{aligned} \tag{5}$$

$$\begin{aligned} \textit{Fats}: 4.7x_1 &+ 6.2x_2 + 7.3x_3 + 4.6x_4 + 0.4x_5 + 0.1x_6 + 0.3x_7 + 0.4x_8 + 0.6x_9 + 0.1x_{10} + 3.1x_{11} + 0.3x_{12} \\ &+ 0.2x_{13} + 1.2x_{14} + 11x_{15} + 0.4x_{16} + 0.4x_{17} + 2.4x_{18} + 0.2x_{19} + 0.1x_{20} + 6.4x_{21} + 16x_{22} \\ &+ 1.5x_{23} + 1.5x_{24} + 23.2x_{25} + 0.9x_{26} + 0.5x_{27} + 1x_{28} + 4.7x_{29} + 2.1x_{30} + 2x_{31} + 4x_{32} \\ &+ 0.2x_{33} + 3.8x_{34} \leq 65 \end{aligned} \tag{6}$$

$$\begin{aligned} \textit{Carbohydrates}: 10.5x_2 &+ 24.5x_3 + 0.7x_4 + 4.5x_5 + 3.8x_6 + 4.8x_7 + 10.5x_8 + 4.6x_{10} + 2.9x_{12} + 2.9x_{13} \\ &+ 36.1x_{14} + 2.8x_{15} + 19.9x_{17} + 16.4x_{19} + 5.1x_{20} + 5x_{24} + 18.6x_{26} + 80.4x_{27} + 13.8x_{28} \\ &+ 0.1x_{29} + 2x_{31} + 4x_{32} + 4.7x_{33} \geq 300 \end{aligned} \tag{7}$$

$$\begin{aligned} \textit{Carbohydrates}: 10.5x_2 &+ 24.5x_3 + 0.7x_4 + 4.5x_5 + 3.8x_6 + 4.8x_7 + 10.5x_8 + 4.6x_{10} + 2.9x_{12} + 2.9x_{13} \\ &+ 36.1x_{14} + 2.8x_{15} + 19.9x_{17} + 16.4x_{19} + 5.1x_{20} + 5x_{24} + 18.6x_{26} + 80.4x_{27} + 13.8x_{28} \\ &+ 0.1x_{29} + 2x_{31} + 4x_{32} + 4.7x_{33} \geq 300 \end{aligned} \tag{8}$$

$$\begin{aligned} \textit{Calcium}: 752x_1 &+ 11x_2 + 17x_3 + 50x_4 + 42x_5 + 24x_6 + 74x_7 + 69x_8 + 46x_9 + 24x_{10} + 86x_{11} + 120x_{12} \\ &+ 22x_{13} + 18x_{14} + 74x_{15} + 55x_{16} + 41x_{17} + 78x_{18} + 10x_{19} + 26x_{20} + 44x_{21} + 65x_{22} \\ &+ 81x_{23} + 34x_{24} + 29x_{25} + 32x_{26} + 27x_{27} + 12x_{28} + 121x_{29} + 51x_{30} + 146x_{31} + 57x_{32} \\ &+ 19x_{33} + 74x_{34} \geq 750 \leq 2500 \end{aligned} \tag{9}$$

$$\begin{aligned} \textit{Calcium}: 752x_1 &+ 11x_2 + 17x_3 + 50x_4 + 42x_5 + 24x_6 + 74x_7 + 69x_8 + 46x_9 + 24x_{10} + 86x_{11} + 120x_{12} \\ &+ 22x_{13} + 18x_{14} + 74x_{15} + 55x_{16} + 41x_{17} + 78x_{18} + 10x_{19} + 26x_{20} + 44x_{21} + 65x_{22} \\ &+ 81x_{23} + 34x_{24} + 29x_{25} + 32x_{26} + 27x_{27} + 12x_{28} + 121x_{29} + 51x_{30} + 146x_{31} + 57x_{32} \\ &+ 19x_{33} + 74x_{34} \geq 700 \leq 2500 \end{aligned} \tag{10}$$

$$\begin{aligned} \textit{Phosphorus}: 482x_1 &+ 27x_2 + 34x_3 + 192x_4 + 38x_5 + 16x_6 + 28x_7 + 38x_8 + 166x_9 + 10x_{10} + 188x_{11} + 33x_{12} \\ &+ 17x_{13} + 56x_{14} + 180x_{15} + 173x_{16} + 51x_{17} + 246x_{18} + 19x_{19} + 15x_{20} + 194x_{21} + 202x_{22} \\ &+ 185x_{23} + 11x_{24} + 142x_{25} + 16x_{26} + 155x_{27} + 21x_{28} + 285x_{29} + 212x_{30} + 210x_{31} \\ &+ 148x_{32} + 24x_{33} + 184x_{34} \geq 700 \end{aligned} \tag{11}$$

$$\begin{aligned} \textit{Phosphorus}: 482x_1 &+ 27x_2 + 34x_3 + 192x_4 + 38x_5 + 16x_6 + 28x_7 + 38x_8 + 166x_9 + 10x_{10} + 188x_{11} + 33x_{12} \\ &+ 17x_{13} + 56x_{14} + 180x_{15} + 173x_{16} + 51x_{17} + 246x_{18} + 19x_{19} + 15x_{20} + 194x_{21} + 202x_{22} \\ &+ 185x_{23} + 11x_{24} + 142x_{25} + 16x_{26} + 155x_{27} + 21x_{28} + 285x_{29} + 212x_{30} + 210x_{31} \\ &+ 148x_{32} + 24x_{33} + 184x_{34} \leq 4000 \end{aligned} \tag{12}$$

(13)

$$Iron: 1.2x_1 + 0.8x_2 + 0.7x_3 + 2.8x_4 + 0.8x_5 + 0.4x_6 + 0.8x_7 + 2.1x_8 + 0.3x_9 + 0.4x_{10} + 1.5x_{11} + 0.5x_{12} + 0.4x_{13} + 1.1x_{14} + 2.8x_{15} + 0.6x_{16} + 0.8x_{17} + 1.3x_{18} + 0.6x_{19} + 0.4x_{20} + 1.2x_{21} + 1.5x_{22} + 0.6x_{23} + 1x_{24} + 1.3x_{25} + 0.4x_{26} + 1x_{27} + 1.2x_{28} + 1.6x_{29} + 1.2x_{30} + 1.4x_{31} + 0.3x_{32} + 0.8x_{33} + 0.8x_{34} \geq 12 \leq 45$$

(14)

$$Iron: 1.2x_1 + 0.8x_2 + 0.7x_3 + 2.8x_4 + 0.8x_5 + 0.4x_6 + 0.8x_7 + 2.1x_8 + 0.3x_9 + 0.4x_{10} + 1.5x_{11} + 0.5x_{12} + 0.4x_{13} + 1.1x_{14} + 2.8x_{15} + 0.6x_{16} + 0.8x_{17} + 1.3x_{18} + 0.6x_{19} + 0.4x_{20} + 1.2x_{21} + 1.5x_{22} + 0.6x_{23} + 1x_{24} + 1.3x_{25} + 0.4x_{26} + 1x_{27} + 1.2x_{28} + 1.6x_{29} + 1.2x_{30} + 1.4x_{31} + 0.3x_{32} + 0.8x_{33} + 0.8x_{34} \geq 10 \leq 45$$

(15)

$$Total\ Vitamin\ A: 68x_1 + 8x_2 + 5x_3 + 16x_4 + 33x_5 + 2x_6 + 3x_7 + 1668x_8 + 196x_9 + 6x_{10} + 20x_{11} + 90x_{12} + 302x_{15} + 20x_{16} + 10x_{17} + 110x_{18} + 194x_{19} + 37x_{20} + 139x_{21} + 50x_{22} + 60x_{23} + 75x_{24} + 150x_{25} + 32x_{29} + 63x_{30} + 58x_{31} + 51x_{32+} 6x_{33} + 67x_{34} \geq 550 \leq 3000$$

(16)

$$Total\ Vitamin\ A: 68x_1 + 8x_2 + 5x_3 + 16x_4 + 33x_5 + 2x_6 + 3x_7 + 1668x_8 + 196x_9 + 6x_{10} + 20x_{11} + 90x_{12} + 302x_{15} + 20x_{16} + 10x_{17} + 110x_{18} + 194x_{19} + 37x_{20} + 139x_{21} + 50x_{22} + 60x_{23} + 75x_{24} + 150x_{25} + 32x_{29} + 63x_{30} + 58x_{31} + 51x_{32+} 6x_{33} + 67x_{34} \geq 500 \leq 3000$$

(17)

$$Thiamine: 0.04x_2 + 0.02x_3 + 0.11x_4 + 0.05x_5 + 0.04x_6 + 0.04x_7 + 0.04x_8 + 0.01x_9 + 0.02x_{10} + 0.02x_{11} + 0.06x_{12} + 0.02x_{13} + 0.32x_{14} + 0.07x_{15} + 0.04x_{16} + 0.04x_{17} + 0.09x_{18} + 0.09x_{19} + 0.02x_{20} + 0.02x_{21} + 0.09x_{22} + 0.02x_{23} + 0.03x_{24} + 0.45x_{25} + 0.02x_{26} + 0.1x_{27} + 0.04x_{28} + 0.01x_{29} + 0.14x_{30} + 0.07x_{31} + 0.02x_{32} + 0.04x_{33} + 0.06x_{34} \geq 1.2$$

(18)

$$Thiamine: 0.04x_2 + 0.02x_3 + 0.11x_4 + 0.05x_5 + 0.04x_6 + 0.04x_7 + 0.04x_8 + 0.01x_9 + 0.02x_{10} + 0.02x_{11} + 0.06x_{12} + 0.02x_{13} + 0.32x_{14} + 0.07x_{15} + 0.04x_{16} + 0.04x_{17} + 0.09x_{18} + 0.09x_{19} + 0.02x_{20} + 0.02x_{21} + 0.09x_{22} + 0.02x_{23} + 0.03x_{24} + 0.45x_{25} + 0.02x_{26} + 0.1x_{27} + 0.04x_{28} + 0.01x_{29} + 0.14x_{30} + 0.07x_{31} + 0.02x_{32} + 0.04x_{33} + 0.06x_{34} \geq 1.1$$

(19)

$$Riboflavin: 0.12x_1 + 0.07x_2 + 0.02x_3 + 0.2x_4 + 0.04x_5 + 0.02x_6 + 0.07x_7 + 0.04x_8 + 0.05x_9 + 0.02x_{10} + 0.06x_{11} + 0.09x_{12} + 0.02x_{13} + 0.28x_{14} + 0.39x_{15} + 0.04x_{16} + 0.09x_{17} + 0.13x_{18} + 0.07x_{19} + 0.02x_{20} + 0.1x_{21} + 0.12x_{22} + 0.03x_{23} + 0.04x_{24} + 0.27x_{25} + 0.1x_{26} + 0.05x_{27} + 0.02x_{28} + 0.17x_{29} + 0.18x_{30} + 0.04x_{31} + 0.09x_{32} + 0.02x_{33} + 0.2x_{34} \geq 1.3$$

(20)

$$Riboflavin: 0.12x_1 + 0.07x_2 + 0.02x_3 + 0.2x_4 + 0.04x_5 + 0.02x_6 + 0.07x_7 + 0.04x_8 + 0.05x_9 + 0.02x_{10} + 0.06x_{11} + 0.09x_{12} + 0.02x_{13} + 0.28x_{14} + 0.39x_{15} + 0.04x_{16} + 0.09x_{17} + 0.13x_{18} + 0.07x_{19} + 0.02x_{20} + 0.1x_{21} + 0.12x_{22} + 0.03x_{23} + 0.04x_{24} + 0.27x_{25} + 0.1x_{26} + 0.05x_{27} + 0.02x_{28} + 0.17x_{29} + 0.18x_{30} + 0.04x_{31} + 0.09x_{32} + 0.02x_{33} + 0.2x_{34} \geq 1.1$$

(21)

$$Niacin: 1.9x_1 + 1.3x_2 + 0.6x_3 + 5.5x_4 + 0.2x_5 + 0.4x_6 + 0.3x_7 + 0.8x_8 + 2.5x_9 + 0.4x_{10} + 6.6x_{11} + 0.8x_{12} + 0.1x_{13} + 1.1x_{14} + 0.1x_{15} + 4.3x_{16} + 0.8x_{17} + 8.4x_{18} + 0.8x_{19} + 0.3x_{20} + 7.8x_{21} + 3.8x_{22} + 2.1x_{23} + 0.5x_{24} + 4.5x_{25} + 0.5x_{26} + 2.8x_{27} + 0.9x_{28} + 8x_{29} + 7.8x_{30} + 3.2x_{31} + 2.1x_{32} + 0.3x_{33} + 4.6x_{34} \geq 16 \leq 35$$

(22)

$$Niacin: 1.9x_1 + 1.3x_2 + 0.6x_3 + 5.5x_4 + 0.2x_5 + 0.4x_6 + 0.3x_7 + 0.8x_8 + 2.5x_9 + 0.4x_{10} + 6.6x_{11} + 0.8x_{12} + 0.1x_{13} + 1.1x_{14} + 0.1x_{15} + 4.3x_{16} + 0.8x_{17} + 8.4x_{18} + 0.8x_{19} + 0.3x_{20} + 7.8x_{21} + 3.8x_{22} + 2.1x_{23} + 0.5x_{24} + 4.5x_{25} + 0.5x_{26} + 2.8x_{27} + 0.9x_{28} + 8x_{29} + 7.8x_{30} + 3.2x_{31} + 2.1x_{32} + 0.3x_{33} + 4.6x_{34} \geq 14 \leq 35$$

(23)

where x_1 is anchovy; x_2 is red avocado, x_3 is banana (latundan), x_4 is beef lean meat, x_5 is bitter gourd, x_6 is bottle gourd, x_7 is green cabbage, x_8 is carrot, x_9 is catfish, x_{10} is chayote, x_{11} white meat chicken, x_{12} is Chinese cabbage, x_{13} is cucumber, x_{14} is durian, x_{15} is chicken egg, x_{16} is spotted grouper, x_{17} is horseradish tree pod, x_{18} is short bodied mackerel, x_{19} is mango (Carabao), x_{20} is musk melon, x_{21} is milkfish, x_{22} is striated mudfish/murrel, x_{23} is ribbon-finned nemipterid, x_{24} is papaya, x_{25} is lean pork Boston butt, x_{26} is rambutan, x_{27} is well-milled rice, x_{28} is santol, x_{29} is Indian sardines, x_{30} is round scad, x_{31} is white shrimp, x_{32} =common slipmouth, x_{33} is sponge gourd, and x_{34} is tilapia.

2.3. Sensitivity Analysis

A technique used to determine how different values of an independent variable will impact a particular dependent variable under a given set of assumptions. This technique is used within specific boundaries that will depend on one or more input variables, such as the effect that changes in interest rates will have on a bond's price. Sensitivity analysis is a way to predict the outcome of a decision if a situation turns out to be different compared to the key prediction(s).

The objective of the model is to minimize food expenditure; Z. Constraints in the model include the nutritional requirements: energy, protein, fats, carbohydrates, calcium, phosphorus, iron, total vitamin A, thiamine and niacin. The Food and Agriculture Organization and the World Health Organization have documented the harmful effects that may arise from excess consumption of some nutrients. Vitamin D and excess calcium are associated with kidney stone formation. High levels of vitamin A are associated with hair loss, bone pain and dry skin. Nutrients including calcium, phosphorus, iron, vitamin A, and niacin have allowable maximum intake levels based on internationally accepted dietary recommended intake.

2.4. Data

The researchers gathered the cheapest combination of foods for a diabetic person and its prices as shown in table 1. These data are gathered from the Philippine Statistic Authority (Bureau of Agricultural Statistics) [8]. The Energy and Nutritional requirement is based on Required Energy and Nutrient Intake (RENI). The nutritional value of each food is based on the Philippine Food Composition Tables (FCT) of Food and Nutrition Research Institute (FNRI) [9].

Table 1. Nutritional content of food items and its cost.

Food and Description	Energy (kcal)	Protein (g)	Fats (g)	Carbohydrates (g)	Calcium (mg)	Phosphorus (mg)	Iron (mg)	Total Vit. A (RE) (μg)	Thiamine (mg)	Rivaflavin (mg)	Niacin (mg)	Price
Anchovy	77	14.1	4.7	0	752	482	1.2	68	0	0.12	1.9	11.31
Avocado, red	100	0.9	6.2	10.5	11	27	0.8	8	0.04	0.07	1.3	6.9
Banana, latundan	105	1.2	7.3	24.5	17	34	0.7	5	0.02	0.02	0.6	4.2
Beef lean meat	137	23.1	4.6	0.7	50	192	2.8	16	0.11	0.2	5.5	26.21
Bitter Gourd	25	0.9	0.4	4.5	42	38	0.8	33	0.05	0.04	0.2	4.82
Bottle gourd	18	0.5	0.1	3.8	24	16	0.4	2	0.04	0.02	0.4	2.66
Cabbage, green	28	1.4	0.3	4.8	74	28	0.8	3	0.04	0.07	0.3	5.44
Carrot	52	1.5	0.4	10.5	69	38	2.1	1668	0.04	0.04	0.8	5.08
Catfish	80	18.7	0.6	0	46	166	0.3	196	0.01	0.05	2.5	14.07
Chayote	21	0.4	0.1	4.6	24	10	0.4	6	0.02	0.02	0.4	2.44
Chicken, white meat	110	20.6	3.1	0	86	188	1.5	20	0.02	0.06	6.6	13.98
Chinese cabbage	21	1.7	0.3	2.9	120	33	0.5	90	0.06	0.09	0.8	5.24
cucumber	16	0.6	0.2	2.9	22	17	0.4	0	0.02	0.02	0.1	4.76
Durian	163	2	1.2	36.1	18	56	1.1	0	0.32	0.28	1.1	15
Egg, chicken	160	12.4	11	2.8	74	180	2.8	302	0.07	0.39	0.1	10.14
Grouper, spotted	63	14.9	0.4	0	55	173	0.6	20	0.04	0.04	4.3	29.63
Horseradish tree pod	93	2.5	0.4	19.9	41	51	0.8	10	0.04	0.09	0.8	4.54
Mackerel, short bodied	108	21.6	2.4	0	78	246	1.3	110	0.09	0.13	8.4	17.09

Mango, carabao	70	0.6	0.2	16.4	10	19	0.6	194	0.09	0.07	0.8	10.91
Melon, musk	22	0.3	0.1	5.1	26	15	0.4	37	0.02	0.02	0.3	5.28
Milkfish	136	19.8	6.4	0	44	194	1.2	139	0.02	0.1	7.8	11
Mudfish/Murrel,striated	99	21.2	16	0	65	202	1.5	50	0.09	0.12	3.8	18.64
Nemipterid,ribbon-finned	90	19	1.5	0	81	185	0.6	60	0.02	0.03	2.1	19.61
Papaya	53	0.5	0	0	34	11	1	75	0.03	0.04	0.5	3.98
Pork boston butt, lean	278	17.3	23.2	0	29	142	1.3	150	0.45	0.27	4.5	19.8
Rambutan	87	1.2	0.9	18.6	32	16	0.4	0	0.02	0.1	0.5	7.64
Rice, well-milled	356	7.4	0.5	80.4	27	155	1	0	0.1	0.05	2.8	4.36
Santol	67	0.7	1	13.8	12	21	1.2	0	0.04	0.02	0.9	5.28
Sardines, indian	121	19.5	4.7	0.1	121	285	1.6	32	0.01	0.17	8	7.67
Scad, round	100	20.4	2.1	0	51	212	1.2	63	0.14	0.18	7.8	13.3
Shrimp, White	91	19	0	0	146	210	1.4	58	0.07	0.04	3.2	36.46
Slipmouth, common	91	18.2	0	0	57	148	0.3	51	0.02	0.09	2.1	24.6
Sponge gourd	23	0.6	0.2	4.7	19	24	0.8	6	0.04	0.02	0.3	3.97
Tilapia	107	18.1	3.8	0	74	184	0.8	67	0.06	0.2	4.6	10.09
REQUIRED NUTRIENTS:	**2000**	**58**	**65**	**300**	800	700	10	500	1.1	1.1	14	
MAXIMUM REQUIRED					**2500**	**4000**	**45**	**3000**			**35**	

Table 1 indicates the nutritional content as well as its price of each food item. Each food is specifically described in the table. The data becomes the input of this research.

3. Results and Discussion

According to the findings, a 50 year old and above diabetic male needs to spend a minimum of 72.22 pesos for foods that satisfy all the nutrients needed per day. in order to attain the minimum cost, the food must consists of 60.49 g of anchovy, 91.24 g of carrot, 121.92 g of durian, 121.41 g of chicken egg, 70.82 g of pork (lean), and 369.70 g of rice (well-milled) shown in table 2. Furthermore, for ages 50 and above, diabetic female's minimum spending for foods is 64.65 pesos per day consisting of 75.87 g of anchovy, 43.38 g of carrot, 160.46 g of durian, 69.66 g of chicken egg, 23.16 g of pork (lean) and 416.19 g of rice (well-milled) as shown in table 3.

In addition, in every 100 g consumption of avocado will have a 4.06 pesos increase in the minimum cost, consumption of beef will increase by 16.36 per 100 g, for a 100 g consumption of bitter gourd, catfish and chicken will increase the cost by 1.25 pesos, 12.25 pesos and 9.81 pesos respectively. A 100 g consumption of any meat or poultry product: beef, chicken would increase food cost by at least 10 pesos to 16 pesos, for a 100 g consumption of other vegetables will increase the cost from 0.50 pesos to 2.00 pesos. Obviously, the consumption of any of the above food items will increase food cost because they are relatively more expensive than the components of the food items suggested by the model.

Table 2. Optimal values of food items for males.

Name	Final Value	Reduced Cost	Objective Coefficient	Allowable Increase	Allowable Decrease
Anchovy	0.604925777	0	11.313	1.952737368	8.245277361
Avocado, red	0	4.061649992	6.9	1E+30	4.061649992
Banana, latundan	0	2.607837089	4.199999999	1E+30	2.607837089
Beef lean meat	0	16.35540355	26.21	1E+30	16.35540355
Bitter Gourd	0	1.248238915	4.82	1E+30	1.248238915
Bottle gourd	0	0.32528019	2.66	1E+30	0.32528019
Cabbage, green	0	1.578952684	5.44	1E+30	1.578952684
Carrot	0.912381187	0	5.08	1.511165648	2.32140067
Catfish	0	12.25130867	14.07	1E+30	12.25130867
Chayote	0	0.832436811	2.44	1E+30	0.832436811
Chicken, white meat	0	9.807756198	13.98	1E+30	9.807756198

Chinese cabbage	0	0.295638773	5.239999999	1E+30	0.295638773
Cucumber	0	3.151441943	4.76	1E+30	3.151441943
Durian	1.219183293	0	15	0.259597939	2.234035492
Egg, chicken	1.214058829	0	10.14	3.349175752	1.009256568
Grouper, spotted	0	26.36521256	29.63	1E+30	26.36521256
Horseradish tree pod	0	1.108040065	4.54	1E+30	1.108040065
Mackerel, short bodied	0	10.09060649	17.09	1E+30	10.09060649
Mango, carabao	0	6.441651033	10.91	1E+30	6.441651033
Melon, musk	0	3.655461493	5.28	1E+30	3.655461493
Milkfish	0	7.442842668	11	1E+30	7.442842668
Mudfish/Murrel, striated	0	11.60291878	18.64	1E+30	11.60291878
Nemipterid, ribbon-finned	0	16.8685933	19.61	1E+30	16.8685933
Papaya	0	1.079688659	3.98	1E+30	1.079688659
Pork Boston butt, lean	0.708186743	0	19.8	3.474837913	0.354966758
Rambutan	0	5.443741097	7.64	1E+30	5.443741097
Rice, well-milled	3.696979465	0	4.36	1.283083203	3.772623586
Santol	0	2.25971609	5.28	1E+30	2.25971609
Sardines, Indian	0	2.373586482	7.669999999	1E+30	2.373586482
Scad, round	0	4.488145104	13.3	1E+30	4.488145104
Shrimp, White	0	30.08335815	36.46	1E+30	30.08335815
Slipmouth, common	0	21.98212	24.6	1E+30	21.98212
Sponge gourd	0	1.203899137	3.97	1E+30	1.203899137
Tilapia	0	4.264337665	10.09	1E+30	4.264337665

Table 3. Optimal values of food items for females.

Name	Final Value	Reduced Cost	Objective Coefficient	Allowable Increase	Allowable Decrease
Anchovy	0.758707454	0	11.313	1.952737372	8.24527736
Avocado, red	0	4.061649992	6.9	1E+30	4.061649992
Banana, latundan	0	2.60783709	4.200000001	1E+30	2.60783709
Beef lean meat	0	16.35540355	26.21	1E+30	16.35540355
Bitter Gourd	0	1.248238915	4.82	1E+30	1.248238915
Bottle gourd	0	0.32528019	2.66	1E+30	0.32528019
Cabbage, green	0	1.578952684	5.44	1E+30	1.578952684
Carrot	0.43379878	0	5.08	1.51116565	2.321400671
Catfish	0	12.25130867	14.07	1E+30	12.25130867
Chayote	0	0.832436811	2.44	1E+30	0.832436811
Chicken, white meat	0	9.807756197	13.98	1E+30	9.807756197
Chinese cabbage	0	0.295638774	5.239999999	1E+30	0.295638774
Cucumber	0	3.151441943	4.76	1E+30	3.151441943
Durian	1.604649335	0	15	0.259597939	2.234035492
Egg, chicken	0.696620418	0	10.14	3.349175753	1.009256568
Grouper, spotted	0	26.36521256	29.63	1E+30	26.36521256
Horseradish tree pod	0	1.108040065	4.54	1E+30	1.108040065
Mackerel, short bodied	0	10.09060649	17.09	1E+30	10.09060649
Mango, carabao	0	6.44165103	10.91	1E+30	6.44165103
Melon, musk	0	3.655461493	5.28	1E+30	3.655461493
Milkfish QUANTITY	0	7.442842669	11	1E+30	7.442842669
Mudfish/Murrel, striated	0	11.60291878	18.64	1E+30	11.60291878
Nemipterid, ribbon-finned	0	16.8685933	19.61	1E+30	16.8685933
Papaya	0	1.079688659	3.98	1E+30	1.079688659
Pork Boston butt, lean	0.23157692	0	19.8	3.474837912	0.354966758
Rambutan	0	5.443741098	7.64	1E+30	5.443741098
Rice, well-milled	4.161872183	0	4.36	1.283083203	3.772623596
Santol	0	2.25971609	5.28	1E+30	2.25971609
Sardines, Indian	0	2.373586482	7.669999999	1E+30	2.373586482

Scad, round	0	4.488145104	13.3	1E+30	4.488145104
Shrimp, White	0	30.08335815	36.46	1E+30	30.08335815
Slipmouth, common	0	21.98212	24.6	1E+30	21.98212
Sponge gourd	0	1.203899137	3.97	1E+30	1.203899137
Tilapia	0	4.264337666	10.09	1E+30	4.264337666

Table 4. Nutrient constraint sensitivity for males.

Name	Final Value	Shadow Price	Constraint R.H. Side	Allowable Increase	Allowable Decrease
minreq: Energy	2000	-0.003951458	2000	462.4217828	165.8061758
minreq: Protein	67	0.021643521	67	12.16898951	7.036338438
minreq: Fats	36.30419287	0	65	1E+30	28.69580713
minreq: Carbohydrates	354.229033	0	300	54.22903303	1E+30
minreq: Calcium	750	0.011545564	750	936.6761509	432.0084098
maxreq: Calcium	750	0	2500	1E+30	1750
maxreq: Phosphorus	1286.643898	0	4000	1E+30	2713.356102
minreq Phosphorus	1286.643898	0	700	586.6438979	1E+30
minreq: Iron	12	0	45	1E+30	33
minreq: Iron	12	1.266754534	12	1.057059658	1.629062317
maxreq: Total Vit A	2035.86055	0	3000	1E+30	964.1394496
minreq: Total Vit A	2035.86055	0	550	1485.86055	1E+30
minreq: Thiamine	1.2	35.65689469	1.2	0.391624655	0.124390224
minreq: Riboflavin	1.3	9.247657976	1.3	0.109035298	0.194986467
maxreq: Niacin	16.88015428	0	35	1E+30	18.11984572
minreq: Niacin	16.88015428	0	16	0.880154276	1E+30

Table 5. Nutrient constraint sensitivity for females.

Name	Final Value	Shadow Price	Constraint R.H. Side	Allowable Increase	Allowable Decrease
minreq: Energy	2000	-0.003951458	2000	151.2118285	302.8402717
minreq: Protein	58	0.021643521	58	19.24657535	3.798000706
minreq: Fats	20.78136898	0	65	1E+30	44.21863101
minreq: Carbohydrates	399.0477889	0	300	99.04778889	1E+30
minreq: Calcium	800	0.011545564	800	306.2929099	541.8317639
maxreq: Calcium	800	0	2500	1E+30	1700
minreq: Phosphorus	1275.407496	0	700	575.4074956	1E+30
maxreq Phosphorus	1275.407496	0	4000	1E+30	2724.592504
minreq: Iron	10	0	45	1E+30	35
minreq: Iron	10	1.266754534	10	2.170513324	0.570427467
maxreq: Total Vit A	1020.284376	0	500	520.2843758	1E+30
minreq: Total Vit A	1020.284376	0	3000	1E+30	1979.715624
minreq: Thiamine	1.1	35.65689469	1.1	0.322787609	0.129530511
minreq: Riboflavin	1.1	9.247657972	1.1	0.105221721	0.288866681
minreq: Niacin	16.31869775	0	14	2.318697751	1E+30
maxreq: Niacin	16.31869775	0	35	1E+30	18.68130225

As shown in table 4 and 5, all the maximum constraints and the energy, fats, carbohydrate, phosphorus, total vitamin A and niacin minimum requirement are not binding in the model. The binding constraints in the model are minimum protein, calcium, iron, thiamine and riboflavin requirements. According to Table 4 and 5 (Shadow Price column), although these constraints are binding, a unit increase in the right hand side of any of them will not increase food expenditure significantly. A unit increase in the minimum protein, calcium, iron, thiamine, and riboflavin constraints will increase food expenditure by 0.021643521 pesos, 0.011545564 pesos, 1.266754534 pesos, 35.65689469 pesos and 9.247657976 pesos respectively. A unit increase in the right hand side of the minimum Energy, on the other hand, will decrease the food expenditure by 0.003951458 pesos.

IOP Conf. Series: Journal of Physics: Conf. Series **820** (2017) 012002 doi:10.1088/1742-6596/820/1/012002

As long as the Right Hand Side Constraint increases or decreases within the allowable ranges (allowable increases and decreases column) we can use shadow price to find out the new objective function value.

4. Conclusion and Recommendations

People with diabetes have specific calorie need based on the recommendation of the dietitian for support in the daily responsibility of managing the rules of conduct regarding diet. A diabetic male need to spend a minimum of 72.22 pesos for a 60.49 g of anchovy, 91.24 g of carrot, 121.92 g of durian, 121.41 g of chicken egg, 70.82 g of pork (lean), and 369.70 g of rice (well-milled), for female, it cost 64.65 pesos for 75.87 g of anchovy, 43.38 g of carrot, 160.46 g of durian, 69.66 g of chicken egg, 23.16 g of pork (lean) and 416.19 g of rice (well-milled). The foods needed by a diabetic male to satisfy the nutrient requirement are the same as a diabetic female based on 2000-calorie diet. The only difference is the quantity required that results to a higher cost for male. Although diabetes is not a curable disease, a healthy diet is important and can control the disease. We have come up to a combination that is consistent in Calorie and fats, carbohydrates, and proteins have been evenly distributed throughout the day. On the other hand, other patients require insulin injections to control the disease. A good understanding of how to meet the individual's needs regarding nutrients is necessary in the development of diabetic diet strategies to ensure a sustainable health. Further research is needed to explore how diet options can be differentiated to meet the diverse needs of the diabetic patient group (50-64 years old, male and female and 65 years old above male and female).

References

[1] *Diabetes: Symptoms, Causes and Treatments. Diabetes Cases Rise From 30 Million to 230 Million in 20 Years* Retrieved from http://www.medicalnewstoday.com/info/diabetes

[2] *Philippine Diabetes Statistics* 2012 Retrieved from http://.allaboutdiabetes.net/Philippine-diabetes-statistics/

[3] Wild S 2004 *Global Prevalence of Diabetes* Retrieved from https://www.ncbi.nlm.nih.gov/pubmed/15111519

[4] *The Truth about the So-Called Diabetes Diet* Retrieved from http://www.joslin.org/info/the-truth-about-the-so-called-diabetes-diet.html

[5] *Linear Programming: An Essential Optimization Technique* Retrieved from http://www-01.ibm.com/software/commerce/optimization/linear-programming/

[6] *Excel Solver* Retrieved from www.ce.memphis.edu/1112/notes/excel/Excel_solver.pdf

[7] Mead R & Nelder J *A Simplex Method for Function Minimization* Retrieved from http://comjnl.oxfordjournals.org/content/7/4/308.short?rss=1&ssource=mfc

[8] *Country Stat Philippines* Retrieved from http://www.countrystat.gov

[9] *Food and Nutrition Research Institute* 1997 The Philippine Food Composition.

[10] *Required Energy and Nutrient Intakes* (RENI)

IOP Conf. Series: Journal of Physics: Conf. Series **820** (2017) 012003 doi:10.1088/1742-6596/820/1/012003

Laser-induced plasma emission enhanced by microwaves in argon gas for potential application of nuclear fuel material analysis

Ali Khumaeni[1*], Motonobu Tampo[2], Masabumi Miyabe[3], Katsuaki Akaoka[3], and Ikuo Wakaida[3]

[1]Department of Physics, Faculty of Science and Mathematics, Diponegoro University, Tembalang, Semarang 50275, Indonesia

[2]Institute of Materials Structure Science, KEK, Tokaimura, Ibaraki 319-1195, Japan

[3]Fuel Debris Analysis Group, Collaborative Laboratories for Advanced Decommisioning Science, Japan Atomic Energy Agency, Tokaimura, Ibaraki 319-1195, Japan

*E-mail: khumaeni@fisika.undip.ac.id

Abstract. To perform quantitative analysis of nuclear fuel materials by using laser-induced breakdown spectroscopy (LIBS), emission spectral lines of atoms should be clearly identified, which is far from other atomic and molecular disturbances, low background emission, and optimum emission intensity. To this end, intensified microwaves were introduced into the laser-induced plasma in LIBS. Gadolinium oxide (Gd_2O_3) was employed for the simulation of nuclear fuel material. Enhancement of laser plasma was obtained when the microwave was introduced into the laser plasma. Total emission intensity of Gd significantly increased for the case of microwave-assisted laser plasma, which revealed that the microwave can effectively be employed to increase the sensitivity. Furthermore, the analytical lines of Gd can be clearly identified, which are far from the molecular bands disturbance and have high emission intensity with low background intensity. The enhancement factors for neutral Gd I at 470.9 nm and ionic Gd II at 458.3 nm were 40 and 45 times for the LIBS with microwaves case. The plasma temperatures were 9500 K and 4800 K for the LIBS with and without microwaves, respectively. The microwave-assisted laser-induced plasma in Ar gas has high potential for the analysis of impurity in nuclear fuel materials.

1. Introduction

Recently, laser-induced breakdown spectroscopy (LIBS) has become well known method for the rapid and direct qualitative and quantitative elements in various samples in science and industries [1-2]. In this conventional method, a pulsed neodymium-doped yttrium aluminium garnet (Nd:YAG) laser is focused onto a sample target to induce a small luminous plasma just above the sample. Elemental composition of the target can be obtained by analyzing the atomic emission from the plasma [3-4]. Compared to other conventional analytical methods such as inductively coupled plasma optical emission spectroscopy (ICP-OES) and atomic absorption spectroscopy (AAS), the LIBS has several advantages including simple sample preparation and rapid and multi-elemental detection [5]. One of the LIBS applications is monitoring of uranium and plutonium in the fuel production process [6]. However, the LIBS has weak points including low sensitivity to perform the analysis of impurity elements in the materials, which have complex spectral lines, such as nuclear fuel elements.

Several methods have been developed to improve the sensitivity including double pulse LIBS [7], combination between LIBS and laser-induced fluorescence (LIF) [8], and resonant and LIBS [9]. However the methods are costly because the other laser system is required. The other method used to improve the sensitivity is microwave-assisted laser-induced breakdown spectroscopy (MA-LIBS) [10]. The use of MA-LIBS has advantages including low-cost instrument and that the sensitivity of analytical measurement can be effectively enhanced. In this method, the microwave-assisted laser-induced plasma was induced in air for nuclear fuel analysis. However, it was found that the molecular bands coming from air clearly appeared disturbing the analytical lines.

In this study, an enhancement of laser-induced plasma emission assisted by microwaves in argon (Ar) gas was performed for the detection of Gd lines in gadolinium oxide (Gd_2O_3) pellet as a simulation of nuclear fuel material. By using Ar gas as a replacement of air, the molecular bands from air such OH bands and N_2 molecules can significantly be suppressed, increasing the signal to noise ratio (S/N) of the analytical lines coming from the analyte. The results certified that the emission lines of Gd can be clearly and separately detected. This method has high prospect to analysis of impurity in materials, which have complex spectral lines such as nuclear fuel materials.

2. Experimental Procedure

The basic experimental setup used in this study is shown in Fig. 1. A second-order harmonic of Q-switched Nd:YAG laser (Quanta-Ray, Spectra Physics, 532 nm, 8 ns) was directly focused onto a sample surface through a quartz window using quartz lens with a focal length of 200 mm. During the experiment, the laser energy was 5 mJ. The laser plasma was enhanced by intensified microwave field induced with an antenna; the antenna was circular shaped with a diameter of 3 and 9 mm. Microwave was generated by using a magnetron at a frequency of 2.45 GHz (Muegge MG0500D-215TC, 0-1 ms). The power of the magnetron was 400 W. The Nd:YAG laser and microwave were operated in the synchronization mode with a delay time of 10 μs for the Nd:YAG laser bombardment relative to the microwave generation.

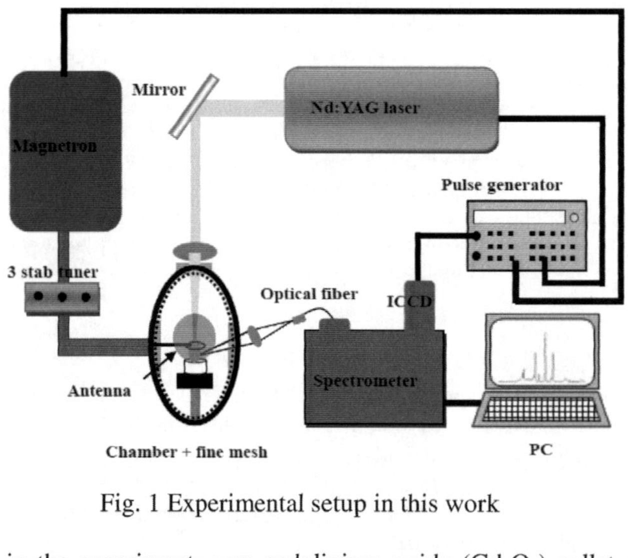

Fig. 1 Experimental setup in this work

The sample used in the experiments was gadolinium oxide (Gd_2O_3) pellet as a simulation of nuclear fuel of uranium oxide (U_2O_3). The Gd_2O_3 has similar physical and chemical characteristics with U_2O_3 and it is not harmful for human being; if the analysis is made by using real nuclear fuel of U_2O_3, special treatment of sample preparation and its experiment should be carried out. The sample was 1 mm thick and 10 mm in diameter. The sample was made by compressing Gd_2O_3 powder of 1 g by the pressure of 1.5 MPa. During the experiment, the sample was placed in a metal chamber equipped by windows, on which the fine gold mesh (lattice constant of 100 μm and wire diameter of 30 μm) was attached. The chamber functioned to block the microwave radiation. The pressure of the surrounding gas in the chamber was set at 0.67 kPa. During the experiment, the Ar gas (purity 99.999%) was flowed.

The emission spectrum was obtained by using an ICCD camera (Andor, iStar) through a high-resolution echelle spectrometer (ARYELLE) with the resolution of 50 pm at the wavelength of 360-460 nm. The light emission of the laser plasma was collected by using an optical fiber, which fed into the spectrometer. The plasma radiation at 2-5 mm from the sample surface was imaged in a ratio of 1:1 onto one end of the fiber by using a quartz lens (f= 100 mm).

3. Results and Discussion

First, the generation of luminous plasma from the Gd_2O_3 sample has been carried out by using different method of standard LIBS and MA-LIBS in Ar gas at 0.67 kPa. Figure 2 shows the photographs of plasma emission in (a) standard LIBS, and (b) MA-LIBS coupled by antenna. In this experiment, the laser energy was 5 mJ and the microwave power was 400 W.

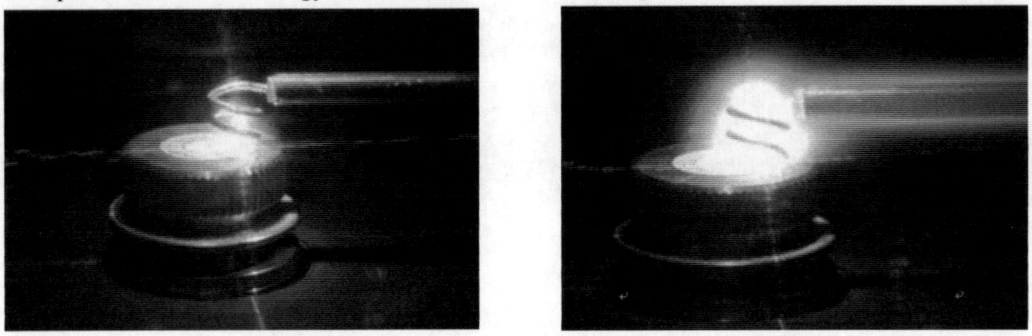

(a) **(b)**

Fig. 2 Photographs of plasma emission taken from the Gd_2O_3 sample by using (a) LIBS, and (b) MA-LIBS

It is clearly seen that bright white color plasma with a diameter of around 3 mm was produced by using LIBS method [Fig. 2(a)]. When a microwave was introduced into the LIBS plasma region via a loop antenna, the plasma diameter was enlarged of around 10 mm and the plasma emission was much brighter [Fig. 2(b)] compared the case of LIBS plasma. This revealed that the microwave can effectively enhance the plasma emission induced by the LIBS method. The mechanism of plasma enhancement by microwave is assumed as follows: when the microwave is radiated into the LIBS plasma, the electrons are accelerated, increasing the kinetic energy. Together with the increment of kinetic energy, the number of collisions between electrons and constituents increase, which affect the plasma temperature and the densities of electrons, ions, and neutral atoms in the plasma, allowing the enhancement of plasma emission.

To ensure that the antenna is never ablated and disturbed the analytical spectrum during the experiment, the emission spectra from the Gd_2O_3 sample in Ar gas were obtained by using only microwave without laser irradiation (Fig. 3). It should be noted that the antenna for transmitting the microwaves from magnetron was made of Cu coaxial cable. It is seen in Fig. 3(a) that completely no neutral Cu lines at 324.7 nm and 327.4 nm were observed, which means that no ablation of antenna material takes place during the microwave irradiation. Thus, the antenna used in this study never disturbs the analytical lines. Furthermore, completely no Gd lines occurred in the spectrum by using microwaves. This certified that microwave irradiation cannot ablate and vaporize the Gd_2O_3 sample and thus, it cannot be employed to perform analysis by using microwave only.

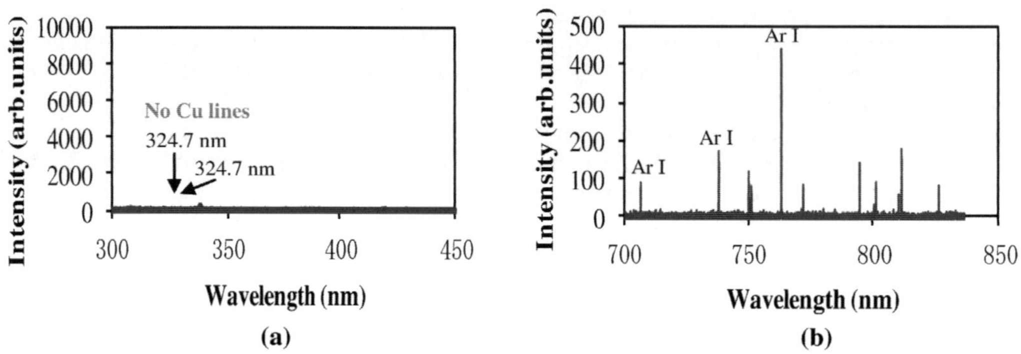

Fig. 3 Emission spectra obtained from the Gd_2O_3 sample by using only microwave radiation at wavelength region (a) 300-450 nm, and (b) 700-850 nm

Figure 3 (b) shows the emission spectrum obtained from the Gd_2O_3 sample at wavelength region of 700-850 nm by using microwave only. Completely no Gd lines occurred in the spectrum. However, many neutral argon (Ar) lines clearly appear with high emission intensity such as Ar I at 706.7 nm, 738.3 nm, and 763.5 nm. From this result, we concluded that the microwave irradiation can excite Ar gas used as an ambient gas, but cannot vaporize and excite the Gd in the Gd_2O_3 sample target. This result certified that the microwave only cannot be employed to perform analysis of Gd because it is not able to ablate the sample target and therefore no atomization and excitation of elements from the sample target take place.

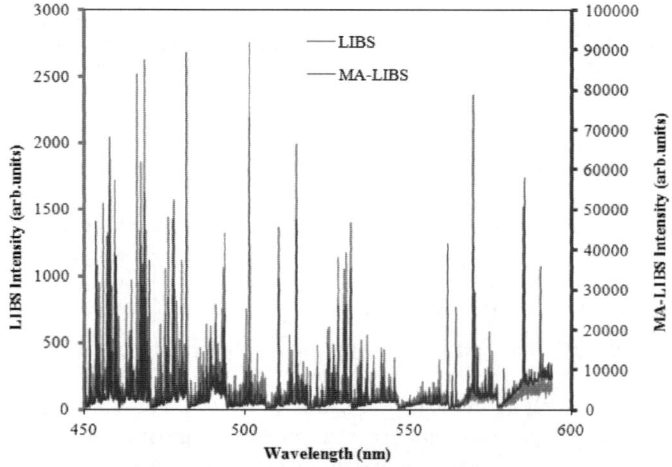

Fig. 4 Emission spectra of Gd obtained from the Gd_2O_3 sample at wavelength region 450-600 nm by using LIBS and MA-LIBS

To examine the effect of the microwave on the enhancement of emission spectrum for nuclear material analysis, the MA-LIBS was then employed to detect the Gd emission obtained from the Gd_2O_3 sample target, as a simulation material of nuclear fuel material. Figure 4 shows the emission spectrum taken from the Gd_2O_3 sample by using LIBS (red line) and MA-LIBS (blue line) at 450-

600 nm. The argon gas was used as an ambient gas at 0.67 kPa. The laser energy was 5 mJ and the laser was irradiated at 10 Hz irradiation. The power of microwaves was 400 W. Many Gd lines clearly appear in the spectrum of LIBS and MA-LIBS; Gd has complex spectral lines as well as nuclear fuel material such as uranium. Based on NIST database, neutral and ionic Gd has 821 lines from 200 to 800 nm [11]. The emission intensities of Gd lines significantly increased when the microwaves were employed (blue line). Total emission intensity enhancement is approximately 30 times higher for the case of MA-LIBS. From this spectrum, we also obtained the enhancement factor of neutral and ionic Gd lines. The enhancement factors for the neutral Gd I at 470.9 nm and ionic Gd II at 458.3 nm were approximately 40 and 45 times as shown in Fig. 5(a) and 5(b), respectively for the case of MA-LIBS. It should also be noted that the hydroxyl (OH) bands and N_2 molecular lines appeared in MA-LIBS method in air [10] completely disappeared. This result indicates that the microwave-assisted laser-induced plasma in Ar gas can be used to clearly and separately detect adjacent analytical lines, which are very complex in nuclear fuel materials.

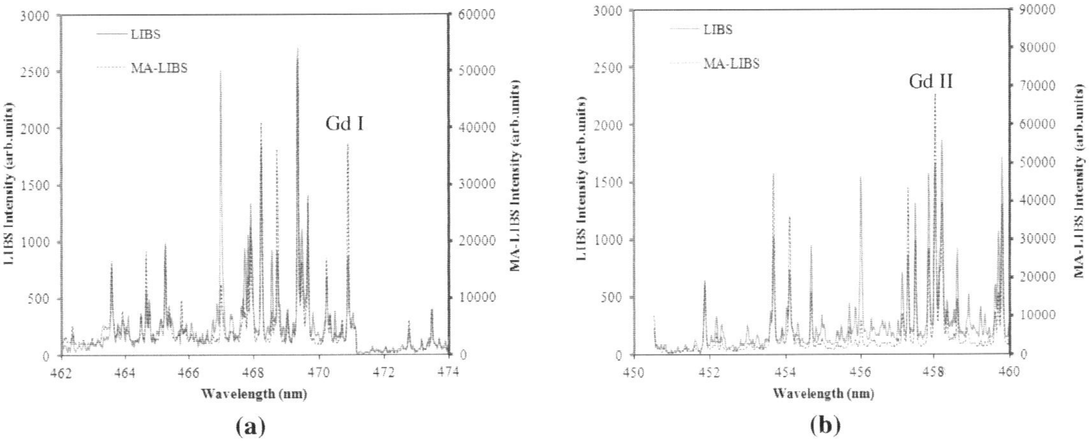

Fig. 5 Emission spectra of the Gd obtained from the Gd_2O_3 sample at wavelength region (a) 462-474 nm) and (b) 450-460 nm by using LIBS and MA-LIBS

Based on our previous experiment [12], it is assumed that the enhancement of laser-induced plasma emission by microwaves is due to the increment of plasma temperature when the microwaves are introduced in the laser plasma. In order to estimate the plasma temperature in LIBS and MA-LIBS a Boltzmann plot method was employed. For the calculation of plasma temperature, selected lines and their spectroscopic data obtained from Kurucz database were used as shown in Table 1 [13].

Table 1. Wavelength, upper level energy, upper level degeneracy, and transition probability for the ionic Gd emission lines used in this study.

Elements	Wavelength (nm)	Upper level energy (eV)	Upper level degeneracy	Transition probability (s^{-1})
Gd II	454.01	5.055	10	8.58×10^7
Gd II	458.24	3.956	8	3.17×10^6
Gd II	460.11	3.251	10	5.15×10^6
Gd II	501.08	3.531	16	1.03×10^6
Gd II	510.01	4.028	14	4.08×10^6
Gd II	515.67	3.973	10	7.28×10^6
Gd II	561.62	3.267	14	3.12×10^5
Gd II	585.52	3.711	8	3.06×10^6

Gd II	590.41	3.721	14	2.34 x 10^6

It is assumed that local termodynamic equilibrium (LTE) is achieved in the plasma LIBS and MA-LIBS. The Boltzmann plot equation in LTE condition is as follows,

$$\ln\left(\frac{I\lambda}{g_k A_k}\right) = -\frac{1}{k_B T} E_k + C \quad (1)$$

Where I is the integrated intensity of spectral lines occurring between the upper energy level k and lower energy level i, λ is the transition line wavelength, g_k is the degeneracy of the upper energy level, A_k is the transition probability, k_B, T, and E_k are Boltzmann constant, the plasma temperature, and energy of the upper energy level, respectively, C is a constant for a given atomic species. Using the Boltzmann plot equation (Eq. 1) and measurements of spectral lines (Table 1) using LIBS and MA-LIBS, Boltzmann plot was made as shown in Fig. 6.

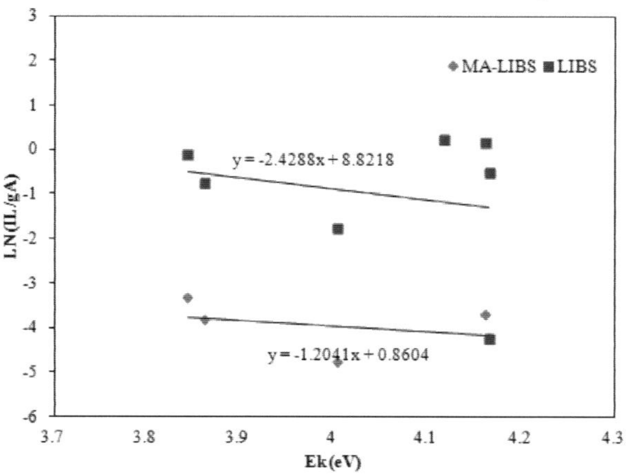

Fig. 6 Boltzmann plot from the analysis of ionic Gd II lines

The slope of the curve results in plasma temperature of approximately 4800 K and 9500 K in LIBS and MA-LIBS, respectively. It is assumed that the increment of the plasma temperature in MA-LIBS is due to the reheating of plasma by microwave as in the case of double pulse LIBS (DP-LIBS) [14]. As is known the plasma temperature is contributed by electron collision in the plasma region. When the microwaves are introduced in the plasma region, the electrons are accelerated and multiple collisions of electrons increase, which affects the plasma temperature and the densities of electrons, ions, and neutral atoms in the plasma, resulting in the improvement of emission intensity. The enhanced plasma temperature in MA-LIBS can effectively increase the intensities of analytical lines, resulting in high signal to intensity (S/N) ratio of Gd lines.

4. Conclusions

Enhancement of laser-induced plasma was carried out using antenna-coupled microwaves in the Gd_2O_3 pellet. In this study, the MWs was generated by magnetron and transferred into the laser-induced plasma via coaxial cable to enhance the plasma emission. The results revealed that the laser plasma emission was enhanced indicated by enlarge in diameter. The analytical lines can clearly be identified with high emission intensity and low background emission and is far from the molecular disturbance. Total emission intensity of Gd significantly increased for the case of MA-LIBS (approximately 30 times) compared to the case of LIBS emission. The enhancement factors for neutral Gd I at 470.9 nm and ionic Gd II at 458.3 nm were approximately 40 and 45 times, respectively, for the MA-LIBS case. The plasma temperature for MA-LIBS is 9500 K and it is

IOP Conf. Series: Journal of Physics: Conf. Series **820** (2017) 012003 doi:10.1088/1742-6596/820/1/012003

4800 K for LIBS case. The MA-LIBS method has high potential to be employed for the sensitive detection of impurity in nuclear fuel materials, which is very difficult to be made using standard LIBS and conventional analytical methods. In this study, the experiment was made at low pressure environment because the MWs power has limitation at 400 W. This is a challenge to study at atmospheric pressure by using higher power of MWs in order to realize in-situ analysis, which is generally carried out at 1 atmosphere.

References

1. Tognoni E, Palleschi V, Corsi M, Cristoforetti G 2002 *Spectrochim. Acta part B* **57** 1115
2. Rusak D A, Castle B C, Smith B W, Winefordner J D 1998 *Trends in Analyt. Chemist.* **17** 453
3. Cremers D A, Barefield J E II, Koskelo A C 1995 *Appl. Spectrosc.* **49** 857
4. Radziemski L J 2002 *Spectrochim. Acta Part B* **57** 1109
5. Cremers D A, Radziemski L J 2006 *Handbook of Laser-Induced Breakdown Spectroscopy* (John Wiley and Sons, Chichester)
6. Cremers D A, Beddingfield A, Smithwick R, Chinni R C, Jones C R, Beardsley B, Karch L 2012 *Appl. Spectrosc.* **66** 250
7. Babushok V I, DeLucia F C Jr., Gottfried J L, Munson C A, Miziolek A W 2006 *Spectrochim. Acta Part B* **61** 999
8. Telle H H, Beddows D C S, Morris G W, Samek O 2001 *Spectrochim. Acta Part B* **56** 947
9. Gouegucl B, Laville S, Vidal F, Sabsabi M, Chaker M 2010 *J. Anal. At. Spectrom.* **25** 635
10. Khumaeni A, Motonobu T, Katsuaki A, Ikuo W 2013 *Optics Express* **21** 29755
11. Kramida A, Ralchenko Y, Reader J, and NIST ASD Team 2015 *NIST Atomic Spectra Database (ver.5.3), [Online]*. Available:http://physics.nist.gov/PhysRefData/ASD/lines_form.html [2016, October 11]. National Institute of Standards and Technology, Gaithersburg, MD.
12. Khumaeni A, Akoka K, Miyabe M, Wakaida I 2016 Front. of Phys. **11** 114209
13. Smith P L, Heise C, Esmond J R, Kurucz R L, 2016 Atomic Spectral Line Database. Available: http://www.pmp.uni-hannover.de/cgi-bin/ssi/test/kurucz/sekur.html [2016, October 11]
14. Giacomo A D, Aglio M D, Bruno D, Gaudiuso R, Pascale O D 2008 *Spectrochim. Acta Part B* **63** 805

IOP Conf. Series: Journal of Physics: Conf. Series **820** (2017) 012004 doi:10.1088/1742-6596/820/1/012004

Time-resolved plasma imaging in microwave-assisted laser-induced breakdown spectroscopy

Ali Khumaeni[1*], Masabumi Miyabe[2], Katsuaki Akaoka[2], and Ikuo Wakaida[2]

[1]Department of Physics, Faculty of Science and Mathematics, Diponegoro University, Tembalang, Semarang 50275, Indonesia
[2]Fuel Debris Analysis Group, Collaborative Laboratories for Advanced Decommisioning Science, Japan Atomic Energy Agency, Tokaimura, Ibaraki 319-1195, Japan

*E-mail: khumaeni@fisika.undip.ac.id

Abstract. To study the dynamics of luminous plasma induced by a pulse laser and microwaves (MWs), time resolved imaging of microwave-assisted laser-induced plasma was carried out. In this study, a luminous plasma was induced by a second harmonic Nd:YAG laser (532 nm, 8 ns, 5 J) on calcium oxide (Ca_2O_3) pellet at low pressure (5 Torr) of Ar gas. The luminous plasma was enhanced by intensified microwaves (MWs) generated from a magnetron. The image of plasma induced by laser has different features from the plasma induced by the laser with MWs. For the case of laser only, the plasma lifetime is approximately 50 μs, the plasma size at 5 μs is approximately 3 mm, and the neutral Ca emission is much brighter than the ionic Ca emission. When the MWs was introduced into the laser-induced plasma, the plasma emission was enhanced, namely the plasma lifetime was elongated of approximately 8 times, the plasma size at 5 μs was enlarged of approximately 3 times, and the ionic Ca emission is much brighter than the case of neutral Ca emission, which indicated that the plasma temperature is much higher than the case of laser only. Therefore, the plasma induced by the microwave-assisted laser can be effectively used for the analysis of elemental composition in materials with high sensitivity.

1. Introduction

Laser-induced plasma spectroscopy (LIPS), which is commonly known laser-induced breakdown spectroscopy (LIBS), has become a popular method for qualitative and quantitative elemental analysis in various kinds of materials including solids, liquids, and gases [1-4]. In standard LIBS, a pulse neodymium yttrium aluminum garnet (Nd:YAG) laser is commonly used as an optical source for generation of a luminous plasma on/in materials [5-6]. Compared to other atomic analytical methods such as X-ray fluorescence (XRF) and inductively coupled plasma optical emission spectroscopy (ICP-OES), the LIBS is much superior due to less sample pretreatment, its capability for rapid and remote analysis, and much lower price of experimental devices. Recently, this method has been employed in various fields including environment [7-8], metal industries [9-10], and nuclear fields [11]. However, it is known that this method has less sensitivity for the quantitative analysis of materials.

Various methods have been suggested to improve the sensitivity of analytical detection. Some scientists used double pulse LIBS (DP-LIBS), in which two pulse lasers with interpulse delay in the range of nanosecond to microsecond are used to induce a luminous plasma [12-13]. Khumaeni et al. developed new method of microwave-assisted LIBS (MA-LIBS) coupled by antenna to enhance the laser plasma emission [14-15]. In this method, microwaves (MWs) generated by a magnetron were introduced into the laser-induced plasma. The results revealed that the MWs can be effectively used to increase the plasma temperature and extend the plasma lifetime, resulting in enhancement of laser-induced plasma emission. Enhancement factors of emission intensity from tens to hundreds were achieved.

In this study, time-resolved imaging of a luminous plasma induced by a pulse laser coupled with an intensified MWs was carried out. The microwave-assisted laser-induced plasma was induced on pelletized calcium oxide (Ca_2O_3) used as a sample target. The results show that the plasma emission can be readily enhanced with enlarged plasma size, increased plasma temperature, and elongated plasma lifetime [15]. It should be noted the present study of time-resolved imaging of luminous plasma proved the assumption of mechanism of plasma enhancement generation by using MWs described in the previous report [15]. Furthermore, this study also displayed the dynamics of neutral and ionic atoms in the laser plasma with- and without MWs.

2. Experimental Procedure

The experimental arrangement in this research was shown in Fig. 1(a). The sample employed in this research was Ca_2O_3 pellet. The sample was placed into a circular metal chamber, on which the fine gold mesh (lattice constant of 80 μm and wire diameter of 30 μm) was attached in the inner chamber. During the experiment, Ar gas was flowed in the chamber at a reduced pressure (5 Torr).

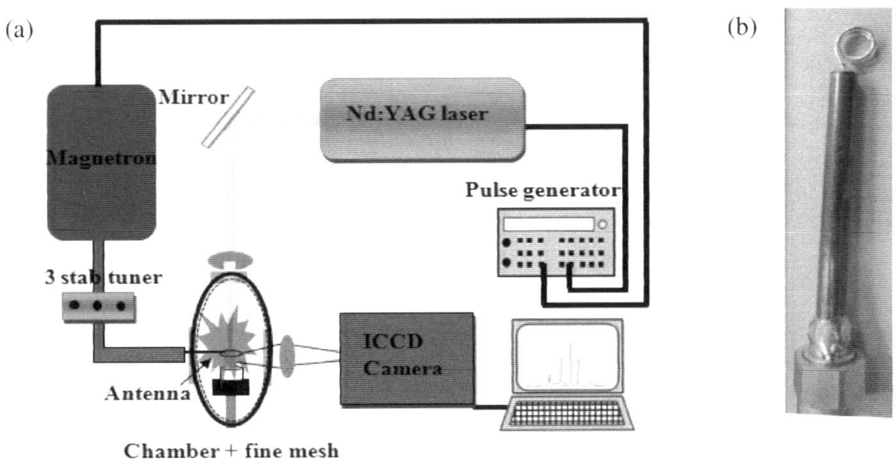

Fig. 1 (a) Experimental setup used in this research, (b) a loop antenna used to transfer the MWs from magnetron to laser plasma region.

A second-harmonic Nd:YAG laser (Quanta-Ray, Spectra Physics, 532 nm, 8 ns) was used as an optical source. To induce a luminous plasma, the laser was directly focused on a material target through a quartz window with a focal length of 200 nm. The laser plasma was enhanced by MWs generated by using a magnetron at a frequency of 2.45 GHz (MUEGGE, 400 W, 1 ms) via a loop antenna with a diameter of 3 mm [Fig. 1(b)]. the Nd:YAG laser was operated 10 μs after the MWs generation. The luminous plasma emission was imaged by an intensified charge coupled device (ICCD) camera (Andor, iStar) through a quartz lens (f = 100 mm).

3. Results and Discussion

The effect of MWs in the enhancement of laser-induced plasma emission obtained from Ca_2O_3 pellet has been studied as reported in the previous work [15]. The plasma lifetime extended and the plasma temperature increased when the MWs was introduced into the laser plasma, resulting in enhancement of laser plasma emission and thus, increasing in signal to noise ratio (S/N) of ionic Ca (II) at 393.3 nm and 396.8 nm and neutral Ca (I) 422.6 nm emission intensities as shown in Fig. 3 in the previous paper [15].

In this study, the effect of the microwave to the enhancement of emission intensity was examined by studying the dynamics of plasma emission in Ca_2O_3 pellet using LIBS and MA-LIBS. To this end, time-resolved emission images of plasma induced by pulse laser without– (LIBS) and with MWs (MA-LIBS) in Ca_2O_3 pellet sample were recorded as shown in Figs. 2 and 3, respectively. Ionic Ca II at 393.3 nm and neutral Ca I at 422.6 nm emission lines were used as a representative of Ca emission spectra. Band-pass interference filters with center wavelengths of 394 nm for Ca ions (Ca II) and 420 nm for Ca neutral (Ca I) were used to study the dynamics of Ca ions (Ca II at 393.3 nm and 396.8 nm) and Ca neutral (Ca I 422.6 nm) in the plasma region, respectively. Both filters have bandwidth of 10 nm.

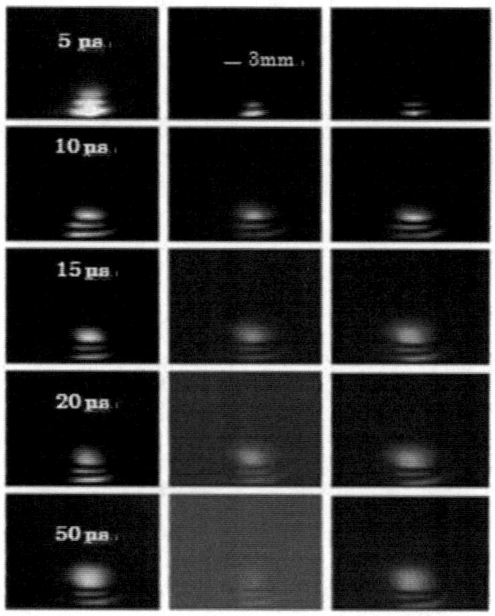

(a) (b) (c)

Fig. 2 Plasma emission images obtained from the Ca_2O_3 pellet sample using LIBS (a) without band-pass filter, (b) with band-pass filter of 394 nm, and (c) with band-pass filter of 420 nm.

Figure 2 represents the plasma emission images in LIBS without the use of band-pass filter [Fig. 2(a)], by using a 394 nm filter [Fig. 2(b)], and by using a 420 nm filter [Fig. 2(c)]. The lifetime of plasma emission was approximately 50 μs. Two black lines in the plasma image are loop antenna used to transfer the MWs from magnetron to the laser plasma region; the plasma emission most often occurred in the inner of loop antenna and therefore the wire of the antenna disturbs the image of the plasma to the ICCD. A laser plasma with a small diameter of approximately 3 mm was observed at 5 μs delay time. The brightest plasma emission clearly occurred at 5 μs. By increasing the delay time, the plasma expands away from the target surface and enlarges the diameter of approximately 5 mm. It should be noticed that the plasma emission lifetime for the Ca ions [Fig. 2(b)] is much shorter (around 20 μs) than the case of Ca neutral of around 50 μs [Fig. 2(c)]. Furthermore, it can be observed that the Ca neutral emission is slightly brighter than the Ca ionic emission. These results indicate that the plasma temperature is not so high enough to ionize the Ca atoms as confirmed by the results shown in Fig. 5 of previous paper [15]. When the temperature is not so high enough, the atoms are excited to the excitation level of neutral atom and it is hard to be ionized due to lack of energy for excitation and ionization.

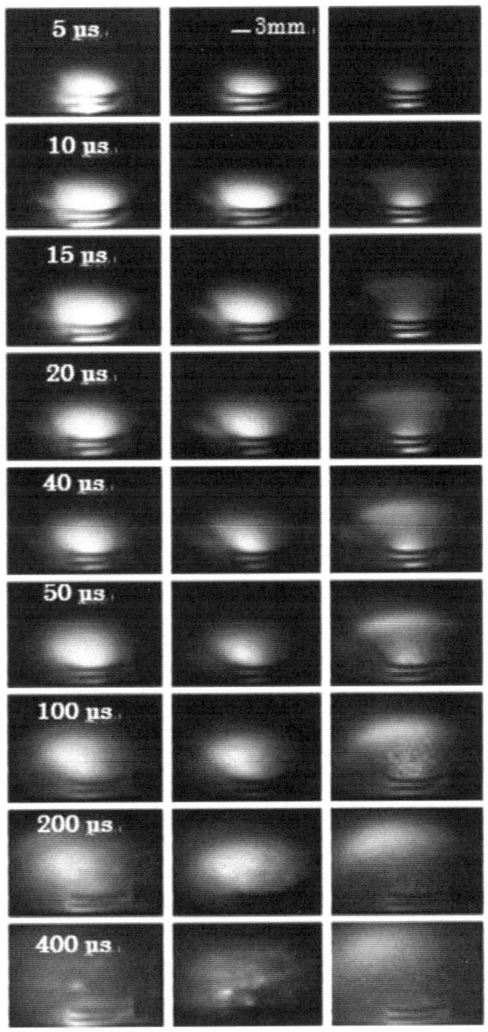

(a) (b) (c)

Fig. 3 Plasma emission images obtained from the Ca_2O_3 pellet sample using MA-LIBS (a) without band-pass filter, (b) with band-pass filter of 394 nm, and (c) with band-pass filter of 420 nm.

In contrary to the laser plasma in LIBS above, the luminous plasma induced by a pulse laser with MWs (MA-LIBS) shows different feature of plasma emission as displayed in Fig. 3. Very high-intense plasma emission with a diameter of approximately 8 mm at 5 μs delay time was clearly observed. With increasing the time, the plasma diameter was enlarged of approximately 15 mm at 200 μs delay time. The intensity of plasma emission slowly decreases with increasing the delay time. It should be noticed that the plasma emission for the MA-LIBS without any filter as shown in Fig. 3(a) was enlarged of approximately 3 times compared to the case of LIBS as displayed in Fig. 2(a). This result revealed that the microwave plays an important role on the enlargement of plasma diameter. It is assumed that in the case of microwave-assisted laser plasma, the plasma consists of two regions, namely main plasma (inner layer) and outer layer plasma. When the laser is focused on the sample target, main plasma with high electron density is induced. Once the microwave is introduced into the laser plasma, the electromagnetic field induced by the microwave is effectively absorbed by the laser plasma especially at outer layer of plasma region, which has lower electron density. The microwave then accelerates the electron in the plasma, increasing its kinetic energy. Along with the increment of kinetic energy, the number of collisions between electrons and constituents increases, which affects the enlargement of plasma diameter and the increase of the plasma temperature. It should be noticed that the plasma lifetime in MA-LIBS extended very long lifetime (approximately 8 times longer than the LIBS case) due to long MW duration.

It is also seen that from initial to later stages, the plasma emission of Ca ions [Fig. 3(b)] is much stronger compared to the case of Ca neutral emission [Fig. 3(c)]. It should be pointed out that stronger emission intensity of Ca ions indicates a high degree of ionization of calcium atoms in the microwave-assisted LIBS. By observing the plasma images induced in LIBS with and without microwave, it can be revealed that the microwave plays an important role on the enlargement of plasma diameter, the extension of plasma lifetime and the increment of plasma temperature, which means that the excitation process of atoms can effectively take place in the plasma region both in the inner and outer layers of microwave-assisted laser plasma region, allowing to the improvement of emission intensity in MA-LIBS. However, the experiment on this study was conducted at low pressure environment, which is due to the limitation of the MWs power at 400 W. By higher power of the MWs, the experiment has high possibility to be carried out at atmospheric pressure, which enables one to study rapid and insitu analysis of materials.

4. Conclusions

Time-resolved imaging of luminous plasma induced on Ca_2O_3 pellet by laser without (LIBS) and with MWs (MA-LIBS) has been conducted. The plasma characteristics are different between LIBS and MA-LIBS. In case of LIBS, the plasma lifetime is approximately 50 μs, the plasma size at 5 μs is approximately 3 mm, and the neutral Ca emission is much brighter than the ionic Ca emission. This indicated that the plasma temperature in LIBS is quiet low because the degree of ionization is low, which is shown by brighter emission of neutral Ca, and therefore, the atomization and excitation of atoms cannot effectively take place. When the MWs were introduced into the laser plasma (MA-LIBS), the plasma lifetime was extended of approximately 8 times, the plasma size was enlarged 3 times longer at 5 μs than the case laser plasma without MWs (LIBS). Also, the ionic Ca emission is much brighter than the neutral Ca emission. The stronger emission intensity of Ca ions indicates a high degree of ionization of calcium atoms in the microwave-assisted LIBS. The results certified that MWs have significant effect in the enhancement of plasma emission. Therefore, LIBS assisted by MWs (MA-LIBS) can be employed to perform analysis of elemental composition in materials with high sensitivity. Furthermore improvement of present method by using higher power of MWs will be made in the near future in order to realize the analysis at atmospheric pressure, which is enable one to conduct rapid and in-situ analysis of materials.

References

1. Cremers D A, Radziemski L J 2006 *Handbook of laser-induced breakdown spectroscopy* (Chichester: Wiley)
2. Miziolek A W, Palleschi V, Schechter I 2006 *Laser Induced Breakdown Spectroscopy* (Cambridge: Cambridge University Press)

3. Tognoni E, Palleschi V, Corsi M, Cristoforetti G 2002 *Spectrochim. Acta Part B* **57** 1115
4. Meinhardt C, Sturm V, Fleige R, Fricke-Begemann C, Noll R 2016 *Spectrochim. Acta Part B* **123** 171
5. Zhang Y, Xiong G, Li S, Dong Z, Buckley S G, Tse S D 2013 Combust. Flame **160** 725
6. Sturm V, Fleige R, Kanter d M, Leitner R, Pilz K, Fischer D, Hubmer G, Noll R 2014 Analyt. Chemis. **86** 9687
7. Wainner R T, Harmon R S, Miziolek A W, McNesby K L, French P D 2001 *Spectrochim. Acta Part B* **56** 777
8. Martin M Z, Labbé N, André N, Harris R, Ebinger M, Wullschleger S D, Vass A A 2007 *Spectrochim. Acta Part B* **62** 1426
9. Noll R, Bette H, Brysch A, Kraushaar M, Mönch I, Peter L, Sturm V 2001 *Spectrochim. Acta Part B* **56** 637
10. Vieitez M O, Hedberg J, Launila O, Berg L 2005 *Spectrochim. Acta Part B* **60** 920
11. Whitehouse A I, Young J, Botheroyd I M, Lawson S, Evans C P, Wright J 2001 *Spectrochim. Acta Part B* **56** 821
12. Uebbing J, Brust J, Sdorra W, Leis F, Niemax K 1991 *Appl. Spectrosc.* **45** 1419
13. Babushok V I, DeLucia F C Jr., Gottfried J L, Munson C A, Miziolek A W 2006 *Spectrochim. Acta Part B* **61** 999
14. Khumaeni A, Tampo M, Akaoka K, Miyabe M, Wakaida I 2013 *Opt. Express* **21** 29768
15. Khumaeni A, Akaoka K, Miyabe M, Wakaida I 2016 *Front. Phys.* **11** 114209

IOP Conf. Series: Journal of Physics: Conf. Series **820** (2017) 012005 doi:10.1088/1742-6596/820/1/012005

Automatic detection of ischemic stroke based on scaling exponent electroencephalogram using extreme learning machine

H A Adhi[1], S K Wijaya[2], Prawito[2], C Badri[1] and M Rezal[1]
[1]Biomedical Engineering, Master Program, Universitas Indonesia, Jakarta, Indonesia
[2] Faculty of Mathematics and Natural Sciences, Universitas Indonesia, West Java, Indonesia

E-mail: habbi.ananto@ui.ac.id, skwijaya@sci.ui.ac.id, prawito@sci.ui.ac.id, cholidbadri@yahoo.co.id and muhammad.rezal@ui.ac.id

Abstract. Stroke is one of cerebrovascular diseases caused by the obstruction of blood flow to the brain. Stroke becomes the leading cause of death in Indonesia and the second in the world. Stroke also causes of the disability. Ischemic stroke accounts for most of all stroke cases. Obstruction of blood flow can cause tissue damage which results the electrical changes in the brain that can be observed through the electroencephalogram (EEG). In this study, we presented the results of automatic detection of ischemic stroke and normal subjects based on the scaling exponent EEG obtained through detrended fluctuation analysis (DFA) using extreme learning machine (ELM) as the classifier. The signal processing was performed with 18 channels of EEG in the range of 0-30 Hz. Scaling exponents of the subjects were used as the input for ELM to classify the ischemic stroke. The performance of detection was observed by the value of accuracy, sensitivity and specificity. The result showed, performance of the proposed method to classify the ischemic stroke was 84 % for accuracy, 82 % for sensitivity and 87 % for specificity with 120 hidden neurons and sine as the activation function of ELM.

1. Introduction

Stroke is one of cerebrovascular diseases which are caused by obstruction of blood flow to the brain [1]. The obstruction causes brain tissue damage. Based on data from WHO in 2012, stroke is the first leading cause of death in Indonesia and the second leading cause in the world [2]. Cardiovascular disease, stroke, and diabetes had become the biggest cause of disability in Indonesia [3] and in some developed countries. Ischemic stroke accounts for most of all stroke cases, up to 87% [4]. Computed Tomography (CT)-Scan was often used to support the diagnosis of stroke. While in ischemic stroke has been established that Diffused / Perfusion-Weighted (D/P-W) MRI as the gold standard in the examination of ischemic stroke [5]. D/P-W MRI is expensive and has limited availability [6]. However, CT scan or MRI, included D/P-W MRI, may take several hours to obtain examination reports while the initial aid is quite limited period for stroke patients [7].

Electroencephalography is a method to record electrical activity of the brain. Nowadays, electroencephalography widely used for evaluation of seizures and epilepsy [8]. In ischemic stroke, the role of electroencephalography as an indicator of disturbed function (locally, regionally, or diffusely) in the event of stroke has not always been duly appreciated [9] especially when CT-Scan and MRI is available. As outlined before that CT-Scan and MRI has limited availability, many research tried to optimize electroencephalography in ischemic stroke detection. Electroencephalography offers a cheaper cost, widely available [10], continuous monitoring [11] and good temporal resolution [12]. Some of the advantages are the basis for optimizing electroencephalography to detect acute ischemic stroke [13]. Tissue damage in ischemic stroke patients causes the changes in electrical activity of brain. Those changes can be known through EEG power [14], variability [15, 16] and functional connectivity of the cortex [17]. Detrended fluctuations analysis (DFA) is one of method for analysing variability of EEG. The result of the DFA is the fluctuation of the EEG indicated by the scaling exponents.

Automatic detection is a technique to analyse signal quantitatively using a computer. Automatic detection offers great promise to assist in interpretation of EEG studies [18]. In automatic detection, there are two main considerations they are feature extraction techniques and classification techniques. Machine learning method is commonly applied in classification problem. Machine learning method based on artificial neural networks were applied in stroke classification problems such as extreme learning machine (ELM) [19] and 1D convolutional neural network (1DCNN) [20]. In this study, we presented the result of automatic detection of ischemic stroke based on the scaling exponent of EEG using ELM.

2. Material and methods

2.1. Data collection

This study was conducted and approved by the Ethics Committees of National Centre for Brain Hospital (RS PON), Jakarta. Until this study was conducted we obtain data from 31 ischemic stroke patients (age 57,6 ± 7,9 years) and 30 normal subjects (age 54,2 ± 8,1 years). The number of normal subjects selected to balance with the amount of ischemic stroke patients data. The EEG recording was conducted for 30 minutes with 32 channels and the electrode were placed with 10-20 international system. The recording was performed using Biologic System and Xltek Netlink EEG 32U with sampling frequency is 256 and 512 Hz and stored in a file in the format of edf (European Data Format).

2.2. Proposed method

The proposed method shows in figure 1. It consists of two stages namely training and testing stages. The method starts from edf file acquisition, to be read as the EEG data. Then, it is followed by the pre-processing; selecting 18 channels of EEG (FP1, F3, F7, C3, T3, P3, T5, T1, O1, FP2, F4, F8, C4, T4, P4, T6, T2 and O2) and bandpass filter the signal in the main frequencies of interest between 0-30 Hz. Channels are selected by the availability of its pair in both hemispheres. The purpose of feature extraction is to get the features of EEG, there are 36 scaling exponents obtained using DFA, two moments and two normalized moments. These features are used as input for ELM algorithm. As outlined before, system divided into two stages there are training and testing. Training stage is intended to get the optimum parameters (input weights, bias hidden neurons and output weights) for the classification. Meanwhile, testing stage is intended to validate the performance of ELM, with parameter

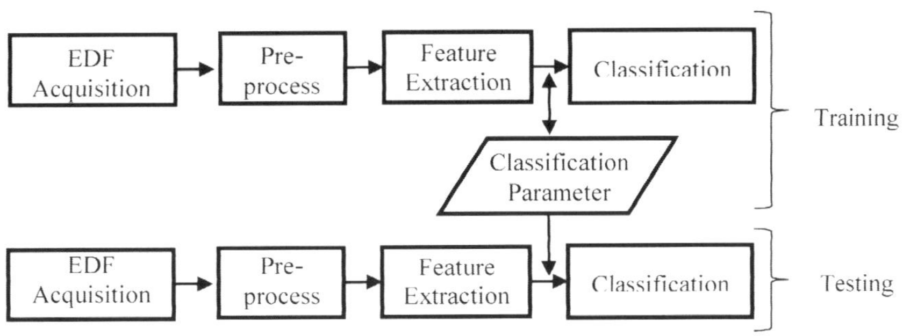

Figure 1. Block diagram of proposed method

was obtained from training stage, as a classifier. Performance of the proposed method were performed using the K-fold cross validation where $K = 10$. Performance of detection was evaluated based on accuracy, sensitivity and specificity.

2.3. Detrended fluctuations analysis

DFA is one of method to analyse the variability of a biomedical signal [21]. DFA is represented by two scaling exponent (α_1 and α_2) [15]. Scaling exponent is used as an indicator of EEG fluctuations that obtained from the slope of a line formed on plots of ln ($F(k)$) versus ln (k). In the spatial domain, the moments of the distributions of α_1 and α_2 summarize the mean and variability across the scalp[15]. Application of DFA with normalize moment follows:

i) Divide the EEG signal, $x(t)$, into B segment with k data point where $1 \leq t \leq T$ and T is the number of data points. Floor is function to round the value of (T/B) to nearest integer less than or equal. k obtained using equation (1).

$$k = floor\ (T/B) \tag{1}$$

ii) Get the trend of signal $x(t)$ on the segment b, $\bar{x}_b(t)$, equation (2). Linear-fit is function to get linear line with one independent and one dependent variable from a group of data point in segment b.

$$\bar{x}_b(t) = linear - fit\ [x(t)]\ ; \quad (b\text{-}1)\ k < t < bk \tag{2}$$

iii) Calculate the average fluctuation $x(t)$ using equation (3).

$$F^2(k) = \frac{1}{kB}\sum_{t=1}^{kB}[x(t) - \bar{x}_b(t)]^2 \tag{3}$$

iv) Get the scaling exponent using equation (4), where i is the order of the scaling exponent, q is a positive integer and N is the number of EEG channel. Then, calculate the moment of the scaling exponent, $G^{(i)}$, using equation (5).

$$\ln(F(k)) \propto \alpha \ln(k) \tag{4}$$

$$G_q^{(i)} = \frac{1}{N}\sum_{j=1}^{N}\alpha_{i,j}^q \tag{5}$$

v) Calculate normalized moments, $M_q^{(i)}$, using equation (6).

$$M_q^{(i)} = \frac{G_q^{(i)}}{\left(G_1^{(i)}\right)^q} \tag{6}$$

2.4. Extreme learning machine

Extreme Learning Machine (ELM) was propounded for single layer hidden feedforward neural networks (SLFNs). ELM has good generalization ability, quick to perform learning and a simple structure [22]. ELM architecture consists of input layer, hidden layer and output layer. Weights input (w_{nk}) connecting n-th node in the input layer and the node-k in the hidden layer. While the weights output (β_k) connecting node-k in the hidden layer and output layer. For each sample of training data (x_i, t_i) where x_i is the value at the input node and t_i is the expected target of a sample. Activation function, $g(x)$, in the standard SLFNs with K hidden node is modelled in equation (7), where b_j is a bias in the k-th hidden neuron. Equation (8) is used to obtain the target value where the input weights and biases hidden neurons randomly determined. Output weights, β, modelled on the equation (10), $H^\dagger$ is the Moore-Penrose's generalized inverse matrix of the matrix H.

In artificial neural network, a parameter in the hidden layer neurons to calculate the value from input layer is called activation function. Thus *g(x)* on ELM will be recalculated using this activation function to pass the value to the output layer. Some activation functions which are often used in the artificial neural network are sigmoid, sine and radial basis. The activation function and the number of neurons in the hidden layer will affect the performance of ELM.

$$\sum_{j=1}^{K} \beta_j g\left(w_{jk} \cdot x_n + b_j\right) = t_n \tag{7}$$

$$H\beta = T \tag{8}$$

where:

$$H(w_1, \dots, w_K, b_i, \dots, b_K, x_1, \dots, x_M) = \begin{bmatrix} g(w_1 \cdot x_1 + b_1) & \cdots & g(w_K \cdot x_1 + b_K) \\ \vdots & \cdots & \vdots \\ g(w_1 \cdot x_M + b_1) & \cdots & g(w_K \cdot x_M + b_K) \end{bmatrix}; \tag{9}$$

$$\boldsymbol{\beta} = \begin{bmatrix} \beta_1^T \\ \beta_2^T \\ \vdots \\ \beta_K^T \end{bmatrix}_{K \times l}, \boldsymbol{T} = \begin{bmatrix} t_1^T \\ t_2^T \\ \vdots \\ t_K^T \end{bmatrix}_{M \times l}$$

$$\beta = H^{\dagger} T \tag{10}$$

2.5. *Evaluation Matrices*

To evaluate the performance of detection, we used the value of accuracy, sensitivity and specificity [23]. Accuracy or success rate of detection is obtained using equation (11), where false positive (FP) is the number the illness subject detected as the normal subjects, false negative (FN) is the number normal subjects detected as the illness subject. Sensitivity or true positive (TP) rate is used to measure the ability of detecting the illness subjects as correctly the illness subjects. Sensitivity is obtained using equation (12). Specificity or true negative (TN) rate is used to measure the ability of detecting the normal subjects as correctly the normal subjects. Specificity was obtained using the equation (13).

$$Accuracy = \frac{TP+TN}{TP+TN+FP+FN} \tag{11}$$

$$Sensitivity = \frac{TP}{TP+FN} \tag{12}$$

$$Specificity = \frac{TN}{TN+FP} \tag{13}$$

3. Experiment results and discussion

3.1. *Scaling exponent analysis*

In this study, we used two parameters k on DFA, because the data has two sampling frequency. The range $3 \le k \le 500$ was applied to data with sampling frequency of 256 Hz and $6 \le k \le 1000$ was applied to data with sampling frequency of 512 Hz. Two scaling exponents are obtained from the slope of a line formed on plots of ln $(F(k))$ versus ln (k) from two regions. First region, to obtained α_1, located between $1 < \ln(k) < 3.5$ for data with sampling frequency 512 Hz and $1 < \ln(k) < 3$ for data with sampling frequency 256 Hz. Second region, to obtained α_2, is located between $4.45 < \ln(k) < 5.5$ for both data.

Figure 2 shows the scatter plot of the moment of scaling exponents ($G_2^{(1)}$ and $G_2^{(2)}$) in ischemic stroke patients and normal subjects. The results showed that ischemic stroke patients generally have $G_2^{(1)}$ and

$G_2^{(2)}$ higher than the normal subject, table 1, and there is correspond to the previous study in [7]. Figure 2 shows that moment of scaling exponent cannot distinguish normal subject and ischemic stroke patient clearly. In [7], normalize moment of scaling exponent can distinguish ischemic stroke patients and normal subjects clearly and shows that ischemic stroke patient have lower normalize moment than normal subject. The result of normalize moment $M_2^{(1)}$and $M_2^{(2)}$ in this study, figure 3, shows that normalize moment still cannot be distinguish between normal subject and ischemic stroke patients clearly. Normal subjects have similar average normalize moment with ischemic stroke but ischemic stroke patients have higher value of standard deviation, table 1. The result also corresponded to the other previous study [24] which generally indicates the acute stroke thalamic ischemic patient has a higher EEG complexity than the normal subjects.

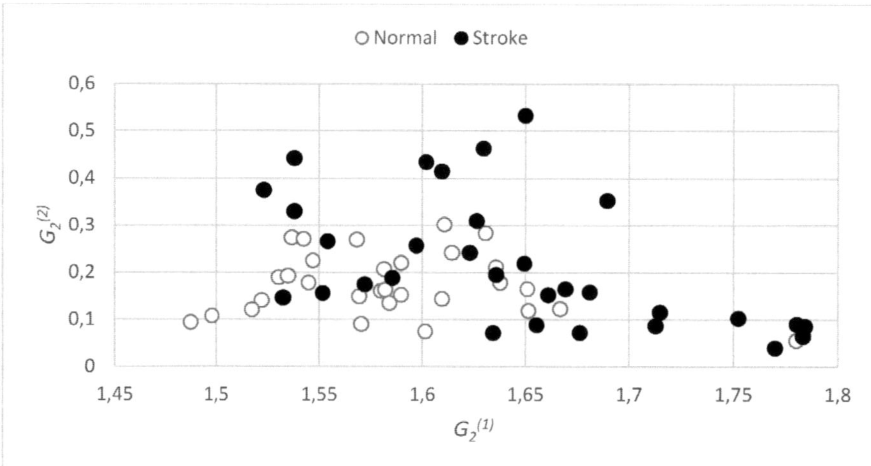

Figure 2. Moment ($G_2^{(1)}$and $G_2^{(2)}$) plot for ischemic stroke patients and normal subjects.

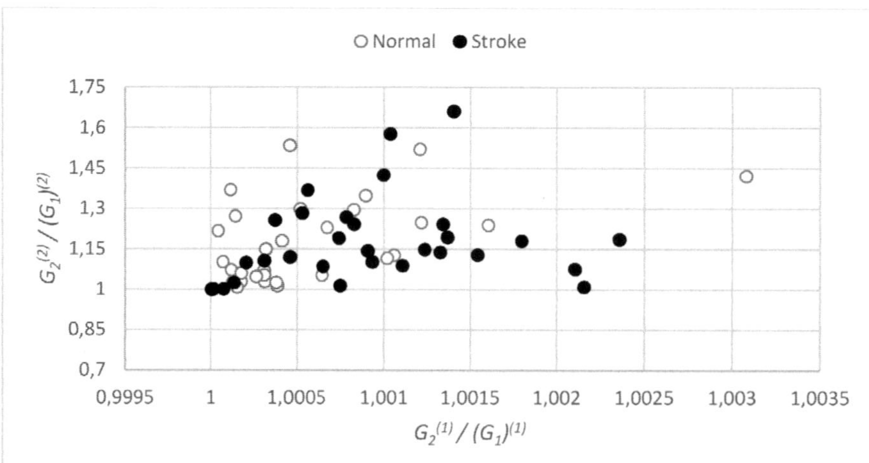

Figure 3. Normalize moment ($M_2^{(1)}$ and $M_2^{(2)}$) plot for ischemic stroke patients and normal subjects.

Table 1. Average moment and normalized moment for each scaling exponent.

Subject	$G_2^{(1)}$	$G_2^{(2)}$	$M_2^{(1)}$	$M_2^{(2)}$
Normal	1.59 ± 0.06	0.17 ± 0.07	1.00 ± 0.00	1.17 ± 0.15
Stroke	**1.64 ± 0.08**	**0.22 ± 0.14**	1.00 ± 0.00	1.17 ± 0.16

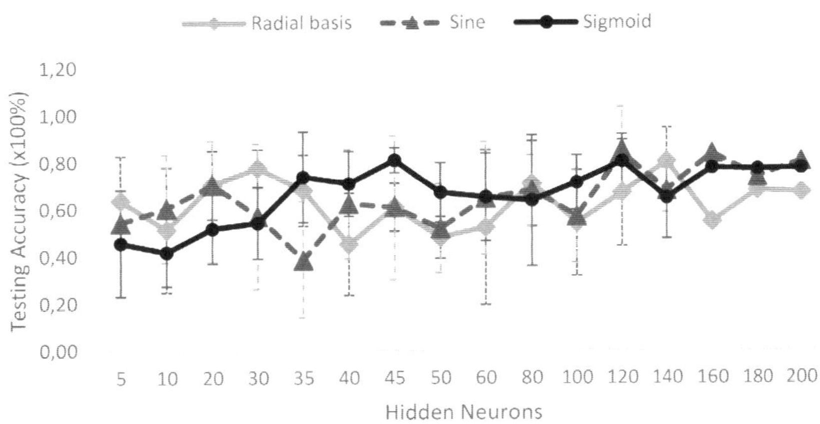

Figure 4. Accuracy of the system using scaling exponent features in testing stage.

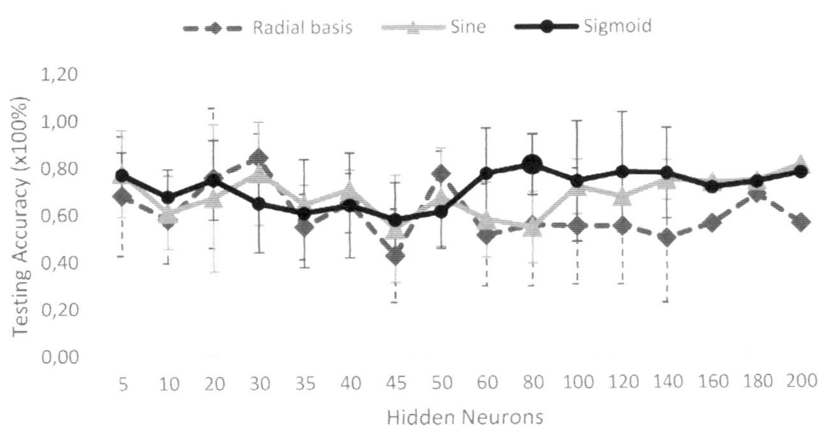

Figure 5. Accuracy of the system using combined features in testing stage.

3.2. Feature and performance analysis

In this study, we divided the features into two groups of features they are scaling exponent features and combined features. The group of scaling exponent features consists of 36 scaling exponents from 18 channels of EEG. Meanwhile, the combined features are the combination of $G_2^{(1)}$, $G_2^{(2)}$, $M_2^{(1)}$, $M_2^{(2)}$and scaling exponent features and will be consist 40 value for each subject. Experiments in those features was conducted using three parameters of activation function (sigmoid, sine and radial basis) and the number of neurons in the hidden layer were varied. The application of some parameter settings is intended to get the optimum settings in detection. Accuracy used as main parameter to measure performance of proposed method.

The accuracy of this experiments with some parameter settings, in best case, are presented in figure 4 and 5. The results obtained in testing stage shows that the addition of neurons in hidden layer was not always followed by the increase of ELM performance. Experiments using three activation function showed, no activation function is dominant over the others. It is showed that in scaling feature using sine activation function get the best accuracy, but in combined feature the best accuracy showed with sigmoid activation function.

Table 2. Best performance for each group of features and functions activation in testing stage.

	Scaling exponent feature			Combined feature		
Parameters	Sigmoid	Sine	Radial Basis	Sigmoid	Sine	Radial Basis
Accuracy	0.81± 0.05	**0.84 ± 0.12***	0.80± 0.14	0.81 ± 0.13	0.80 ± 0.08	0.77 ± 0.10
Sensitivity	0.70± 0.18	**0.82 ± 0.17**	0.68± 0.24	0.70 ± 0.30	0.88 ± 0.16	0.68± 0.24
Specificity	0.93 ± 0.15	**0.87± 0.30**	0.93± 0.15	0.93 ± 0.15	0.73 ± 0.28	0.87 ± 0.18

*Optimum parameter settings.

Table 3. Best performance for each optimum parameter in the training stage.

	Scaling Feature			Combined Feature		
Parameters	Sigmoid	Sine	Radial Basis	Sigmoid	Sine	Radial Basis
Accuracy	1	1	1	1	1	1
Sensitivity	1	1	1	1	1	1
Specificity	1	1	1	1	1	1

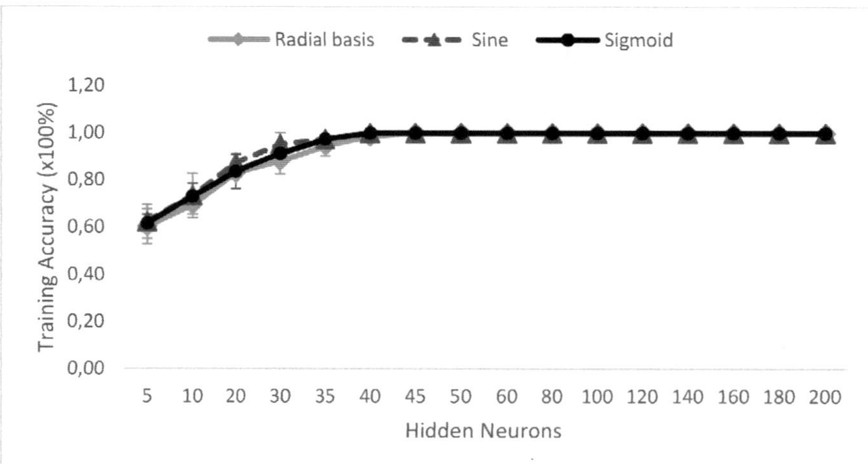

Figure 6. Accuracy using scaling exponent features in training stage.

Table 2 shows the summary of the best performance in testing stage obtained from several experiments in this study. The optimum parameters in this study was obtained with 84% as the highest accuracy using sine activation function, 120 neurons and scaling exponent features. These parameter settings gave the performance 82% for sensitivity and 87% for specificity in testing stage. The system performance indicates that the proposed method performed better to detect the normal subject than ischemic stroke patient. The reason for this situation might be because in this study there were several ischemic stroke patients in the mild ischemic stroke (NIHSS ≤ 4) thus more difficult to distinguish from normal subjects. Table 3 shows the summary of performance in the training stage by the same parameters with table 2. In the previous discussion, has been outlined that the addition of neurons is not always followed by the increase of accuracy in testing stage, but in this training stage, the addition of neurons was able to improve training accuracy as shown in figure 6. To reach 100 % accuracy in the training stage the ELM need minimum 45 neurons in the hidden layer. In the training stage, 100 % accuracy can be interpreted that the ELM was able to form a good classification model.

4. Conclusion

The results of this study shows EEG in the ischemic stroke subjects have a higher scaling exponent than the normal subjects. The best performance was achieved for the detection using sine activation function, 120 neurons and scaling exponent features as the parameters. Using these parameters, we obtained 84% for accuracy, 82% for sensitivity and 87% of specificity. In our future work, we will further explore other method of feature extraction and classifier to optimize and get better ischemic stroke detection system.

Acknowledgments

This study was supported by the Department of Education and Culture of Republic of Indonesia (0402/UN2.R12/HKP.05.00/2015) and Universitas Indonesia.

References

[1] Brass L M 1992 Stroke *Yale University School of Medicine Heart Book* ed Zaret B L, Moser M and Cohen L S (New York: Hearst Books) chapter 18 pp 215–233

[2] World Health Organization 2014 *The top 10 causes of death*

[3] World Health Organization 2012 *Indonesia: WHO statistical profile*

[4] American Stroke Association 2016 *Ischemic Strokes (Clots)*

[5] Birenbaum D, Bancroft LW and Felsberg G J 2011 Imaging in Acute Stroke *West J Emerg Med* **12** 67–76

[6] Grunwald I Q, Fassbender K and Wakhloo A K 2012 *How to set up an acute stroke service* (Berlin: Springer)

[7] Hwa R C and Ferree T C 2004 Stroke detection based on the scaling properties of human EEG *Phys A Stat Mech its Appl* **338** 246–254

[8] Acharya U R, Vinitha S S, Swapna G, Martis R J, Suri J S 2013 Automated EEG analysis of epilepsy: A review *Knowledge-Based Syst* **45** 147–165

[9] Niedermeyer E 2005 Cerebrovascular Disorders and EEG *Electroencephalography: Basic Principle, Clinical Applicationand Related fields,* vol 5 ed Niedermeyer E and Silva F L (Philadelphia: Lippincott Williams & Wilkins) chapter 17 pp 339–362

[10] Sheorajpanday R V A, Nagels G, Weeren A J T M, Surgeloose D De and Deyn P P De 2010 Additional value of quantitative EEG in acute anterior circulation syndrome of presumed ischemic origin *Clin Neurophysiol* **121** 1719–1725

[11] Schneider A L and Jordan K G 2005 Regional attenuation without delta (RAWOD): A distinctive EEG pattern that can aid in the diagnosis and management of severe acute ischemic stroke *Am J Electroneurodiagnostic Technol* **45** 102–117

[12] Rijsdijk M, Leijten F S S and Slooter A J C 2008 Continuous EEG monitoring in the Intensive Care Unit *Netherlands J Crit Care* **12** 157–162

[13] Wijaya S K, Badri C, Misbach J, Soemardi T P and Sutarmo V 2014 Optimization Electroencephalography (EEG) for Detecting Acute Ischemic Stroke: A Preliminary Study *Int. Conf. Biomed. Eng. Technol. Appl. 2014 (ICBETA 2014)* pp 165–169

[14] Foreman B and Claassen J 2012 Quantitative EEG for the detection of brain ischemia *Annual Update in Intensive Care and Emergency Medicine* ed Vincent J (Berlin: Springer) pp 746–758

[15] Ferree T C and Hwa R C 2005 Electrophysiological measures of acute cerebral ischaemia *Phys Med Biol* **50** 3927–3939

[16] Seely A J and Macklem P T 2004 Complex systems and the technology of variability analysis *Crit Care* **8** R367–R384

[17] Wu W, Jin Z, Qiu Y, Zhu Y, Li Y, Tong S 2009 Influences of bilateral ischemic stroke on the cortical synchronization *2009 Annu. Int. Conf. IEEE Eng. Med. Biol. Soc. IEEE* pp 2216–2219

[18] Hogan R E 2011 Automated EEG detection algorithms and clinical semiology in epilepsy: Importance of correlations *Epilepsy Behav* **22** S4–S6

[19] Rahma O N, Wijaya S K, Prawito and Badri C 2016 Electroencephalogram Analysis With Wavelet Transform and Neural Network As a Tool for Acute *IASTEM -The 9th Int. Conf. Medical, Biol. Pharm. Sci.* pp 108–113

[20] Giri E P, Fanany M I and Arymurthy A M 2016 Ischemic Stroke Identification Based on EEG and EOG using 1D Convolutional Neural Network and Batch Normalization

[21] Golińska A K 2012 Detrended Fluctuation Analysis (DFA) in biomedical signal processing: selected examples *Stud Logic, Gramm Rheotic* **29** 107–116

[22] Huang G-B, Zhu Q and Siew C 2006 Extreme learning machine: Theory and applications *Neurocomputing* **70** 489–501

[23] Castellino R A 2005 Computer aided detection (CAD): an overview *Cancer Imaging* **5** 17–19

[24] Liu S, Guo J, Meng J, Wang Z, Yao Y, Yang J, Qi H and Ming D 2016 Abnormal EEG Complexity and Functional Connectivity of Brain in Patients with Acute Thalamic Ischemic Stroke *Comput Math Methods Med* **2016** 1–9

The Effect of Rare Earth Dopants in Crystal Structure of Bi-2212 Superconductor

W.G. Suharta[1], IGA. Widagda[2], K. Putra[3], H. Suyanto[4]
Department of Physics, Faculty of Mathematics and Natural Sciences, Udayana University

E-mail: wgsuharta@gmail.com

Abstract. $Bi_2Sr_2CaCu_2O_{8+\partial}$ samples have been successfully synthesized by doping rare earth (RE) variations using wet-mixing method. Samples calcined at 600°C for 3 hours and sintered at 850°C for 10 hours. The purpose of research is to determine the effect of the RE dopant on the microscopic structure of BSCRECO superconductors. Therefore, the research was conducted characterization by X-ray diffraction (XRD), Fourier transform infrared spectroscopy (FTIR) and scanning electron microscope (SEM). Measurements with XRD could be carried out and crystal system of $Bi_2Sr_2CaCu_2O_{8+\partial}$ with rare earth (RE) dopants could be determined clearly. Generally, crystallization has occurred very well demonstrated by the diffraction peaks are sharp, which is dominated by the emergence of Bi-2212 phase. Search match results of XRD spectrum showed $Bi_2Sr_2CuO_x$ (2201) and Ca_2CuO_3 (21) as an impurity phase with small intensity. Also, that is showing volume fraction from 85 to 92% and orthorombic value for all samples from 5 to 7%. The effect of RE dopants resulted a shift angle 2θ and changes in the volume of the unit cells of each sample. The value of the unit cell volume of the largest to smallest is BS(CN)CO, BS(CNG)CO, BS(CNEG)CO, BS(CNE)CO, BS(CG)CO, BS(CEG)CO and BS(CE)CO. Measurement with FTIR showed the bending vibration absorption by CO_3^{2-} in the wavelength range between 1500 and 1520 cm^{-1}, vibration of M-O between 420 and 650 cm^{-1}, the complex formation of BSCCO in the wavelength range between 1690 and 1700 cm^{-1}. Measurement with SEM showed rod shape with particle size in hundreds nanometer.

Keywords : superconductor, wet-mixing, wavelength

1. Introduction

Superconductor of the Bismuth system (BSCCO) is one of the high-temperature superconductors as candidate in applications. Superconductivity symptoms of BSCCO system occur at Cu-O layer. While Sr-O and Bi-O layers, each are an insulator and semiconductor. Superconductor of BSCCO system hase the general formula $Bi_2Sr_2Ca_{n-1}Cu_nO_{2n+4}$ and has three phases associated with n = 1, 2, 3, namely: Bi-2201, Bi-2212 and Bi-2223. Phase Bi-2201 with a critical temperature of 30 K has one layer of Cu-O, phase Bi-2212 with a critical temperature of 80 K has two layers of Cu-O and phase Bi-2223 with a critical temperature of 110 K has three layers of Cu-O [1].

Provision of doping for growth of superconductor crystals have been affecting the properties of the superconductor material. The addition of Ag doping on Bi-2212 phases have been capable of producing the critical current density of 2750 A/mm^2 at a temperature of 4.2 K and zero field Tesla [2]. While the addition of Ag doping on Bi-2223 superconductor resulted in the critical current density (Jc) $3x10^4$ A/cm^2 at 77 K [3]. The addition of Mn on Bi-2212 superconductor produces a small pure

IOP Conf. Series: Journal of Physics: Conf. Series **820** (2017) 012006 doi:10.1088/1742-6596/820/1/012006

crystal and the sample does not melt [4]. The substitution of Nb on Bi/Pb-2223 superconductor has been studied by Marek Pekala et al [5], while the addition of Mo, Cd and Y have been also reported by Beables [6].

In this paper, we report the effect of RE dopants with one, two and three combinations on the crystal structure of $Bi_2Sr_2(RE_xCa_{1-x})Cu_3O_{8+\partial}$ superconductors. The contribution of this paper is the knowledge about the influence of RE dopant on the shift 2θ angle of XRD spectrum, changes of lattice parameter, volume fraction and unit cell volume of $Bi_2Sr_2(RE_xCa_{1-x})Cu_3O_{8+\partial}$ superconductors. We also report the infrared absorption and showed shape and particle size of the samples.

2. Experiment

Samples $Bi_2Sr_2(RE_xCa_{1-x})Cu_3O_{8+\partial}$ synthesized using a wet-mixing method [7] with one, two and three of RE combinations. The raw material using Bi_2O_3 powder compound (99.9%), $SrCO_3$ (99.9%), Nd_2O_3 (99.9%), Gd_2O_3 (99.9%), Eu_2O_3 (99.9%), $CaCO_3$ (99.9%) and CuO (99.9%). The each raw materials are dissolved with HNO_3 and stirred about 1 hour to obtain a clear solution. Then all solutions mixed into one and stirred to obtain a homogeneous solution. That solution heated at 100°C until crust. The crust calcined at 600°C for 3 hours and sintered at 850°C for 10 hours. Steps of synthesis process is also done by W.G. Suharta et.al [8]. In this research has been conducted characterization using XRD, FTIR and SEM.

3. Results and Discussion

Samples $Bi_2Sr_2(Ca_{0.5}RE)Cu_3O_{8+\partial}$ has been successfully synthesized by using wet-mixing methods and variations of RE, and then calcinated at 600°C for three hours and sintered at 850°C for ten hours. Samples that have been synthesized are $Bi_2Sr_2(Ca_{1-x}RE_x)Cu_3O_{8+\partial}$ with one, two and three combinations of RE. Sample $Bi_2Sr_2(Ca_{0.5}Nd_{0.5})Cu_3O_{8+\partial}$ with symbol BS(CN)CO, sample $Bi_2Sr_2(Ca_{0.5}Eu_{0.5})Cu_3O_{8+\partial}$ with symbol BS(CE)CO, sample $Bi_2Sr_2(Ca_{0.5}Gd_{0.5})Cu_3O_{8+\partial}$ with symbol BS(CG)CO, sample $Bi_2Sr_2(Ca_{0.5}Nd_{0.25}Eu_{0.25})Cu_3O_{8+\partial}$ with symbol BS(CNE)CO, sample $Bi_2Sr_2(Ca_{0.5}Gd_{0.25}Eu_{0.25})Cu_3O_{8+\partial}$ with symbol BS(CGE)CO, sample $Bi_2Sr_2(Ca_{0.5}Nd_{0.25}Gd_{0.25})Cu_3O_{8+\partial}$ with symbol BS(CNG)CO, sample $Bi_2Sr_2(Ca_{0.5}Nd_{0.166}Eu_{0.166}Gd_{0.166})Cu_3O_{8+\partial}$ with symbol BS(CNEG)CO.

The results of the XRD characterization of $Bi_2Sr_2(Ca_{0.5}RE)Cu_3O_{8+\partial}$ samples that was calcinated at 600°C for 3 hours and sintered at 850°C for 10 hours is shown in Fig. 1. In general, all of the samples have shown the occurrence of crystallization very well, which is characterized by a sharp spectrum. All samples showed the same pattern spectrum. On some peak also shows that there are still overlapped peaks. This happens because the sintering time was too short, so we need the addition of sintering time, so that the overlaps peaks can be separated as well.

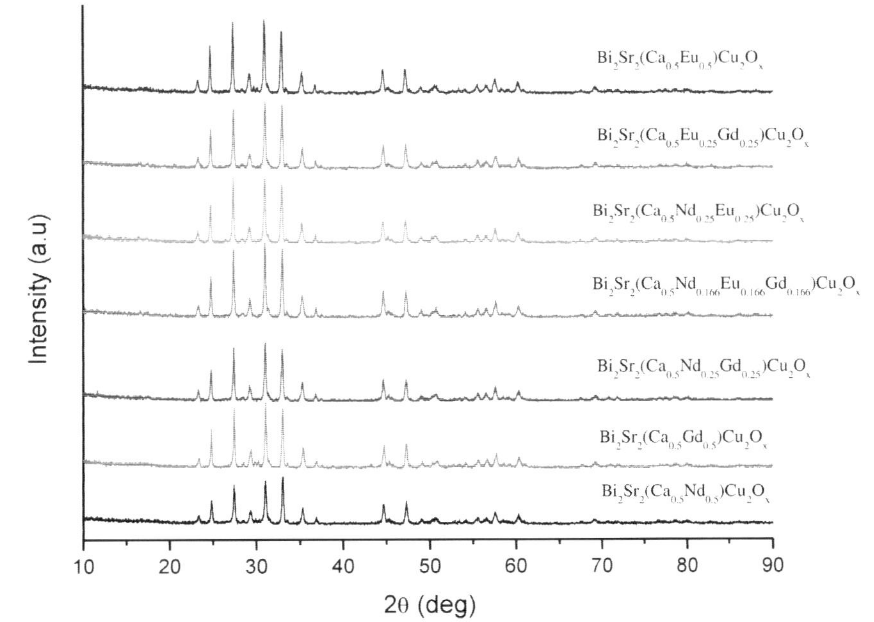

Figure 1. Spectrum of BS(CRE)CO samples with a variation of RE that was calcinated at 600°C for 3 hours and sintered at 850°C for 10 hours

The results of search match of BS(CRE)CO at 2θ from 5° to 30° are shown in Fig. 2. While the results of search match of BS(CRE)CO at 2θ from 30° to 60° are shown in Fig. 3.

In general, the appearance of the diffraction peaks dominated by the phase BS(CRE)CO, but in some peaks showed an impurity phase with a very small intensity. That impurity is 2201 phases ($Bi_2Sr_2CuO_{3+\delta}$) and 21 phases (Ca_2CuO_3).

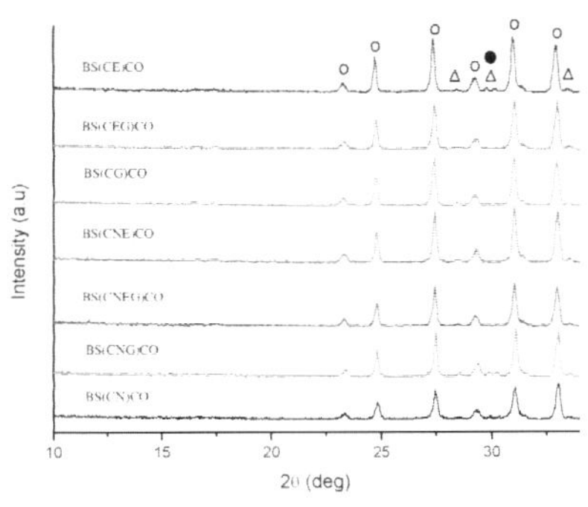

Figure 2. The results of search match of BS(CRE)CO samples at 2θ from 10° to 35°. Symbol o (Bi-2212 phase), Δ (Bi-2201 phase), ● (Bi-21 phase)

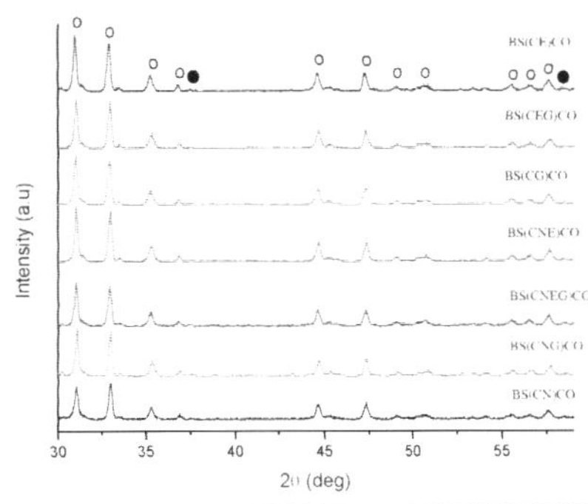

Figure 3. The results of search match of BS(CRE)CO samples at 2θ from 30° to 60°. Symbol o (Bi-2212 phase), Δ (Bi-2201 phase), ● (Bi-21 phase)

The addition of the element RE (rare earth) on the superconductor BSCCO (Bi-Sr-Ca-Cu-O) resulted the angle shift of X-ray diffraction peak. The angle shift of X-ray diffraction peaks at an angle 2θ between 20 and 26° are shown in Fig. 4. While the angle shift at 2θ from 26 to 30° are shown in Fig. 5. The shift is caused by differences in ionic radii of Ca and RE. The angle shift of samples from the largest angle to smallest angle is BS(CN)CO, BS(CNG)CO, BS(CNEG)CO, BS(CNE)CO, BS(CNE)CO, BS(CEG)CO and BS(CE)CO.

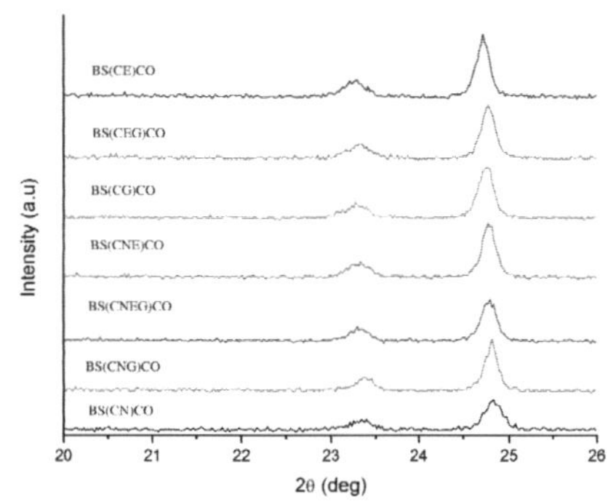

Figure 4. The angle shift at 2θ from 20 to 26°

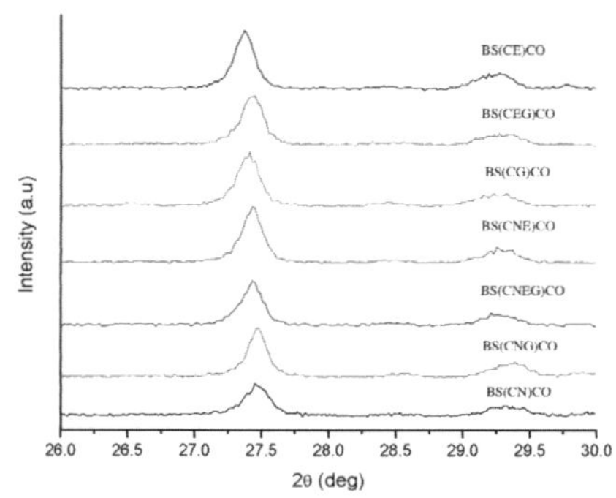

Figure 5. The angle shift at 2θ from 26 to 30°

Rietveld analysis is a method of refinement or matching of the diffraction pattern of models with the observed diffraction pattern using the least squares method. The degree of match between the diffraction pattern calculation results and observations were expressed by R_{WP} (weighted profile factor), Rp (profile factor), Re (expected weighted profile factor), S (goodness-of-fit), R_B (Bragg factor).

The value of lattice parameter, reliability and fraction volume of BS(CRE)CO samples are shown in Table 1. Refinement results was showed changes on lattice parameters and unit cell volume changes at each sample. The value of the lattice parameters (a, b, c) and the volume of the unit cell from the largest to smallest is BS(CN)CO, BS(CNG)CO, BS(CNEG)CO, BS(CNE)CO, BS(CG)CO, BS(CEG)CO and BS(CE)CO.

Table 1. The value of lattice parameter, releability and fraction volume of BS(CRE)CO samples

Sample	Lattice Parameter			Releability				FV
	a (Å)	b (Å)	c (Å)	R_p	R_{wp}	R_e	GoF	(%)
BS(CN)CO	5.4318(2)	5.4601(3)	30.6070(1)	17.90	23.05	17.20	1.79	87
BS(CNG)CO	5.4385(4)	5.4463(1)	30.5835(2)	17.90	23.37	17.25	1.84	85
BS(CNEG)CO	5.4377(1)	5.4398(4)	30.5585(2)	18.03	23.45	16.27	2.08	89
BS(CNE)CO	5.4348(6)	5.4408(8)	30.5502(5)	18.48	24.08	16.79	2.06	88
BS(CG)CO	5.4433(2)	5.4382(1)	30.5102(3)	18.19	23.89	16.66	2.06	92
BS(CEG)CO	5.4345(3)	5.4403(3)	30.5425(7)	17.94	23.40	16.56	1.99	90
BS(CE)CO	5.4341(7)	5.4422(4)	30.5144(2)	17.80	23.53	16.10	2.14	89

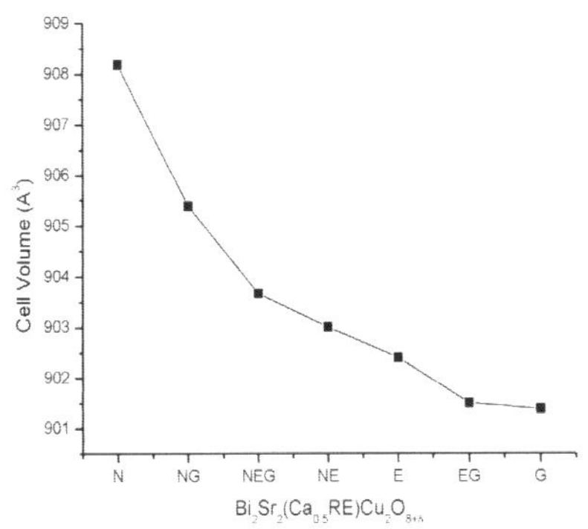

Figure 6. The value of the volume of unit cell BS(CRE)CO samples

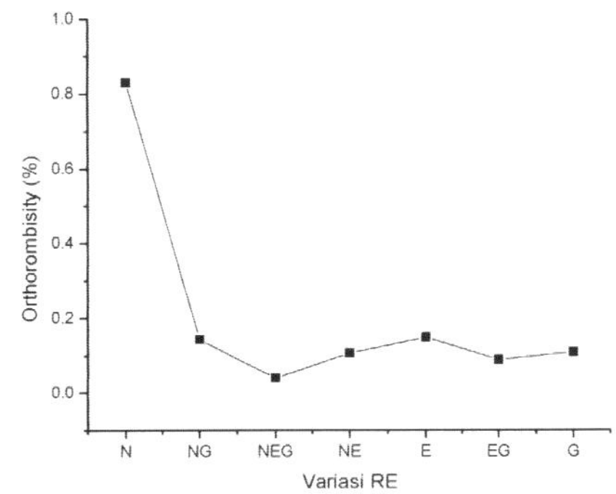

Figure 7. The value of the orthorombicity of BS(CRE)CO samples

Figure 8 shows the presence of a broad absorption peak at around 4000-450 cm^{-1} from the FTIR spectrum. Peak around 3012.81-3055.24 cm^{-1}, 2872.01-2883.58 cm^{-1}, 1695.43-1701.22 cm^{-1} and 1500.62-1514.12 cm^{-1} are corresponding to the stretching vibrations of carbon-hydrogen (C-H). The bending vibrations of carbon-hydrogen (C-H) is detected at 1413.82-1416.21 cm^{-1}. Peak around 1691.57-1701.22 cm^{-1} and 1287.34-1344.38 cm^{-1} are corresponding to the stretching vibrations of carbon -oxygen (C-O), while the absorption of CO_3^{2-} is presented around 1500.62-1514.12 cm^{-1}. Peak around 503.42-522.71 cm^{-1} and 426.27-439.77 cm^{-1} are corresponding to the stretching vibrations of metal-oxide (M-O) [9, 10, 11]. The absorption band of $SrCO_3$ is detected at 929.69-952.84 cm^{-1}, while peak around 3734.19-3747.69 cm^{-1} is corresponding to the stretching vibrations O-H. The value of the absorption bands in the FTIR spectra of BS(REC)CO sample is shown in Table 2.

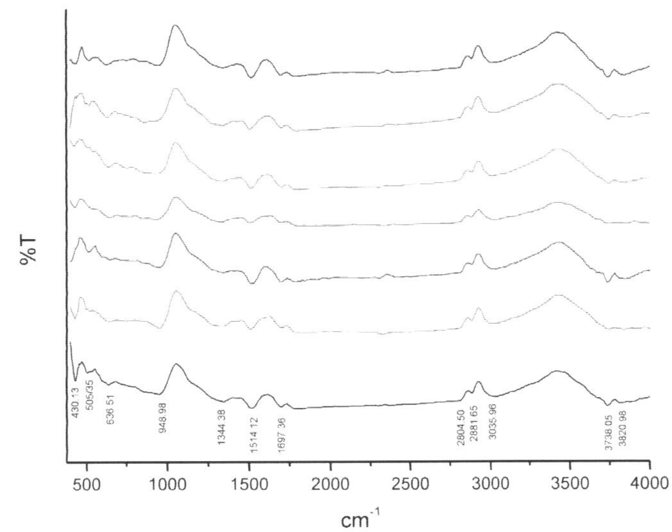

Figure 8. The results of FTIR characteritations of BS(CRE)CO samples

IOP Conf. Series: Journal of Physics: Conf. Series **820** (2017) 012006 doi:10.1088/1742-6596/820/1/012006

Table 2. The value of the absorption bands in the FTIR spectra of BS(REC)CO samples

Assignment (ions)	**EG**	**NEG**	**NE**	**NG**	**N**	**G**	**E**
M-O	432.05	431.98	-	428.20	426.27	428.43	439.77
M-O	509.21	506.32	511.14	-	503.42	520.78	522.71
M-O	648.08	636.51	644.22	646.15	640.37	632.65	626.87
C-H (streching)	1697.36	1695.43	1691.57	1697.36	1691.57	1701.22	1697.36
C-H (streching)	2881.65	2872.01	2881.32	2877.79	2883.58	2881.25	2881.65
C-H (streching)	3051.39	3055.24	3030.17	3026.31	3012.81	3037.89	-
C-H (bending)	1413.82	1416.21	-	-	-	1414.65	-
$SrCO_3$	950.91	952.84	947.05	929.69	927.76	932.12	943.19
C-O/-COO-	1697.36	1695.43	1691.57	1697.36	1691.57	1701.22	1697.36
C-O/-COO-	1288.45	1287.34	1296.16	1328.95	1319.31	1319.31	1344.38
CO_3^{2-}	1500.62	1510.26	1514.12	1506.41	1514.12	1506.41	1514.12
O-H	3739.97	3737.90	3734.19	3734.19	3739.97	3747.69	3736.12

The morphological characterization results of BS(CRE)CO samples by using SEM was shown in Fig. 9, 10, 11 and 12. These pictures show the rod-shaped particles and plate-shaped particles. Rod-shaped particles in accordance with the orthorhombic of BSCCO crystal structure. While a plate-shaped particles indicates that agglomeration process is already happening. This happens in all samples. The crystal size calculation was done using two ways, first by Scherrer calculation and the second using ImageJ program. The result of the calculation both ways are shown in Table 3. There are differences in the results of calculations for crystal size. Scherrer calculations utilizing the characterization results of XRD involving the whole sample information, while the calculation with ImageJ uses the results of SEM or TEM characterization involving information on the material surface.

Value crystal size between 92.81 and 202.85 nm is obtained from Scherrer calculation, while the value of crystal size of between 63.99 and 182.48 nm is obtained by utilizing the ImageJ program. Crystal size calculation results for all samples using both methods are shown in Table 3.

Figure 9. SEM image of BS(CNG)CO samples

Figure 10. SEM image of BS(CNE)CO samples

IOP Conf. Series: Journal of Physics: Conf. Series **820** (2017) 012006 doi:10.1088/1742-6596/820/1/012006

Figure 11. SEM image of BS(CEG)CO samples

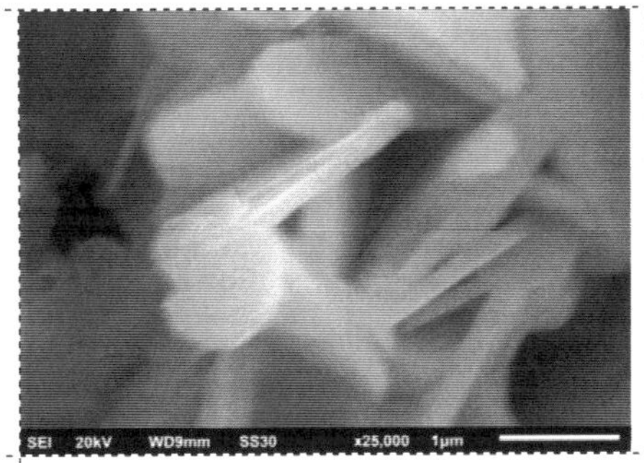

Figure 12. SEM image of BS(CNEG)CO samples

Table 3. The value of particle size of BS(CRE)CO samples

Sample	**Particle size (nm)**	
	Scherrer equation	**ImageJ**
BS(CN)CO	192.81	118.19
BS(CNG)CO	153.94	93.99
BS(CNEG)CO	196.01	124.03
BS(CG)CO	202.52	182.42
BS(CNE)CO	192.47	85.88
BS(CEG)CO	172.21	89.19
BS(CE)CO	183.54	93.75

Conclusions

XRD characterization results showed that the addition of RE doping on the superconducting BSCCO result shift of angle that caused by differences in ionic radii RE with Ca. Also, RE dopants resulting in a change in the unit cell volume. The volume of the unit cell from the largest to smallest is BS(CN)CO, BS(CNG)CO, BS(CNEG)CO, BS(CNE)CO, BS(CG)CO, BS(CEG)CO and BS(CE)CO. FTIR characterization results showed the characteristic vibrations of M-O, C-H (stretching), C-H (bending), C-O/-COO- and O-H. Crystal size was obtained in this study between 90-200 nm.

References

[1] Anderson P W 1998 *Science* 279-282
[2] Kim S C, Ha D W, Oh S S, Han I Y, Bruzek C E, Oh J G, Sohn H S 2007 *Cryogenics* **47** 437-441
[3] Zhang P X, Zhou L, Li C S, Shao H M, Zheng H L, Ji P, Wu X Z, Feng Y, Liu X H 2003 *Physica C* **392-396** 1011-1014
[4] Uthayakumar S, Srinivasan E, Jayavel R, Subramanian C, Endo T 2003 *Physica C* **392-396** 463-467

[5] Marek Pekala, Jan Mucha, Rudi Cloots, Jean-Francois Fagnard, Philippe Vanderbemden, Marcel Ausloos 2003 *Physica C* **387** 191-197
[6] Beables P V, Parberry J M 1997 *Phys. Stat. Sol* **162** 689
[7] Suharta W G, Suasmoro S, Pratapa S, Darminto 2013 *AIP Conference Proceeding*
[8] Suharta W G, Mugirahardjo H, Pratapa S, Darminto D, Suasmoro S 2013 *Journal of Superconductivity and Novel Magnetism* **26** 11 3209-3214
[9] Arshad M, Qureshi A H, Masud K and Qasi N K 2007 *Journal of Thermal Analysis and Calorimetry* **89** 595-600
[10] Singh H S, Rohitash Kumar 2016 *International Journal of Scientific & Engineering Research* **7** 1464-1467
[11] Stuart B H 2004 Infrared Spectroscopy: Fundamentals and Applications, John-Wiey & Sons.

Forecasting the Quarterly Production of Rice and Corn in the Philippines: A Time Series Analysis

Jackie D. Urrutia[1], John Lean B. Diaz[2] and Francis Leo T. Mingo[3]

[1]Director, Intellectual Property Management Office
Chief, Center for Statistical Studies, Institute for Data and Statistical Analysis
Office of the Vice President for Research, Extension, Planning and Development
Polytechnic University of the Philippines-Sta. Mesa, Manila, Philippines
[2]Polytechnic University of the Philippines – Parañaque Campus, Philippines
[2]Faculty Member, Polytechnic University of the Philippines – Commonwealth Branch

E-mail: jdurrutia@pup.edu.ph, jackieurrutia20@gmail.com, math_urrutia@yahoo.com, djlean1998@gmail.com, tententenstranger@gmail.com, franz_kyo@yahoo.com, fltmingo@pup.edu.ph

Abstract. Rice and corn are crops that are produced on most of the agricultural areas in the Philippine geographic territory. Gathered from the Philippine Statistics Authority (PSA), the quarterly volume of total production of the mentioned crops will be predicted using Seasonal Autoregressive Integrated Moving Average (SARIMA) Modelling through a Box-Jerkins method of forecasting. Investigators aims to forecast rice production and Corn production actual observed in the next 5 years of data, particularly in years mentioned. Obtaining models through SARIMA method, for rice production is SARIMA (2 1 8) (1 1 0)4 and for Corn production would be SARIMA (3 1 8) (0 1 1)4. Results suggest that two models have satisfactorily fitted their specified series and diagnostic checking for models' accuracy. The dynamically forecasted values of both series were following the trend pattern of the past values of their allotted observed values which were going on an upward direction indicating favourable outcome for the study. This may help people for better understanding the possible changes in values of the data series gathered and may even help the government for any reforms they can make for the upward mobility of production of crops, in agriculture and especially, economy of the country.

1. Introduction

The Philippines has been one of the world's largest suppliers in the global market since it has two season which could sustain the needs of the plants, forestry and fisheries. On that note, rice production gives off the supply for the demand of the local people and also makes good trading internationally.
Rice is one most produced of the crops because it is most needed by people for each day, and is mainly consumed as the main part of the Filipino meals and for almost three times a day, moreover, it sustains proteins and calories that are essential for the body. However, the harvested areas are still

very small compared to the consumption. This leads to more importation of rice products from other countries like Vietnam and Thailand making the Philippines the biggest importer in Asia. On the other hand, the Philippine production of rice increased for approximately 70% which came from the irrigated areas from 1995 to 2010. [1]

Corn also is one utmost massively produced crops in the Philippines and in this country is known as the substitute food whenever there is a shortage for the production of rice. Various classification of corn like yellow corn which is mostly being locally utilized for the livestock and poultry since it is most commonly primary source of food in the animal industry, white corn as the staple food for almost 20 percent of the population primarily in the southern regions of Visayas and Mindanao and makes a major ingredients for producing raw material and industrial products like starch oil and organic liquids, a mere corn having as most of use like the two mentioned and others. [2] From 1980, the Philippines produces for about 3 million metric tons of corn crops from almost 3.2 million hectares of land which surged up to approximately 5 million after the land tilled were increased for about 00.6 million hectares for ten years. In year 2000, the production reduced by 0.3 million metric tons of total corn crops harvested from the number of areas declined which was about 1.6% annually. [3] It was signposted in an article by Louise Maureen Simeon (2016) that the production of corn in the Philippines at present has been 7% less than the earlier forecast of 4.37 million metric tons of corn products, by the Philippine Statistics Authority in 2015. [4]

On that note, the researchers plan on forecasting the rice and corn production for the year 2016-2020. The aim of this study is to provide information for the government's use, and to those that are authorized to make decisions regarding the topics discussed to ensure the betterment of however they can improve the trading statuses of the crops mentioned in the study. This may even help the common people to attain information about the process regarding the production, the system of trading, and how the possible financial aspect is being assessed through the processes involved.

1.1. Objective of the Study

The main purpose of this study is to forecast the production of rice and corn utilizing the Seasonal Autoregressive Integrated Moving Average (SARIMA) through the Box-Jenkins Approach. With the use of SARIMA we can determine rice and corn can be produced for the year 2016-2020. Forecasting the amount of rice and corn production can support the government for their further decisions in the near future.

1.2. Statement of the Problem

This research aim to encourage the government to engross in rice and corn production by answering the following questions:

1) *What are the behavior of the graphs for:*
    a) Rice Production
    b) Corn Production
2) *What are the models that will be used in forecasting the Rice and Corn production in the Philippines?*
3) *What will be the predicted values of each of the Rice production and Corn Production?*
4) *What will be the Forecasted Values for the next five years of each of the two mentioned series?*
5) *Is there a significant difference between the values of:*
    a) Predicted and the actual values of Rice production?
    b) Predicted and the actual values of Corn production?

1.3. Scope and Limitations

The researchers limited the crops as rice and corn with a quarterly volume of production from year 1987-2015. The data were gathered from the Philippine Statistics Authority (PSA). The study will forecast the quarterly volume of production for the next five years (2016-2020) with the same location and unit using the chosen methods of forecasting: Seasonal Autoregressive Integrated Moving Average (SARIMA) Modelling through a Box-Jerkins Approach.

1.4. Significance of the Study
This approach was already used as the statistical basis of analysis the huge amount data by the investigators for predicting the future values in which might be a benefit for the observation of a particular series, especially in economics like financials, vast units of goods, prices changes, and many others that are recorded through time.
The study may help the decision makers to generate any agricultural reforms that could support the production of the crops mentioned in the study, and for so, can help the people for better understanding of trading of the goods and to be more mobilized after knowing the results, the future values in the next 5 years, if ever might become favourable after the systematic changes in the production.

1.5. Research Paradigm

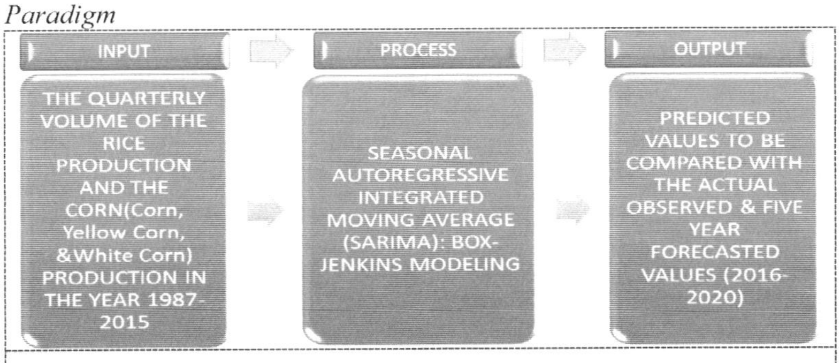

Figure 1. Input, Process, and Output

2. Methodology
The quarterly data of import and export trades Rice and corn of the Philippines from 2016-2020 were gathered to produces mathematical models using Seasonal Autoregressive Integrated Moving Average (SARIMA) with the same location and unit using the chosen methods of forecasting: Seasonal Autoregressive Integrated Moving Average (SARIMA) Modelling through a Box-Jerkins Approach.[14]

2.1. Box-Jenkins
This approach comprises four steps in analyzing time based data. A step by step process introduced by the statisticians, George Box and Gwilym Jenkins, fitting well identified model through the Autoregressive Integrated Moving average, a time series analysis represented by *ARIMA* $(p, d, q), (P, D, Q)s$ where p, d, and q indicates the non-seasonal part of the model, and P, D, and Q the are seasonal part of the model for s seasonality interval of the time series.[15] The approach uses four stages of modeling:
The *model identification* contains certain assumptions to be well thought-out in a time series analysis. On the first part of the model identification, supposing that the series is normally distributed after testing the normal distribution of the data through the Jarque-Bera test for normality, making certain classification of the model is primarily considered, whether the model to be used in the series has seasonally based or not, by which the series may have been seasonally adjusted to moderately deduce the seasonal component, if present within the data. Seasonality, along with the trend & cycle, is one of the seasonal components that are assumed to be present in a time series. It is the recurrence of a pattern in the same length of period after it happened. Another assumption is the stationarity of the

data. Assumption if there is a presence of a unit root within the series. Unit root is a random walk component that makes the data series be less possible to predict. In other words, unit root is a feature or a characteristic of a time series data having non-standard and non-normal asymptotic distributions that is not predictable. There is an augmented version of the test used to determine the stationarity of the data that is developed by the statisticians, David Dickey and Wayne Fuller, called the Dickey-Fuller Test applied to the equation

$$\Delta y_t = \alpha + \beta t + \gamma y_{t-1} + \delta_1 \Delta y_{t-1} + \ldots + \delta_{p-1} \Delta y_{t-p+1} + \boldsymbol{\varepsilon}_t$$ [16]

Where *α is constant, β* is the coefficient on a time trend, for a lag order of *p* in an autoregressive process by which it tests the assumption of the presence of a unit root, preferably used by researchers for much larger amount of data and a lot more complicated set.[17]

The second part will be the *model estimation* which comprises the estimation of the model where a series will be choosing Autoregressive terms and Moving Average terms out of the transformed series, to be regressed against the actual observed data series in which the series may have been differenced non-seasonally, undergone seasonal adjustment and logarithmically transformed based on the identification results on the first part of the approach in generating the model. It is where the possible combinations of the AR terms and the MA terms, as well as the seasonal terms that may possibly be included as part of the model. The plotted autocorrelation and partial-autocorrelation of the transformed series may show significant spikes indicates the AR and MA terms in specific lags.

In the *diagnostic checking* of the series, among the possible models that will be listed, one with favourable diagnostics will be the model to be used in forecasting the actual series. Diagnostic checking comprises the generation of the expected values that will be compared with the actual observed values so that the difference can be check with the mean zero. It also includes the R-square value that is that basis of the accuracy of the model and the Durbin-Watson test statistics for the goodness-of-fit-test, and the AR and MA Processes where to test if the autoregressive terms are stationary and if the moving average terms are invertible or not. Last part of the approach is regressing the model chosen in the process to forecast the series. *Forecasting* the series will be held to represent the possible outcomes in the near future. The graphical results as well as the listed values will be generated.

3. Results and Discussions

3.1. Rice

3.1.1. Actual Data and Logarithmic Transformation

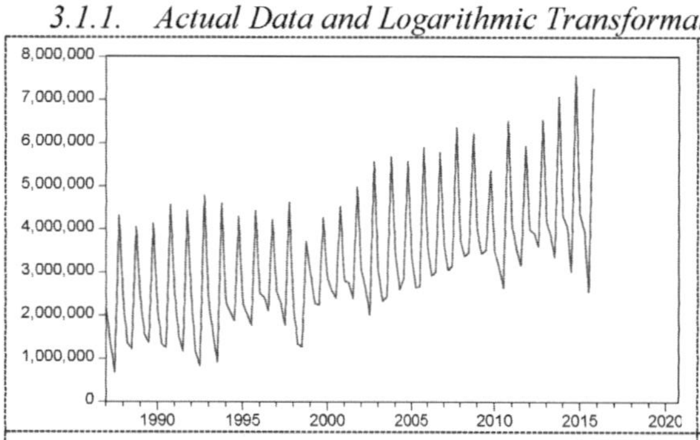

Figure 2. Plotted Graph of the Actual Rice Production Series

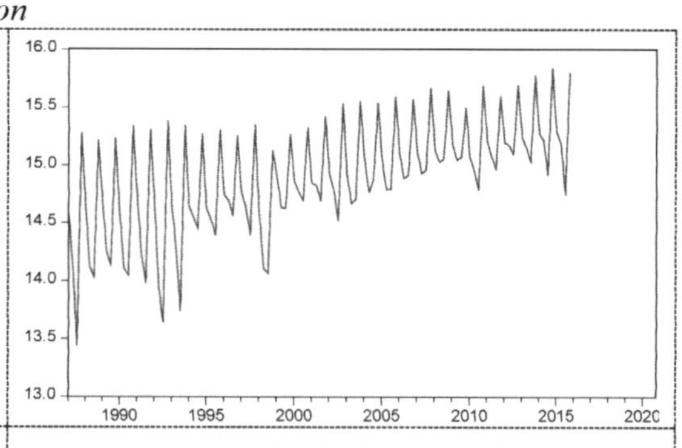

Figure 3. Logarithmically Transformed Rice Production Series.

The data shows an obvious trend started between the year 1995 and 2000, and the data series patterns exhibits which is an obvious seasonal recurrence of a pattern. At every 4th quarter, the series exceeds the preceding values on an instantaneous upsurge. This indicates that the total production in the last three months, specifically: October, November and December, always having the highest production

IOP Conf. Series: Journal of Physics: Conf. Series **820** (2017) 012007 doi:10.1088/1742-6596/820/1/012007

each year. This may be possible because the farmers are more likely to work on less warmer or more humid seasons, because it tends to have an occurrence of longer dry season called El Niño on the earlier part of the year due to climate change.

Also, the correlogram displays possible occurrence of a unit root in the series since the autocorrelation displays significant spikes that are recurring at every fourth lags in the autocorrelation which slowly turns to zero. This indicates the presence of unit root and seasonality of the data. However, it is necessary to have the presence of seasonality further tested in statistical basis.

The results of seasonality test suggest that there is a presence of seasonality failing to reject the null hypotheses which propose that seasonal adjustment is necessary for the data series. However, the autocorrelation has significant spikes at every after 3rd lags that are having values which were slowly turning to zero. This indicates that there might be a presence of a unit root within the series and it indicates that the series is not stationary.

To further test the stationarity of the data series, the augmented version of the Dickey-Fuller Test is performed. The results show that there exists a unit root in the actual observations. Though, it also shows that if the series undergoes an order of non-seasonal differencing, it becomes stationary.

It can be seen at the graph of the logarithmic transformed Rice production data series that the values has been reduced that utterly follows the same pattern, a recurrence of an upsurge value at every 4th quarter. Also, the variance has been minimized, the linear variations has been compressed which makes a minimal fluctuations in the preceding years. The significance level has reduced giving of more significant spikes at specific lags in the Correlogram of the series. At the fourth lag, both the autocorrelation and partial autocorrelation displayed significant spikes which mean a seasonal component is still present. Furthermore, it still follows the same outline of significant spikes of the correlogram in the actual observed values where at every after 3rd lags there occurs significant spikes that are slowly turning to zero value. These simply suggest that there might still be a presence of unit root and it means an order of non-seasonal differencing is still necessary.

3.1.2. Differencing Seasonally and Non-seasonally

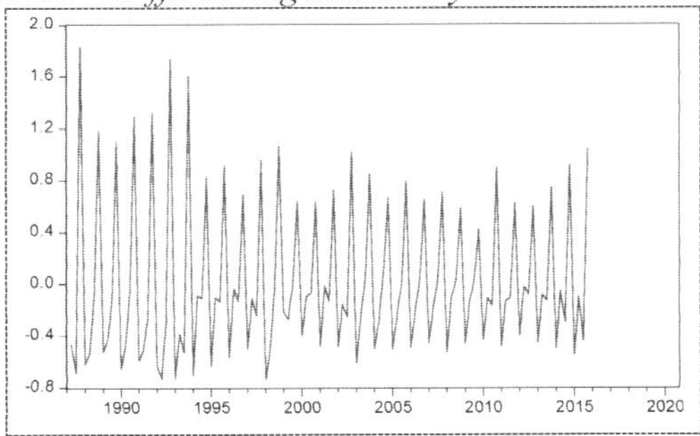

Figure 4. Non-seasonally Differenced Rice Production Series

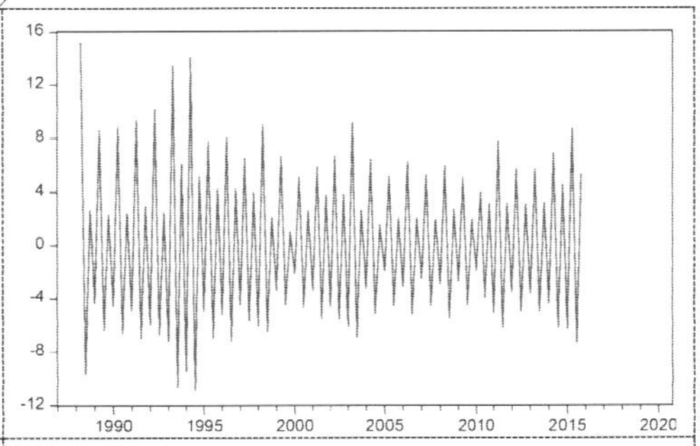

Figure 5. Seasonally Differenced Rice Production Series

Differencing the series reduced negative values and made the linear fluctuation having the mean zero, reduced its trend as well as the vertical values where it have now its units smaller than the preceding graph of the transformed data series. The correlogram have more spikes became significant on the negative side of the autocorrelation. The correlogram also states that the spikes at the 4th lag of autocorrelation and partial autocorrelation are significant having a seasonal component present. Moreover, the data series has already rejected the null hypothesis in the augmented Dickey-Fuller test on the first difference. It means that the data series is statistically tested to be stationary on the first order of differencing. It suggests that it no longer need of differencing the series non-seasonally.

The linear graph of the seasonally differenced series of the Rice production after it was transformed logarithmically and differenced on the first order of non-seasonal differencing, the linear plot shows that the line wavers vigorously while following zero trend and it also indicates that the mean now is much closer to zero. It can now be inferred the number of terms that can be used to create a model for the actual data series.

Estimating the number of terms to be included in the model, the possible combinations of the terms were regressed against the actual data series and produced the R-square values, Durbin-Watson statistics of each of the combinations, as well as their AR processes whether they are stationary or not, and the MA processes if it can be inverted or not. Among the listed combinations, 6 combinations with the highest R-square values were included to the selection of the possible models, however, it seems that their AR processes were not stationary, and one of them has its MA process appears not to be invertible. Regressing it again to the actual values of the Rice production, by which a constant now is included so that the AR terms would minimize its effect without adding more moving average terms, some combinations appeared stationary. One with highest R-square value is chosen to model the actual data series of the Rice production.

The model obtained has a total of 2 AR terms, 1 seasonal AR, and 8 MA terms including the constant. It was having the highest r-square value, which is about 86% probability, among the rest of the selection with acceptable Durbin-Watson test Statistics, invertible Moving Average processes, and stationary Autoregressive processes. The model will be SARIMA (2 1 8) $(1\ 1\ 0)_4$, by which will be used in forecasting the actual series.

3.1.3. Static and Dynamic Forecasting of the Rice Production Series

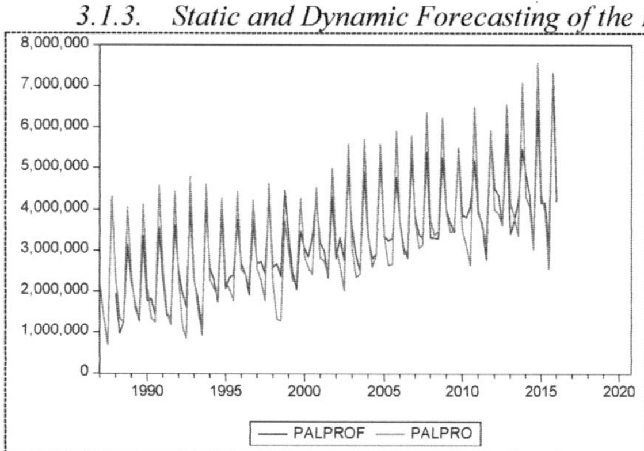

Figure 6. Predicted and Actual (Static Forecast)

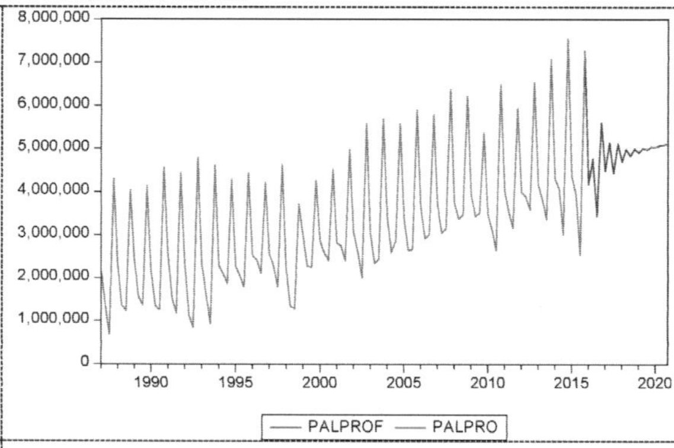

Figure 7. Predicted 5 years and Actual (Dynamic Forecast)

The plotted graph of the predicted values in Figure 6, generated through static forecasting followed the same trend when compared with the actual series. Replicating the same pattern of the actual values, where on every 4th quarter, it has sudden up surged value then makes a downfall for the next three quarters after it. This may indicate that there might be no significant difference between the expected values from the observed values of the Rice production.

To test this assumption, paired t-test was performed to statistically infer that the two compared series have not much of a difference with each other. Paired t-test gives a 0.89335022 probability value which indicates the rejection of the null hypothesis which signifies that there is no significant difference between the compared values. Thus, we can say that the model is actually fitted to the data series.

Hence, in Figure 7, the forecasted values gives off a plotted graph that follows the same trend component, an upward direction of the line, where the line wavers on the same seasonal component vigorously. It shows an upsurge value at every 4th quarter of the year. However, it seems that the line reduces its variation year after year. This indicates that the model could forecast effectively for the next few years which is fairly favourable for predicting the data series.

3.2. Corn Production

3.2.1. Actual Data and Logarithmically Transformed

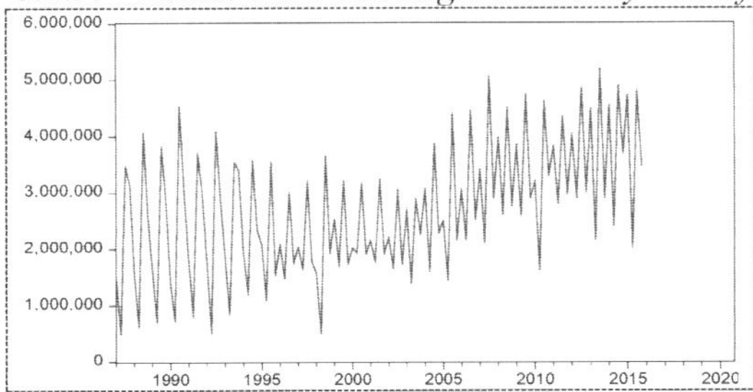

Figure 8. Predicted and Actual (Static Forecast)

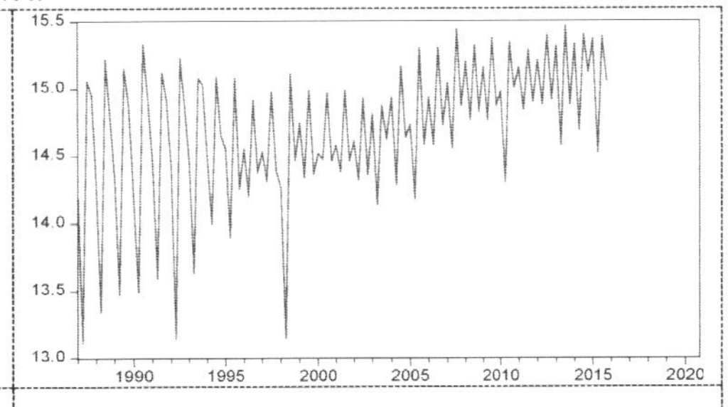

Figure 9. Predicted 5 years and Actual (Dynamic Forecast)

On the graph of the actual values of the total corn production, which includes the corn, white corn and yellow corn, it can be seen that the values follows a pattern where it has bigger changes of values that somehow it seemed that the values are almost the same at two consecutive quarters and makes a sudden huge increase in the 3rd quarter and a slight drop on the 4th quarter in a year. This some sort of pattern may also indicate a seasonal component which means it might have seasonality. However, bring about on the beginning of the data until the after 1995 where it has wavered in to some extent, different than the previous form and procured into a trend going upward. Dropping into lesser amount except from a particular quarter in 1998, and continues to increase values after 2002. The plotted correlogram of the actual series of values of corn production has its significant spikes on both autocorrelation and partial autocorrelation at 4th lag which means it has a recurring pattern at every 4th quarter of the actual data series of the corn production. It indicates seasonality within the series of the actual observed values of Corn Production. Then, to statistically infer the assumed presence of seasonal component, seasonality test is established where the results indicates less than hypothesized probability F-stats value of 0.001 which means it has failed to reject the null hypothesis that there is a seasonal component within the series. Hence, the data series of the actual observed values of corn production have a presence of seasonality. Also, the autocorrelation displays significant spike at every after 3rd lags slowly decreasing in length until it reaches zero on a particular lag. This indicates that there might be a presence of a unit root which means the data might not be stationary.

Testing the stationarity of the data, augmented dickey-fuller test results suggests that the actual values has a unit root since the value greater than 0.01 hypothesized probability, it has failed to reject the null hypothesis even when intercept has included, both trend and intercept, or none. The test result shows that the data needs to be differenced at least once with non-seasonal differencing.

The graph of logarithmic transformed data series of corn production displays the same pattern which is having two consecutive quarters having nearly similar values and makes an unexpected hug increase on the 3rd quarter and a slight drop on the 4th quarter. Then after 1995, it starts to waver on a slightly different way from the preceding term. The logarithmic transformed values slightly differ from the actual observed values in terms of trend component. It seems that the trend of the logarithmic transformed data has continuous trend from the very beginning of the series. On the other hand, the actual values are having a few changes with its trend on a particular group of years. Also, the correlogram have significant spikes on both 4th lags of autocorrelation and partial autocorrelation. Thus, indicating the presence of seasonality. Moreover, on every after 3rd lags significant spikes appears on the plotted autocorrelation that are slowly turning to zero. It means that there might still be a presence of unit root within the series and it signifies that transformed data is not stationary.

3.2.2. Differencing Seasonally and Non-seasonally

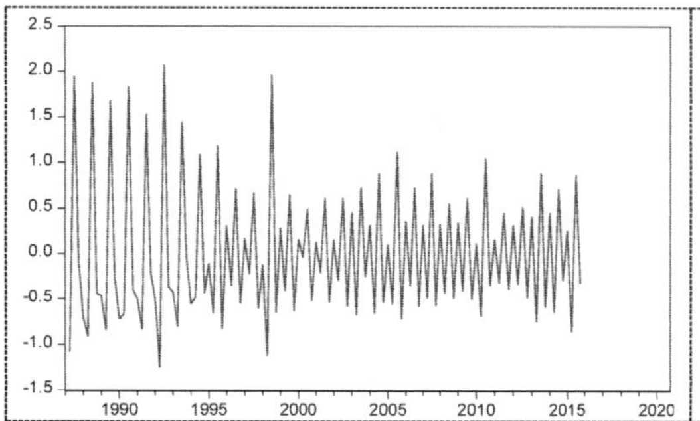

Figure 10. Predicted and Actual (Static Forecast)

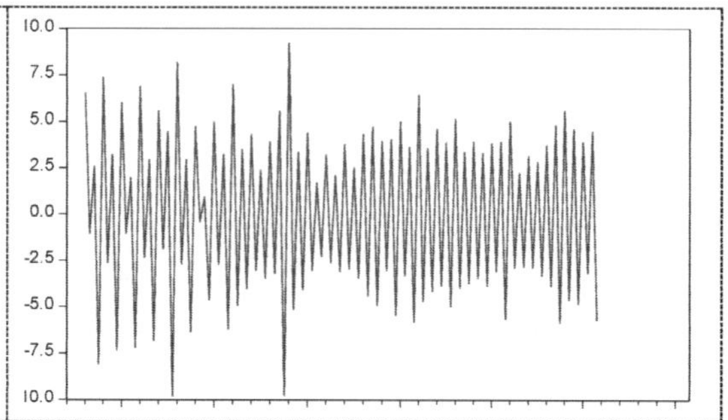

Figure 11. Predicted 5 years and Actual (Dynamic Forecast)

Differencing the series, the values are distributed and having negative values acquire the same distribution with the positive values. It means that the different logarithmic series are nearly having zero mean value. And also, the correlogram of the data series has reduced its significant level making more significant spikes on the left side of the plotted autocorrelation. The correlogram also suggest that there still might be a presence of seasonality within the series given that both autocorrelation and partial autocorrelation have significant spikes on the 4th lags.

The linear graph of the seasonal adjusted differenced logarithmically transformed data of corn production wavers on a vigorous manner despite the fact of having no trend at present, which means it has now zero mean value. On the plotted correlogram, autocorrelation and partial autocorrelation exhibits significant spikes at specific lags which can be concluded as autoregressive and moving average terms to be used to create a list of combinations. These combinations will be used to estimate the model, utilizing it to produce expected predicted values. Additionally, forecasted values for the next five years of the data series of corn production will be generated as an output of the study.

It can now be inferred in the correlogram, the number of terms indicated by the significant spikes on the specific lags in the autocorrelation and partial autocorrelation, to be include to create combinations of autoregressive terms and moving average terms to be regressed with the actual data series of the corn production and thus, to produce the diagnostic tests results – the R-square values of each combinations, as well as their Durbin Watson statistics, and the AR and MA Processes.

Checking the diagnostics, there were few of the combinations that are having stationary AR processes and invertible MA processes. Hence, choosing the model to be used from these combinations, the one with the highest r-square value of is having 3 AR terms, 8 MA terms and 1 seasonal MA term, has its AR processes to be stationary, MA processes were invertible and the Durbin-Watson stats was very satisfactory at 73% R-square Probability. The model will be SARIMA (3 1 8) (0 1 1)$_4$ which will be used to forecast the actual data series of the corn production.

3.2.3. Static and Dynamic Forecasting of the Corn Production Series

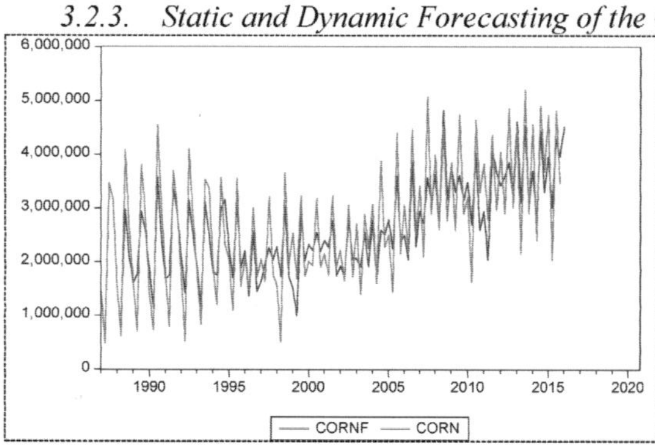

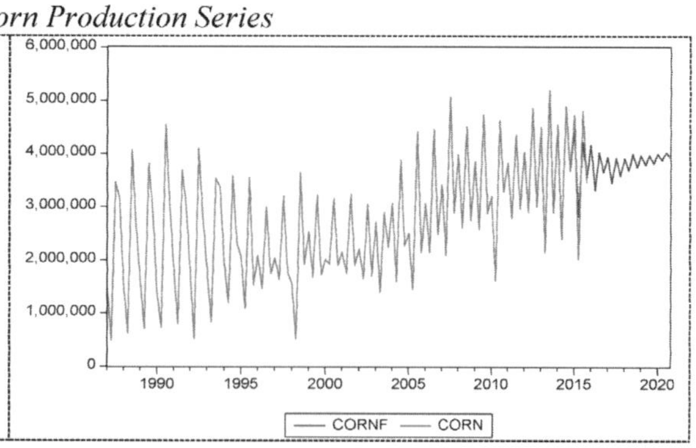

Figure 12. Predicted and Actual (Static Forecast)

Figure 13. Predicted 5 years and Actual (Dynamic Forecast)

The graph of the Predicted values generated using static forecasting, emulating similar trend when plotting it along with the actual series. The dropping values every after two consecutive quarters and upturns to a huge change of value on the 3rd quarter and smaller dropping value each year starting from earlier than 1990 until after year 1995. Then, it wavers on a different manner on a slow upward trend until the end of the actual observed series. The predicted series differs on the variance since the actual observed values emits bigger variations. But to test if there is a significant difference between the two series, paired t-test was perform. Thus, the test result indicates that there is no significant difference between the two compared series. Since the probability value 0.955, it suggests that the null hypothesis was rejected.

Moving on to forecasting in Figure 2.5, the actual data series of the corn production, the forecasted values also gives off a plotted graph that follows the same upward trend linear wavering slowly reducing in variance. Therefore, it means the model can forecast fairly enough for the next few years. The values increases slowly upon having the same fluctuating pattern from the preceding years, however, having less and less variance.

4. Conclusions and Recommendation

The study utterly submits the models of the Rice and Corn production to be used in forecasting the two mentioned data series. Coming up with these models through SARIMA method, the investigators obtained the first model, for the Rice production which is SARIMA (2 1 8) (1 1 0)4 and the second model, for the Corn production would be SARIMA (3 1 8) (0 1 10)4. The corn production predicted values, as well as the predicted values of the Rice production, were statistically inferred that there are no significant differences with their particular actual observed values. Results suggesting that the two models has satisfactorily fitted their specified series and satisfied the diagnostic checking for the models. The dynamically forecasted values of both series were following the trend pattern of the past values of their allotted observed values which were going on an upward direction indicating favorable outcome for the study.

Showing results that are favourable enough, such as having the expected future values on both observations to continue to rise up time after time, the investigators suggests to further include the actual observed values in the near future to be compared with the forecasted values of the Rice and corn production in the study. This may be even of use for the decision making that the government has to make in the future instances, to make the possible changes and reforms for any possible improvement of the production of both corn and Rice.

5. References

[1] GRiSP (Global Rice Science Partnership). 2013. Rice almanac, 4th edition. Los Baños (Philippines): *International Rice Research Institute*. 283 p.

[2] Ocampo, D. M., Ph. D., Pelegrina, W., Benosa-Llorin, V., Soriano, Z., & Baconguis, B. (2013, November 21). White Corn in the Philippines: Contaminated with Genetically Modified Corn Varieties. Retrieved April 21, 2016, from http://www.greenpeace.org/seasia/ph/PageFiles/566478/White-Corn-In-The-Philippines.pdf

[3] Gerpacio, R. V., J. D. Labios, R. V. Labios, and E. I. Diangkinay. 2004. Maize in the Philippines: Production Systems, Constraints, and Research Priorities. Mexico, D.F.: CIMMYT.

[4] Simmeon, L. (2016, March 23). Rice, corn production down in Q1. *The Philippine Star*. Retrieved April 26, 2016, from http://www.philstar.com/business/2016/03/23/1565636/rice-corn-production-down-q1

[5] Agriculture. (2011). Retrieved April 26, 2016, from http://education.nationalgeographic.org/encyclopedia/agriculture/
[6] Reyes, C., Mina, C., Crean, J., De Guzman, R., & Parton, K. (2009). Incorporating Regional Rice Production Models in a Simulation Model of Rice Importation: A Discrete Stochastic Programming Approach [PDF]. *Philippine Institute for Development Studies*.
[7] Delos Reyes, M. F., & David, W. P. (2009). The Effect of El Niño on Rice Production in the Philippines. *The Philippine Agricultural Scientist*, 92(2). Retrieved September 24, 2015.
[8] Tahir, A., & Habib, N. (2013). Forecasting of Maize Area and Production in Pakistan [PDF]. *Islamabad, Pakistan: ESci Journal of Crop Production*.
[9] Awal, M. A., & Siddique, M. (2011, March). Rice Production in Bangladesh Employing by ARIMA Model [PDF]. Bangladesh J. Agril. Res.
[10] Lansigan, F. B., De los Santos, W. L., & Hansen, J. (2007, January). Delivering Climate Forecast Products to Farmers: Ex Post Assessment of Impacts of Climate Information on Corn Production Systems in Isabela, Philippines. Retrieved May 6, 2016, from https://www.researchgate.net/publication/278013847_Delivering_Climate_Forecast_Products_to_Farmers_Ex_Post_Assessment_of_Impacts_of_Climate_Information_on_Corn_Production_Systems_in_Isabela_Philippines
[11] Nasiru, S., &Sarpong, S. (2012, July 15). Production and Consumption of Corn in Ghana: Forecasting Using ARIMA Models. Retrieved May 11, 2016, from https://www.researchgate.net/publication/266507667_Production_and_Consumption_of_Corn_in_Ghana_Forecasting_Using_ARIMA_Models
[12] Siopongco, J. C., Wassman, R., & Sander, B. (n.d.). International Rice Research Institue. Retrieved April 26, 2016, from http://irri.org/resources/publications/serials/technical-bulletins/alternate-wetting-and-drying-in-philippine-rice-production-feasibility-study-for-a-clean-development-mechanism
[13] Socio-economic Impacts of Genetically Modified Corn in the Philippines. (2013). *MASIPAG: Magsasaka at Siyentipiko Para Sa Pag-unlad Ng Agrikultura*, 1-94. Retrieved April 26, 2016, from http://masipag.org/wp-content/uploads/2013/05/MASIPAG-Book-low-res.pdf
[14] Tripathi, R. (2014, September 30). Forecasting Rice Productivity and Production of Odisha, India, Using Autoregressive Integrated Moving Average Models. Retrieved April 28, 2016, from http://www.hindawi.com/journals/aag/2014/621313/
[15] Stellwagen, E. (n.d.). Forecasting 101: Box-Jenkins Forecasting. Retrieved April 27, 2016, from http://www.forecastpro.com/Trends/forecasting101June2012.html
[16] Zivot, E. (2005, April 5). Unit Root Tests. Retrieved April April 26, 2016, from http://faculty.washington.edu/ezivot/econ584/notes/unitroot.pdf
[17] Fomby, T. B., Ph. D. (2006, September 12). Augmented Dickey-Fuller Unit Root Tests. Retrieved July 27, 2016, from http://faculty.smu.edu/tfomby/eco6375/bj notes/adf notes.pdf
[18] Ngo, T. (2013). The Box-Jenkins Methodology for Time Series Models. SAS Global Forum 2013: Statistics and Data Analysis, (454). Retrieved from http://support.sas.com/resources/papers/proceedings13/454-2013.pdf
[19] Section 3: Transformations of Exponential and Logarithmic Functions. (2006, August 28). Retrieved June/July, 2016, from http://faculty.uoit.ca/kay/precalc/pdf/4.3_Explog_transforms.pdf
[20] Nau, R. (2016, May 1). Statistical forecasting: Notes on regression and time series analysis. Retrieved June/July, 2016, from http://people.duke.edu/~rnau/411home.htm
[21] SHier, R. (2004). Statistics: 1.1 Paired t-tests. Mathematics Learning Support Centre. Retrieved from http://www.statstutor.ac.uk/resources/uploaded/paired-t-test.pdf

An Analysis on the Unemployment Rate in the Philippines: A Time Series Data Approach

J D Urrutia[1], **R L Tampis**[2] and **JB E Atienza**[3]

[1]Director, Intellectual Property Management Office
Chief, Center for Statistical Studies, Institute for Data and Statistical Analysis
Office of the Vice President for Research, Extension, Planning and Development
Polytechnic University of the Philippines – Sta. Mesa, Manila, Philippines
[2]Polytechnic University of the Philippines – Parañaque Campus, Philippines
[3]Department of Mathematics and Statistics, College of Science
Polytechnic University of the Philippines – Sta. Mesa Manila, Philippines

E-mail: jackieurrutia20@gmail.com, math_urrutia@yahoo.com.ph, jdurrutia@pup.edu.ph, razzcelletampis@gmail.com, jackyboyatiena@gmail.com

Abstract. This study aims to formulate a mathematical model for forecasting and estimating unemployment rate in the Philippines. Also, factors which can predict the unemployment is to be determined among the considered variables namely Labor Force Rate, Population, Inflation Rate, Gross Domestic Product, and Gross National Income. Granger-causal relationship and integration among the dependent and independent variables are also examined using Pairwise Granger-causality test and Johansen Cointegration Test. The data used were acquired from the Philippine Statistics Authority, National Statistics Office, and Bangko Sentral ng Pilipinas. Following the Box-Jenkins method, the formulated model for forecasting the unemployment rate is SARIMA (6, 1, 5) x $(0, 1, 1)_4$ with a coefficient of determination of 0.79. The actual values are 99 percent identical to the predicted values obtained through the model, and are 72 percent closely relative to the forecasted ones. According to the results of the regression analysis, Labor Force Rate and Population are the significant factors of unemployment rate. Among the independent variables, Population, GDP, and GNI showed to have a granger-causal relationship with unemployment. It is also found that there are at least four cointegrating relations between the dependent and independent variables.

1. Introduction

In a country with almost a 100 million people living in it like the Philippines, having a 9.1 million unemployed citizens is a huge problem. According to the latest Social Weather Stations (SWS) survey, 21.4 percent of the population declares themselves as unemployed. The unemployed Filipinos are considered unemployed based on two criteria, either they don't have a job, or they are currently looking for a job, except for housewives, students, retired workers and disabled person who are considered to be not part of the labor force [1].

Despite of the fact that employment in the Philippines has been growing fast for the past decade, still, many Filipinos are jobless. Employment growth rate was recorded to be almost 50 percent at the beginning of the twentieth century mainly on industrial, agricultural and services sectors. However, as

2000 enters, the country's population was quickly rising as it reached 76 million. It proves that employment growth is not enough to decrease the number of unemployed due to the rapid growth in population, a rise in the labor force participation, and the slow formation of jobs [2].

The country's uneven employment market has traditionally led millions of Filipinos to seek better-paying jobs overseas. As year 2000 enters, overseas Filipino workers almost reached 2.9 million, about nine percent of the labor force [2]. One out of every ten Filipinos works abroad, sending billions of dollars in remittances home, helping to drive the country's consumption-driven domestic economy — but doing little to promote employment [3]. The Philippines has long been a labor exporting market. About 2,500 Filipinos leave the country on a daily basis to seek for greater opportunities abroad and to provide better for the needs of the family [4].

This study aims to generate a mathematical model that can project the future values of unemployment rate in the Philippines from 2015 to 2020 on a quarterly basis. Aside from the model, factors which relatively affect the Unemployment Rate are to be determined among the considered variables, namely Labor Force Rate, Population, Inflation Rate, Gross Domestic Product (GDP), and Gross National Income (GNI). The relationship between the dependent and independent variables are also to be identified, whether they affect each other, or have a long run relationship, or a linear one. This paper aids to provide insights and information regarding the condition of the unemployment rate in the Philippines to be able to suggest preventive measures in solving the unemployment in the country.

1.1. Research Paradigm

A research paradigm was charted by the researchers with the intention of attaining the purpose of the study by following a set of procedures. The research paradigm includes the inputs used in conducting the study, the process or method applied, and the expected outputs.

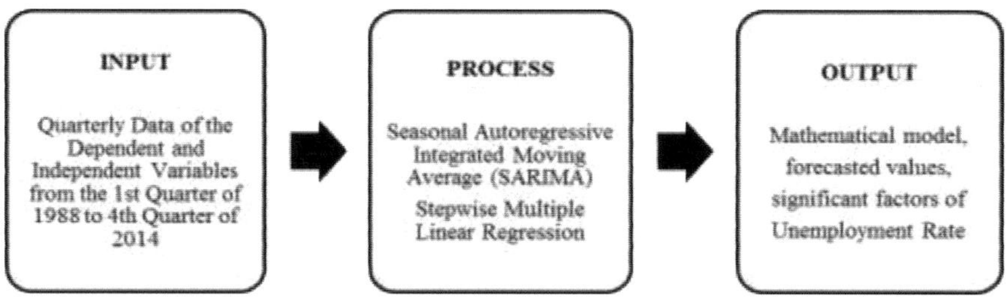

Figure 1. Research paradigm.

Using the quarterly data of the dependent and independent variables, the researchers aim to construct a mathematical model while following the Box-Jenkins methodology. Furthermore, the relationships of the dependent and independent variables, whether causal, cointegration, or linear, are to be identified using several statistical tests.

1.2. Statement of the Problem

So as to meet the objective of the study, the researchers intend to answer the following question which specifically leads to the formulation of the model in forecasting the unemployment rate in the Philippines.

1. What is the trend of the variables starting from the first quarter of 1988 to the fourth quarter of 2014?
   a. Unemployment Rate (y)
   b. Labor Force Rate (x_1)
   c. Population (x_2)
   d. Inflation Rate (x_3)

   e. Gross Domestic Product (x_4)
   f. Gross National Income (x_5)
2. Is there a significant relationship between the dependent and independent variables by means of Pearson correlation?
3. Which among the five independent variables can actually influence the dependent variable using Stepwise Multiple Linear Regression?
4. What is the mathematical model in projecting the unemployment rate of the Philippines generated through Box-Jenkins method?
5. Are the actual values of unemployment rate obtained from BSP, and the predicted ones obtained using the formulated model different from each other?
6. Is there a long run relationship between the dependent and independent variables using Johansen Cointegration test?
7. Can the independent variables affect the dependent variable by means of Pairwise Granger Causality test?

1.3. Scope and Limitations
The study covers the data of the dependent and independent variables starting from the first quarter of 1988 to the fourth quarter of 2014. There are six variables involved in the study namely Unemployment Rate, Labor Force Rate, Population, Inflation Rate, GDP, and GNI. These were obtained in three different institutions: the Philippine Statistics Authority (PSA), National Statistics Office (NSO), and the Bangko Sentral ng Pilipinas (BSP).

2. Methodology
This chapter presents the statistical treatments used by the researchers to be able to come up with the expected outputs. Moreover,

2.1. Multiple Linear Regression
Multiple linear regression is applied to determine which among the five given independent variables can influence the unemployment rate in the country. The model created expresses the value of the unemployment rate as a linear function of the five predictor variables and an error term which is written in a general form of [5]:

$$y_i = b_0 + b_1 x_{1i} + b_2 x_{2i} + \ldots + b_k x_{ki} + e_i \qquad (1)$$

where y_i is the predictand in year i, b_0 is the regression coefficient, b_k is the coefficient on the kth predictor, x_{ik} is the value of kth predictor in year i, e_i is the error term, and k is the total number of predictors.

2.2. Johansen Cointegration Test
The Johansen cointegration test is seen as a multivariate generalization of the augmented Dickey-Fuller test. The generalization is the examination of linear combinations of variables for unit roots. The Johansen test and estimation strategy – maximum likelihood – makes it possible to estimate all cointegrating vectors when there are more than two variables. The Johansen cointegration test provides estimates of all cointegrating vectors [6]. The researchers applied this statistical test to the data to determine if there exist a long run relationship between the unemployment rate and independent variables.

2.3. Pairwise Granger Causality Test
Granger causality test examines whether the independent variables presented affects the unemployment rate. The idea of Granger causality test is that an independent variable which Granger-causes the dependent variable can be better predicted using the histories of both variables than using the history of the dependent variable alone. In this paper, the researchers examined whether the six

IOP Conf. Series: Journal of Physics: Conf. Series **820** (2017) 012008 doi:10.1088/1742-6596/820/1/012008

presented independent variables (labor force rate, population, inflation rate, GDP, GNI) have a causal relationship with unemployment. This is done to better analyze the unemployment rate and determine the reasons behind its fluctuating trend. Granger causality can be assessed by regressing the function:

$$Y_t = \sum_{j=1}^{m} \alpha_j Y_{t-j} + \sum_{i=1}^{m} \beta_i X_{t-i} + D_t + \varepsilon_t \quad (2)$$

where D_t are the deterministics, ε_t is the random error term, α_j is the coefficient on the lagged y values, and β_t is the coefficient on the lagged x values [7].

2.4. *Seasonal Autoregressive Integrated Moving Average (SARIMA)*

The researchers used Seasonal Autoregressive Integrated Moving Average (SARIMA) in forecasting the Unemployment Rate for the next six years starting from 2015 to 2020. SARIMA is used when seasonal behavior is present in the time series. The seasonal ARIMA model incorporates both non-seasonal and seasonal factors in a multiplicative model. One shorthand notation for the model is ARIMA (p, d, q) x $(P, D, Q)_s$ where p is the non-seasonal autoregressive (AR) terms, d is the non-seasonal differencing, q is the non-seasonal moving average (MA) terms, P is the seasonal AR terms, D is the seasonal differencing, Q is the seasonal MA terms, and s is the time span of repeating seasonal pattern. The general model is written formally as:

$$\Phi(B^S)\varphi(B)(x_t - \mu) = \Theta(B^S)\theta(B)w_t \quad (3)$$

where $\varphi(B)$ is the non-seasonal autoregressive terms, $\theta(B)$ is the non-seasonal moving average terms, $\Phi(B^S)$ is the seasonal AR terms, $\Theta(B^S)$ is the seasonal MA terms [8].

2.5. *Box-Jenkins Methodology*

Box - Jenkins Analysis refers to a systematic method of identifying, fitting, checking, and using autoregressive integrated, moving average (ARIMA) time series models [9]. Box-Jenkins method follows a set of procedures in formulating the model [10]:

1. Model Identification – involves determining the order of the model required (p, d, q) in order to capture the salient dynamic features of the data. This mainly leads to use graphical procedures such as plotting the series, the autocorrelation function, and the partial ACF.
2. Model Estimation and Selection – involves estimation of the parameters of the different models identified and proceeds to a first selection of models using information criteria.
3. Model Checking – involves determining whether the model specified and estimated is adequate. Notably, residuals undergo diagnostic tests.

2.5.1. Augmented-Dickey Fuller Test. A Dickey-Fuller test is an econometric test for whether a certain kind of time series data has an autoregressive unit root, and has a general form of [11]:

$$y_t = \alpha + \rho y_{t-1} + \delta t + u_t \quad (4)$$

where u_t is an independent and identically distributed zero-mean error term, α is a constant term, δt is the time trend, and y_t follows a unit root process. This unit root test is used to satisfy the assumption of stationarity in the data of unemployment rate. Before formulating the model, the data for unemployment rate have to be differenced to be able to satisfy the said condition.

2.5.2. Correlogram. The plot of the autocorrelation function as a function of lag is also called the correlogram. Autocorrelation refers to the correlation of a time series with its own past and future values. An important guide to the persistence in a time series is given by the series of quantities called the sample autocorrelation coefficients, which measure the correlation between observations at different times. In this research, correlogram is used for the first step of Box-Jenkins methodology of identifying seasonality within the time series data of unemployment rate. This also dictates what time series model to use in forecasting, whether ARIMA or SARIMA. It is given an equation of [12]:

$$r_k = \frac{\sum_{i=1}^{N-k}(y_i - \bar{y})(y_{i+k} - \bar{y})}{\sum_{i=1}^{N}(y_i - \bar{y})^2} \quad (5)$$

where r_k is the autocorrelation coefficient at lag k, i is the amount of observed periods, x_i is the observation in i period, $\bar{y}$ is the mean.

2.5.3. Paired T-test. A paired t-test is used to compare two population means where you have two samples in which observations in one sample can be paired with observations in the other sample. The predicted values of the unemployment rate obtained using the formulated model was tested against its actual values using this test. This determines if there is a significant difference between the two and tells if the model is good enough to estimate the unemployment rate. The assumptions must be that the data are normally distributed, and the sample is randomly selected [13]. The t-statistics is obtained from the formula [14]:

$$t = \frac{\bar{u}}{\sqrt{s^2 / n}} \quad (6)$$

where $\bar{u}$ is the mean difference, s^2 is the sample variance, n is the sample size, and t is a student t-quantile with $n - 1$ degrees of freedom.

3. Results and Discussion

This section provides the interpretation of data and analysis of the results.

3.1. Trend of the Variables

Within the span of 26 years, the six variables shows fluctuating trends, either upward or downward trend with seasonal patterns because of some reasons.

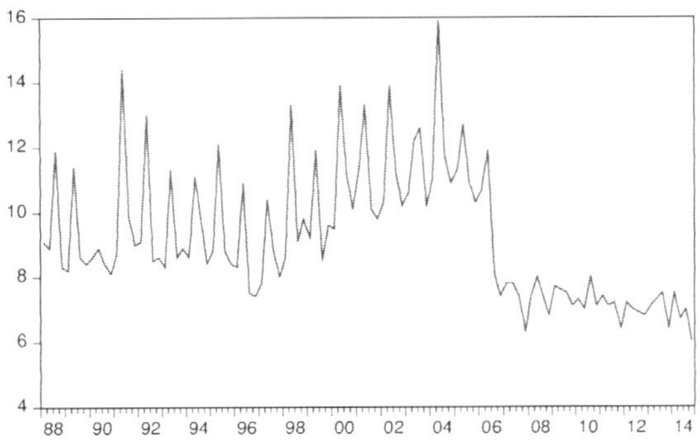

Figure 2. Graph of the unemployment rate.

3.1.1. Unemployment rate. For the past 26 years, unemployment rate reveals a downward trend with seasonal patterns with a major fall between the years 2005 to 2006. An all-time high of 5,002 persons, accounting for 15.9 person of the total unemployed population was recorded in the second quarter of 2004 brought about by 1.9 million new entrants that joined the labor force—including the seasonal influx of new graduates and vacationing students seeking work—that exceeded the 3.6 percent or 1.1 million employment generated during the quarter 2004. This was also due partly to the slump in employment in the agriculture sector and the rise in underemployment rate. The lowest unemployment rate was noted in the fourth quarter of 2014 with 2,482 persons unemployed or six percent because of the millions of jobs created within that year [15].

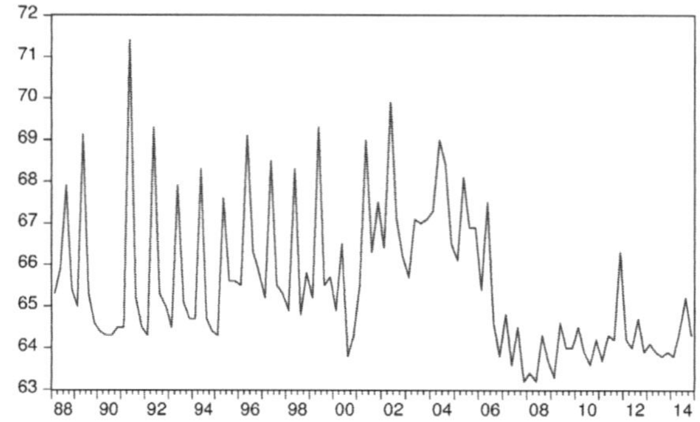

Figure 3. Graph of the labor force rate.

3.1.2. Labor force rate. Labor force participation rate is the proportion of the population ages 15 and older that is economically active which includes all people who supply labor for the production of goods and services during a specified period [16]. The movement of the labor force rate continues to decline with the unemployment from 1988 to 2014. On the second quarter of 1991, it reached its peak of 71.4 percent and on the fourth quarter of 2007, a record low of 63.2 percent was noted. These fluctuations were caused by the ageing of the population, cyclical effects of the Great Recession, and an unexplained portion, which might be due to pre-existing trends unrelated to the first two [17].

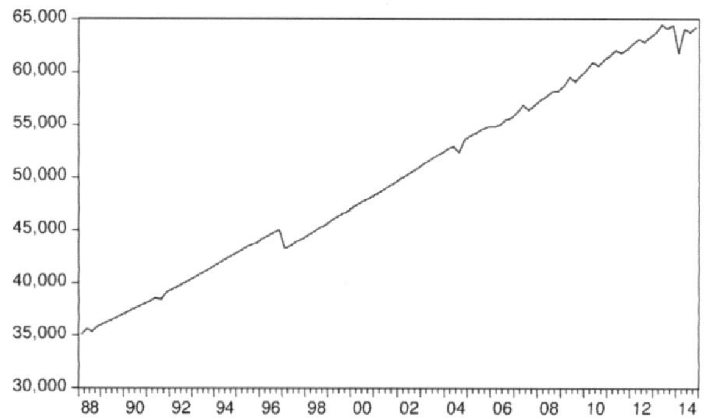

Figure 4. Graph of the population.

3.1.3. Population. For almost three decades, the Philippines has emerged into a country with a hundred million people living in it with continuous increase starting from 1988 to 2014. As of 2014, there is approximately 98 million Filipinos are living in the country, making the Philippines as the 12th most populated nations in the world. This rapid growth in population is caused by the unbalanced condition of the natality and mortality rate. Many are born, but only few dies. Natality rate continues to rise because of an increase in life span and lack of family planning [18].

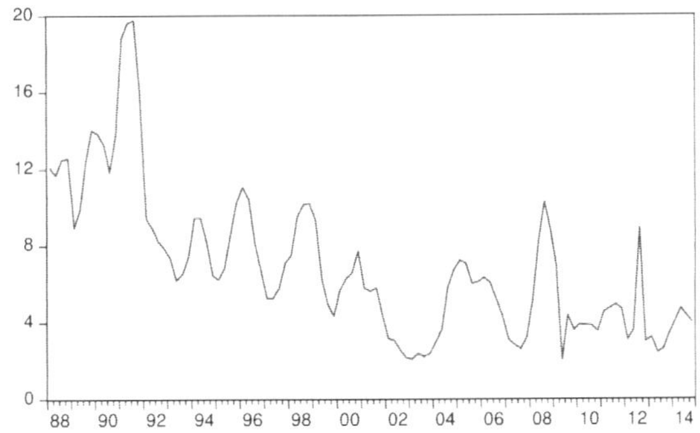

Figure 5. Graph of the inflation rate.

3.1.4. Inflation rate. Inflation rate reveals to have a decreasing pattern with fluctuations from the year 1988 to 2014. The country experienced an inflation rate of 19.77 percent in the third quarter of 1991 due to economic contraction in the early 1990s, which was caused by underlying macroeconomic imbalances, compounded by supply bottlenecks, natural disasters, political instability, a global recession and the Persian Gulf crisis of 1990 to 1991. It was the rebounded in the third quarter of 2009 having the lowest rate record of 2.07 % during the recovery stage of the Philippines after being hit by the Asian Financial Crisis on 2008 [19].

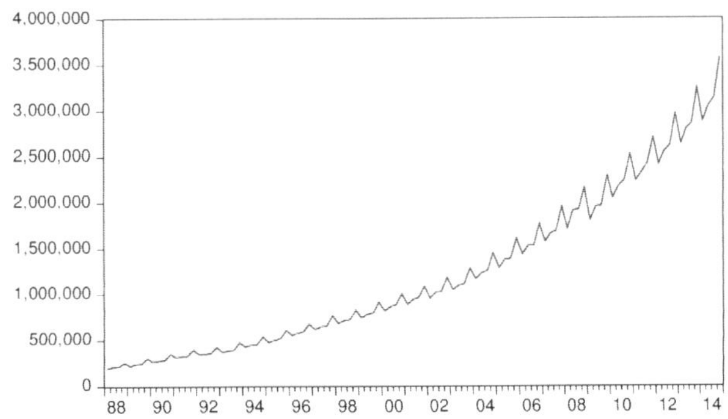

Figure 6. Graph of the GDP.

3.1.5. Gross Domestic Product. Gross Domestic Product exhibits an elevating trend fluctuating seasonally from 1988 to 2014. The Philippines' growing middle class, strong domestic demand, and stable political environment emphasize job creation and inclusive economic growth. Under the administration of President Benigno Aquino, the Philippines implemented reforms to improve the investment climate, making strides in good governance, transparency, and accountability. This result to the continuous rise of GDP, placing the country behind China, which had the fastest economic growth for the year at 7.4 percent, with a record of 6.1 percent in 2014 [19].

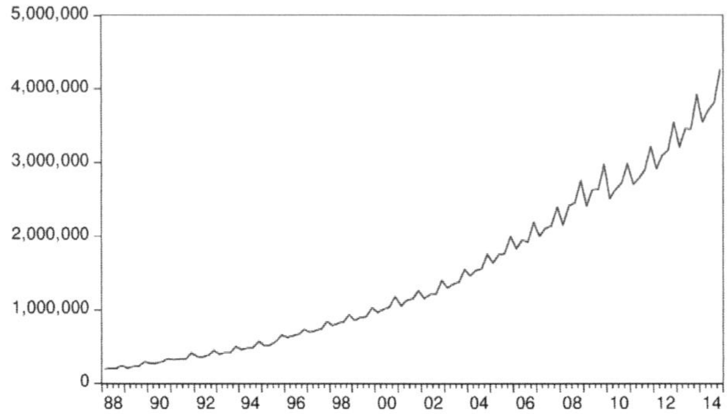

Figure 7. Graph of the GNI.

3.1.6. Gross National Income. Gross National Income exhibits an escalating trend with seasonal patterns for the past 26 years which identical movements with GDP. The government of President Corazon Aquino, embarked on a stabilization program aimed at preventing an upsurge in inflation, controlling the fiscal deficit and improving the external current account position. The economy responded favourably to these measures, posting increases in GNI, investments, private consumption and imports. Successive administrations also intervened in domestic economic affairs by imposing quantitative trade barriers, price controls and subsidies. Initially, the economy grew rapidly, with GNI growing largely due to increased exports and Government investments [19].

3.2. Significant Relationship between Dependent and Independent Variables

To be able to determine the significant relationship between the dependent and independent variables, Pearson correlation was used. The scatterplot diagram was also ascertained to show the linear relationship between the two types of variables.

Table 1. Correlation table.

	Pearson r	p-value
Labor Force Rate (x_1)	0.830	0.000
Population (x_2)	-0.381	0.000
Inflation Rate (x_3)	0.131	0.178
GDP (x_4)	-0.488	0.000
GNI (x_5)	0.480	0.000

Table 1 shows that among the five independent variables, only Inflation Rate reveals to have no significant relationship with the dependent variable having a p-value of 0.178 which is greater than the level of significance of 0.01 accepting the null hypothesis (there is no significant relationship between the dependent and independent variable). Conversely, Population and GDP reveals to have a weak negative correlation with Unemployment Rate based on their Pearson coefficient of determination, but shown to have a significant linear relationship with the dependent variable.

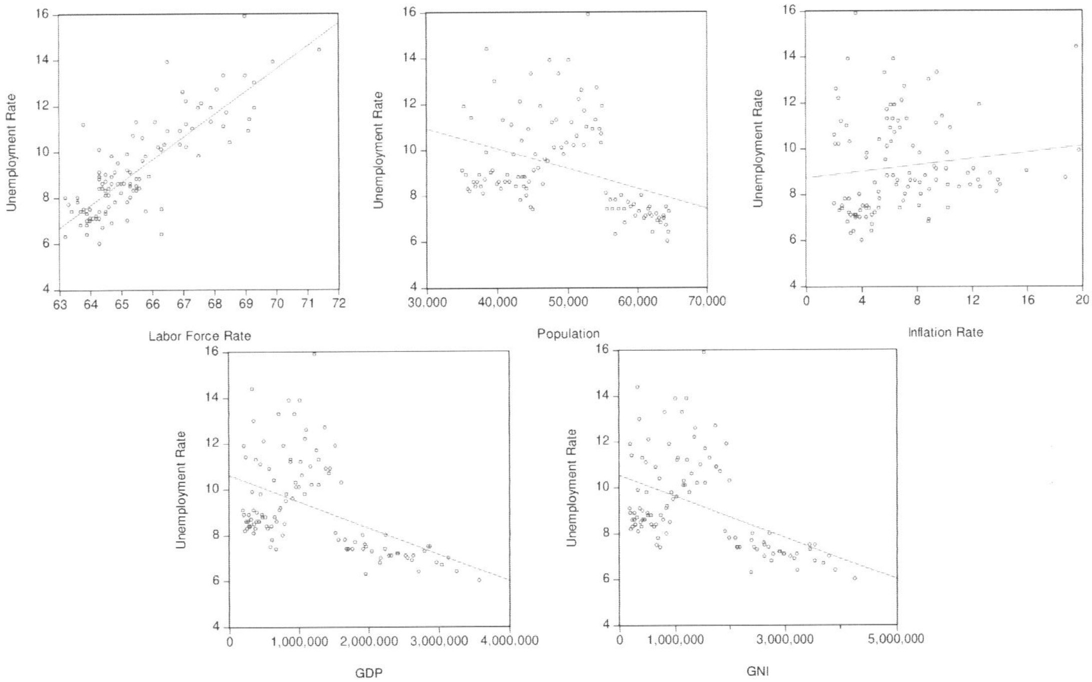

Figure 8. Scatterplot diagram between the dependent and independent variables.

Figure 3 shows the scatterplot diagram revealing the linear relationship between the dependent and independent variables. It can be seen that Population, GDP and GNI have downward trends indicating an inverse relationship with unemployment, while Labor Force Rate and Inflation Rate indicates a direct relationship since the regression line is inclined upward.

3.3. Significant Factors of Unemployment Rate

After conducting a multiple linear regression using the six variables with the Unemployment Rate as the dependent variable, the researchers were able to identify the significant factors of unemployment rate among the five independent variables considered.

Table 2. Regression analysis.

	p-value
Labor Force Rate (x_1)	0.000
Population (x_2)	0.004
Inflation Rate (x_3)	0.571
GDP (x_4)	0.072
GNI (x_5)	0.474

Based on the results obtained, Labor Force Rate and Population revealed to be a significant factor that can actually influence the movement of Unemployment Rate with a p-value of 0.000 and 0.004 correspondingly, with 0.01 level of significance. Since the p-values are less than the level of significance, the null hypothesis was rejected concluding that the variables are significant factors of unemployment rate. This indicates that any movement in the two variables can affect the unemployment in the country. Any increase in the labor force rate and the population can cause the unemployment rate to rise. On the other side, the p-value of Inflation Rate, GDP, and GNI, which is greater than 0.01, accepts the null hypothesis. Therefore, those three remaining variables does not influence unemployment rate.

3.4. *Mathematical Model for Forecasting Unemployment Rate*

3.4.1. Model identification. By examining the correlogram of the unemployment rate, it is found out that there exist seasonality in the time series, and it was non-stationary. So as to satisfy the assumptions of time series that the data must be stationary, the researchers conducted a regular differencing. To minimalize the errors, a logarithmic transformation was applied to the data series of unemployment rate, being stationary at the first regular differencing.

3.4.2. Model estimation. The formulated model was written in the form of SARIMA (6, 1, 5) x $(0, 1, 1)_4$ which contains six non-seasonal AR terms and five non-seasonal MA terms with one regular differencing, and one seasonal MA term with one seasonal differencing, and a time span of repeating seasonal patterns of four as the data used is quarterly.

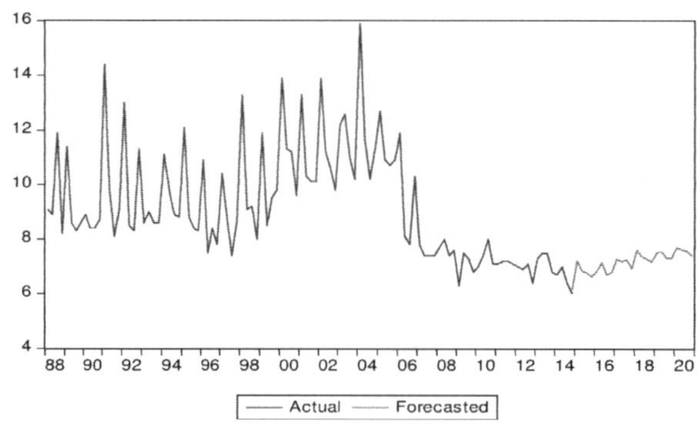

Figure 9. Graph of the actual and forecasted values of unemployment rate.

Using this model, the researchers are able to forecast the unemployment rate from 2016 to 2020 in a quarterly basis, and result shows that unemployment rate for the next five years ranges from six to eight percent. As shown in figure 5, unemployment is stable within the range of six to eight percent for the next five years. This indicates that the unemployment rate in the country will be constant for the following years.

3.4.3. Model checking. R-squared, Durbin-Watson statistics, Mean Absolute Error, and Akaike Information Criterion were considered in choosing the model that can forecast the unemployment rate accurately. The model reveals to have an r-squared of 0.787, a Durbin-Watson of 2.000, an MAE of 0.721, and an AIC of 3.048 indicating that the model is fitted to forecast the unemployment, that the errors are uncorrelated, that the values of the actual and forecasted unemployment rates are closely related, and that it is parsimonious. Furthermore, the residuals are normally distributed as stated by the result of the Jarque-Bera Test having a p-value of 0.01, and the Breusch-Pagan-Godfrey test reveals that the variances are equal with a p-value of 0.0510.

3.5. *Predicted Values*

Through the chosen model, the predicted values of unemployment rate was obtained. To test whether there is a difference between the actual and predicted values of the data series, the researchers used the Paired T-test.

Table 3. Paired t-test.

	p-value

t-statistic	0.9942

Based on the results, it is found out that there is no significant difference between the two values having a p-value of 0.9942. This indicates that the model is actually good enough in estimating the unemployment rate since the actual and predicted values are 99 percent identical.

3.6. Cointegration between Dependent and Independent Variables

Using E-Views, the researchers determine whether there exist a long run relationship between the dependent and independent variables while applying the Johansen Cointegration Test, accounting the Trace Test and the Maximum Eigenvalue Test.

The test shows that there are four cointegrating equations rejecting the null hypothesis of no cointegration from 'none' until 'at most 3' cointegrating vectors with a p-value less than the level of significance of 0.05. This leads to a conclusion that there are at least four cointegrating relations between the dependent and independent variables. This means that there is a long run relationship between the unemployment rate and the dependent variables.

3.7. Granger-causality between Dependent and Independent Variables

By means of Pairwise Granger Causality test, the researchers identify which among the five independent variables Granger-cause the dependent variable.

Table 4. Granger-causality.

	p-value
Labor Force Rate (x_1)	0.1654
Population (x_2)	0.0321
Inflation Rate (x_3)	0.7539
GDP (x_4)	0.0093
GNI (x_5)	0.0079

After conducting the test with the aid of E-Views, it is concluded that Population, GDP, and GNI have a Granger-causal relationship with the Unemployment Rate in the Philippines with a p-value of 0.0321, 0.0093, and 0.0079 correspondingly with a level of significance of 0.05 as shown in table 4. Population was also shown to be one of the significant factors of unemployment rate. This findings only means that the population has a great influence in the movements of the unemployment rate.

4. Conclusion

This paper aims to forecast the unemployment rate in the Philippines using a time series model. The formulated model for estimating and forecasting the unemployment rate in the Philippines is written as SARIMA (6, 1, 5) x $(0, 1, 1)_4$. Forecasted values are within the range of six to eight percent, and is found to be 72 percent closely relative to the actual values. Significant factors of unemployment rate are found to be Labor Force Rate and Population. Also, Population, GDP and GNI are shown to Granger-cause the dependent variable. These variables can affect the movement of the unemployment rate. Any increase on those variables can cause the unemployment rate to ascend or descend.

5. Recommendations

The government must find a way to maintain or reduce the unemployment rate in the Philippines. With over a hundred million people living in it, the government must prevent the rapid growth in the population since it is found that population affects the unemployment rate in the country, or better yet create millions of jobs enough to provide those unemployed a suitable occupation. The researchers suggest investigating on other factors that can affect unemployment rate, as well as adding more observations to the time series data for further studies.

References

[1] Flores H 2016 *9.1 Million Pinoys remained Unemployed – SWS* Retrieved from www.philstar.com/headlines/2016/02/9.1-million-pinoys-remain-unemployed-sws

[2] Brooks R 2002 *Why is Unemployment High in the Philippines?* Retrieved from www.unpan1.un.org/intradoc/groups/public/documents/APCITY/UNPAN027367.pdf

[3] Salvosa F 2015 *Philippine Struggles with Unemployment despite Economic Growth* Retrieved from www.cnbc.com/2015/09/01/unemployment-in-philippines.html

[4] *Six Reasons why Philippines Jobless Rate is High* 2009 Retrieved from www.pinoy-ofw.com/news/224-reasons-why-philippine-unemployment-rate-is-high.html

[5] *Multiple Linear Regression* Retrieved from www.ltrr.arizona.edu/~dmeko/notes_11.pdf

[6] Dwyer G 2015 *The Johansen Tests for Cointegration* Retrieved from http://www.jerrydwyer.com/pdf/Clemson/Cointegration.pdf

[7] *Granger-causality testing within the Context of the Bivariate Analysis of Stationary Macroeconomic Time Series* 2012 Retrieved from http://www.slideshare.net/ThomasReader/granger-causality-testing

[8] *Seasonal ARIMA Models* Retrieved from onlinecourses.science.psu.edu/stat510/node/67

[9] *The Box-Jenkins Method* Retrieved from www.researchgate.net/file.PostFileLoader.html?id=401442285922155

[10] Pelgrin F (2011) *Box-Jenkins Methodology* Retrieved from http://math.unice.fr/~frapetti/CorsoP/Chapitre_5_IMEA_1.pdf

[11] *What's Augmented Dickey-Fuller Test*? Retrieved from http://economics.about.com/cs/economicsglossary/g/augmented.htm

[12] *Autocorrelation* Retrieved from www.ltrr.arizona.edu/~dmeko/notes_3.pdf

[13] *Paired t-tests* Retrieved from www.statstutor.ac.uk/resources/uploaded/paired-t-test.pdf

[14] *Paired t-test.* Retrieved from www.statsdirect.com/help/parametric_methods/paired_t.htm

[15] *Report on Economic and Financial Developments* 2004 Retrieved from www.bsp.gov.ph/downloads/publications/2004/2qltp2004.pdf

[16] *Labor Force Participation Rate* Retrieved from data.worldbank.org/indicator/SL.TLF.CACT.ZS/countries?display=map

[17] Braun S Coglianes J Furman J Stevenson B & Stock J 2014 *Understanding the Decline in the Labor Force Participation Rate in the United States* Retrieved from voxeu.org/article/decline-labour-force-participation-us

[18] Cheung A 2014 *Rapid Population Growth: Its Causes and Effects on Health* Retrieved from https://prezi.com/snbjzejf1yuz/rapid-population-growth-its-causes-and-effects-on-health/

[19] *Philippine Economy* Retrieved from www.globalsecurity.org/world/philippines/economy.htm

Application of Method of Special Series for Representation of Solutions Describing Stationary Gas Flow

M.Yu. Filimonov[1,2]
[1]Ural Federal University, Yekaterinburg, Russia
[2]Institute of Mathematics and Mechanics of UrB RAS, Yekaterinburg, Russia
E-mail: fmy@imm.uran.ru

Abstract. A method of special series with recurrently computed coefficients is considered for solutions of nonlinear partial differential equations. To represent solutions of equation of velocity potential for stationary gas flows the series by the powers of special functions, which provide the recurrent coefficients determining from a sequence of linear differential equations. This makes possible to investigate the convergence of the series, and the functional arbitrariness allows to solve problem of internal flows in a nozzles.

1. Introduction

At present, a prospective direction to obtain approximate solutions of nonlinear partial differential equations is a combination of numerical and analytical methods. This paper considers one of the areas of creation of effective methods for constructing solutions of nonlinear equations in the form of special series with recurrently computed coefficients. This method is developed in the papers [1, 2, 3, 4, 5]. The method is an expansion of the solution of the equation in powers of specially choosed functions that allow to find the coefficients recurrently.

Defenition 1. *If we will find a solution of an initial problem for nonlinear equation in the form of series by the powers of a special function, which allows the series coefficients to be calculated recurrently, then this function we shall call a* basic function *(BF).*

The systematic application of the method of special series investigation of the basic functions were started with [2] in consideration of the following Cauchy problem:

$$u_t = F\left(t, u, \frac{\partial u}{\partial x}, \cdots, \frac{\partial^m u}{\partial x^m}\right), \quad u(0, x) = u^0(x), \tag{1}$$

where F is a polynomial of the unknown function $u(t, x)$ and its derivatives with respect to the space variable.

The proposed series constructions used only the mentioned property of the function F and may contain a functional arbitrariness that was used to prove the solvability of the initial boundary value problem for the generalized Korteweg-de Vries equation [6, 7].

In some cases it is possible to prove global convergence of the constructed series, including in unlimited domains, where the use of numerical methods meets essential difficulties [5]. In some cases it is possible to exactly satisfy zero boundary conditions by the choice of the basic functions (for example, for description of nonlinear vibrating string with fixed end points [1], or membrane with fixed edges [8]).

The solutions in the form of special convergent series are practically important for nonlinear evolution equations such as nonlinear equation of filtration [9, 10], nonlinear wave equations [3, 4] and Lin-Reissner-Tsien equation which describes unsteady transonic flows under the transonic approximation [11, 12, 13]. Convergence of such series is proved, as a rule, in an unbounded domain.

Constructed series converge fast, that allows to use it for testing of numerical methods and for creation new numerical-analytical methods. For example, by special series a nonstationary transonic flow around a wedge was calculated, which describes the transition from non-stationary gas flow to the steady, and it has been shown analytically that this transition is exponential [11]. From a formal point of view, a characteristic feature of the special series, is independency of its form from the type and construction of the equations to be solved (usually, the type of the equation is closely related to study of convergence of the series). Convergence of special series is proved for a broad class of non-linear equations and systems of continuum mechanics. Another area in the theory of the method of special series are constructions of series, taking into account the specificity of the considered equations [14]. Accounting for the specifics of the considered nonlinear equations leads to expansion of classes of basic functions that allow to solve initial boundary value problems and investigate equations with pecularities.

In some cases special series are transformed into finite sums and then generate exact solutions [15]. For real problems, in which satisfaction of various boundary conditions is required it is difficult to manage using of special series, or exact solutions, for example, for long-term problems of forecasting of degradation of permafrost [16, 17, 18]. However, the presence of an exact solution allows sometimes to simplify original mathematical models and to choose numerical methods, as it was done for a geothermal problems [19, 18].

In this paper new results related with with construction and study of special series to the representation of solutions of nonlinear partial differential equations, which have pecularities.

2. General scheme of construction of series with recurrently calculated coefficients in powers of consistent functions

Let find a solution of equation (1) in the form of the series

$$u(t,x) = \sum_{n=0}^{\infty} u_n(t) P^n(t,x) \tag{2}$$

by the powers of basic function $P(t,x)$ satisfying the overdetermined system

$$P_x = A(t,P), \quad P_t = B(t,P) \tag{3}$$

with functions $A(t,P)$ and $B(t,P)$ being analytic respect to P and such that $A(t,0)\equiv 0$ and $B(t,0)\equiv 0$. It was shown that if the initial conditions are written in the form

$$u^0(x) = \sum_{n=0}^{\infty} u_n^0 P^n(0,x), \tag{4}$$

then substituting series (2) into equation (1), collecting similar terms, and taking into account relations (3), we obtain the sequence of first-order ordinary differential equations for the coefficients $u_n(t)$

$$u_n' = F_n(t, u_n, \ldots, u_0), \qquad u_n(0) = u_n^0, \quad n = 1, 2, \ldots \tag{5}$$

where the right-hand sides F_n include only u_j with $j \leq n$ and the coefficients u_n may be linearly contained only. Thus, the coefficients of (2) are calculated recurrently. Examples of basic functions (3) are presented, for example, in [2].

Defenition 2. *Let the functions satisfying relations* (3) *be called universal basic functions.*

Let consider nonlinear equation

$$L_t u = F\left(t, x, \frac{\partial u}{\partial x}, \dots, \frac{\partial^m u}{\partial x^m}\right), \tag{6}$$

where

$$L_t = d_1(t)\frac{\partial}{\partial t} + \ldots + d_l(t)\frac{\partial^q}{\partial t^q}, \qquad d_k(t) \in C[0, \infty), \quad k = \overline{1, l}, \quad d_l(t) \neq 0, \quad t \geq 0,$$

and F_1 is a nonlinear function, which form will be defined in the following. Note, that function F_1 may depend on the spatial variable. The solution of equation (6) will be find in the form of the series

$$u(x,t) = \sum_{n=0}^{\infty} u_n(t) R^n(x,t). \tag{7}$$

Let function $R(x,t)$ satisfy the differential relations

$$\begin{aligned} \left(\frac{\partial R}{\partial x}\right)^2 &= \sum_{k=k_0}^{\infty} a_k(t) R^k, \\ \frac{\partial R}{\partial t} &= \sum_{k=1}^{\infty} b_k(t) R^k. \end{aligned} \tag{8}$$

Here $k_0 \geq 2$ and functions $a_k(t)$, $b_k(t) \in C^1[0, \infty)$ such that series in the right-hand sides of relations (8) are assumed to be absolutely convergent for $|R| \leq R_0$, $R_0 > 0$ and $t \geq 0$. It is easy to show that the system (8) is consistent, if the functions $a_k(t)$ and $b_k(t)$ are fined from differential equations

$$a'_k = \sum_{l+n=k+1} a_l b_n (2n - l).$$

Defenition 3. *Functions, which satisfy* (8), *we will call consistent basic functions, and the special series by the powers of these functions will call consistent series.*

If a consistent basic function (8) depends from x only, then instead of relation (8) the following differential equation should be valid

$$\left(\frac{dR}{dx}\right)^2 = \sum_{k=k_0}^{\infty} c_k R^k, \qquad c_k = \text{const}. \tag{9}$$

Examples of consistent basic functions (9) are

$$R_1(x) = \frac{1}{x^2 + A}, \qquad (R')^2 = 4R^3 - 4AR^4 \quad A = \text{const};$$

$$R_2(x) = a_2^2 a_3^{-2}\left(tg^2 0.5a_2 x + 1\right), \quad (R_2')^2 = -a_2^2 R_2^2 + a_3^2 R_2^3, \quad a_2, a_3 = \text{const};$$

$$R_3(x) = \frac{1}{\cos^2 x + A}, \quad (R_3')^2 = -4R_3^2 + (4 + 8A)R_3^3 + 4A(A-1)R_3^4;$$

$$R_4(x) = \frac{4}{(e^x + e^{-x})^2}, \quad (R_4')^2 = 8R_4^2 - R_4^3.$$

Examples of consistent basic functions with functional arbitrariness are $R_1(x,t) = (x^2 + f(t))^{-1}$, $R_3(x,t) = (\cos^2 x + f(t))^{-1}$, for which differential relations (8) are valid.

We describe a class of equations that allow the construction of solutions in a special series in powers of both universal and consistent basic functions. Let consider a class of nonlinear equations

$$L_t u = \sum_{j=0}^{\infty} A_{p_j}(t) u^{k_{0j}} \left(\frac{\partial u}{\partial x}\right)^{k_{1j}} \left(\frac{\partial^2 u}{\partial x^2}\right)^{l_{1j}} \cdots \left(\frac{\partial^{2m-1} u}{\partial x^{2m-1}}\right)^{k_{mj}} \left(\frac{\partial^{2m} u}{\partial x^{2m}}\right)^{l_{mj}}, \tag{10}$$

$A_{p_j}(t) \in C[0,\infty)$, $p_j = (k_{oj}, k_{1j}, l_{1j}, \cdots, k_{mj}, l_{mj})$, $k_{0j} + \sum_{i=1}^{m}(k_{ij} + l_{ij}) = j$, the right-hand side of equation (10) is assumed to be absolutely convergent in a some domain. It is obvious that the special series by the powers of universal basic functions is a formal solution for equation (10). We check that the series by the powers of consistent basic function is a formal solution for class of equations (10).

The following theorem is valid:

Theorem 1. *Consistent series* (7), (8) *is a formal solution of equation* (10), *if for the powers* k_{ij}, *for any* j *the conditions* $\sum_{i=1}^{m} k_{ij} = 2N_j$, *are satisfied, where* N_j *are integers.*

Proof is carried out by substitution of series (7), (8) into equation (10). Consider the even derivatives of u by x and its powers in the right hand side of equation (10). It is easy to verify that its may be represented by consistent series (7), (8). Indeed, we calculate, for example, u_{xx}

$$u_{xx} = \sum_{n\geq 2} 2n(n-1)u_n(t)R^{n-2}R_x^2 + \sum_{n\geq 1} nu_n(t)R^{n-1}R_{xx}.$$

From the first relation of (8) it follows that R_{xx} is also expanded into a series in powers of consistent basic function R

$$R_{xx} = \sum_{k=k_0-1}^{\infty} \frac{(k+1)a_{k+1}(t)}{2} R^k,$$

therefore

$$u_{xx} = \sum_{m=2}^{\infty} \sum_{k=k_0}^{\infty} 2m(m-1)u_m(t)a_k(t)R^{m-2+k}$$
$$+ \sum_{m=1}^{\infty} \sum_{k=k_0-1}^{\infty} mu_m(t)\frac{(k+1)a_{k+1}(t)}{2} R^{m-1+k}$$
$$+ \sum_{n=2}^{\infty} \Big[\sum_{m+k=n+2, k\geq k_0} 2m(m-1)u_m(t)a_k(t) + \sum_{m+k=n+1, k\geq k_0-1} mu_m(t)\frac{(k+1)a_{k+1}(t)}{2} \Big] R^n$$

Therefore, the derivative of u_{xx} and its powers are represented in a power series by R. Similarly it is shown, that the even derivatives of u by x can also be written in a series by the powers of R. Consequently,

$$\frac{\partial^{2m} u}{\partial x^{2m}} = \sum_{k=0}^{\infty} g_{mk}(t) R^k. \tag{11}$$

Let us now consider the odd derivatives of u by x, included in the right hand side of equation (10). Formula (11) implies that the product of two odd derivatives can be written as a series in the powers of consistent basic function R. Indeed,

$$\frac{\partial^{2m_1+1} u}{\partial x^{2m_1+1}} \frac{\partial^{2m_2+1} u}{\partial x^{2m_2+1}} = \sum_{k_1=1}^{\infty} g_{m_1 k_1}(t) R^{k_1-1} R_x \sum_{k_2=1}^{\infty} g_{m_2 k_2}(t) R^{k_2-1} R_x$$

$$= \sum_{k_1=1}^{\infty}\sum_{k_2=1}^{\infty} g_{m_1k_1}(t)g_{m_2k_2}(t)R^{k_1+k_2-2}R_x^2$$

$$= \sum_{k_1=1}^{\infty}\sum_{k_2=1}^{\infty}\sum_{k_3=k_0}^{\infty} g_{m_1k_1}(t)g_{m_2k_2}(t)a_{k_3}(t)R^{k_1+k_2+k_3-2}.$$

Thus, to make it possible to write the right hand side of equation (10), after substituting in the series (7), (8), also in the form of a series in the powers (8), it is sufficient to require that the terms, which include the odd derivatives should consist of an even number of odd derivatives, including their powers. That is, if the condition $\sum_{i=1}^{m} k_{ij}{=}2N_j$ is valid, the right hand side of equation (10) can be represented as a series by the powers of consistent basic function R. The even derivatives and its powers possess this property, if its are in the right hand side of original equation.

In order to ensure that series (7) is a formal solution of equation (10), we need to check that the coefficients of the consistent series $u_n(t)$ may be found recurrently. Note that the recurrence of finding the coefficients $u_n(t)$ achieved by the specificity of the first equation (8), namely, the fact that the series in the right hand side of the first equation starts no less than the second power of R, $(k \geq k_0 \geq 2)$. Theorem 1 is proved.

Thus, as function F_1 is possible to take the right hand side of equation (10) with the restrictions of Theorem 1.

The consistent series are also useful for constructing solution of nonlinear equations, for which the special series in powers of universal basic functions are not applicable.

3. Investigation of convergence of consistent series

Consider the equation of the velocity potential, describing the stationary gas flow [20],

$$\Phi_z^2\Phi_{zz} + \Phi_r^2\Phi_{rr} + 2\Phi_r\Phi_z\Phi_{rz} - \Theta(\Phi_{rr} + \Phi_{zz} + N\Phi_r/r) = 0. \tag{12}$$

Here $\Theta = (\gamma - 1)[K - 0{,}5(\Phi_r^2 + \Phi_z^2)]$, $\gamma =$ const is the gas adiabatic index, $K =$ const, $\Phi_r(r, z)$, $\Phi_z(r, z)$ are the gas velocity in r and z, respectively, $N = 0$ for a plane case, $N = 1$ for axisymmetric case. Consider the following problem for equation (12):

$$\Phi(0, z) = \Phi_0(z); \quad \Phi_r(0, z) = 0, \quad -\infty < z < \infty, \tag{13}$$

t.i. in the axis $r = 0$ velocity potential distribution $\Phi_0(z)$ is given, for which we will find the current lines (nozzle walls), which create the distribution. Review of numerical and analytical methods for solving the inverse problem (12), (13) is presented in [20].

Solution of equation (12) we will find in the form of consistent series

$$\begin{aligned} &u_i(z)R^i(r, z), \quad u_0 = \text{const}, \\ &R(r, z) = \frac{1}{r^2 + \varphi(z)}, \quad \varphi(z) \in C^2(-\infty, \infty). \end{aligned} \tag{14}$$

Let series (12) substitute into equation (12) and we get an equation for $u_i(z)$

$$\begin{aligned} &u_i''(M_\infty^2{-}1){=}(1{-}i)[4i{-}2{-}2N{-}(1{+}M_\infty^2)\varphi'']u_{i-1}{-}(i{-}1)(i{-}2) \\ &\cdot[(1{+}M_\infty^2)(\varphi')^2 - 4\varphi]u_{i-2} + 2M_\infty^2(i-1)u_{i-1}'\varphi' + R_i(z), \quad i \geq 1; \end{aligned} \tag{15}$$

here

$$R_i(z){=}(\gamma{-}3)M_\infty \sum_{m+n=i} G_1(m)G_2(n){-}M_\infty^2[8{+}2N(\gamma{-}1)] \sum_{m+n=i} u_{n-1}(n{-}1)G_1(m)$$

$$+\frac{\gamma-3}{2}\sum_{m+n+k=i} G_1(m)G_1(n)G_2(k) - 8\sum_{n+l=i} G_3(n)[(l-1)(2l-1)u_{l-1}$$
$$-2\varphi(l-2)(l-1)u_{l-2} - [8+N(\gamma-1)]\sum_{m+n+k=i} u_{m-1}(m-1)G_1(n)G_1(k)$$
$$+8\varphi\sum_{m+n+k=i}(m-1)u_{m-1}G_1(n)(k-1)G_1(k-1)$$
$$+8\varphi\sum_{m+k=i} u_{m-1}(k-1)G_1(k-1) + 2(\gamma-1)\sum_{m+n=i} G_3(n)[G_2(l) - 2N(l-1)u_{l-1}],$$
$$G_1(n) = u_n' - (n-1)\varphi' u_{n-1}; \quad G_2(n) = \frac{dG_1(n)}{dz},$$
$$G_3(n) = \sum_{m+k=n}[m(n-1)u_m u_{n-1} - \varphi(m-1)(n-1)n_{m-1}u_{n-1}].$$

Constants, which describe gas parameters, are for $u_0 = M_\infty$ (M_∞ is the Mach number of the incident flow). So, the solutions of equation (15) are representable in the form

$$\begin{aligned} u_i(z) = \frac{1}{M_\infty^2-1}\int\int &\{(1-i)[4i-2-2N-(1+M_\infty^2)\varphi''(\tau)]u_{i-1}(\tau) \\ &-(i-1)(i-2)[(1+M_\infty^2)(\varphi')^2 - 4\varphi]u_{i-2}(\tau) \\ &+2M_\infty^2(i-1)u_{i-1}'(\tau)\varphi'(\tau) + R_i(\tau)\}d\tau dz + C_i z + D_i, \end{aligned} \tag{16}$$

where $C_i, D_i = \text{const}$, $i \geq 1$.

The theorem is valid.

Theorem 2. *Let the following conditions be fulfilled:*

1. an arbitrary function $\varphi(z) \in C^2(-\infty, \infty)$, which

$$1<\varphi_0 \leq \varphi(z) \leq \varphi_1, \quad |\varphi'|, |\varphi''| \leq M\exp(b|z|), \quad 0<M\leq M_0 \quad 0<b_0\leq b;$$

2. for the constant C_n, D_n the inequalities

$$|C_n|, |D_n| \leq \frac{M}{3n^4}, \quad n \geq 1, \quad \varphi_0, \varphi_1, M, b, M_0, b_0 = \text{const}, \quad \text{are valid.}$$

Then, $L > 0$ exists, that for $M_\infty \neq 1$ series (14) *uniformly converges to a solution of equation* (12) *for all $0 \leq r < \infty$, $|z| \leq L$.*

The proof of Theorem 2 is obtained after estimations for coefficients $u_n(z)$ with $|z| \leq L$

$$|u_n(z)| \leq \frac{M}{n^4}\exp(bn|z|), \quad n\geq 1,$$

$$|u'_n| \leq \frac{M_1}{n^3}\exp(bn|z|), \quad |u_n''| \leq \frac{M_2}{n^2}\exp(bn|z|), \quad M_1, M_2 = \text{const},$$

which are established by induction for direct estimates of expressions (16).

Remark. The second condition (13) is fulfilled out automatically due to the specifics of the basic function $R(r,z)$. In addition, the arbitrariness in the basic function $\varphi(z)$ generates a velocity distribution on the axis $r = 0$. Therefore, if we set $r = h(z)$, then on this line the impermeability condition $h'\Phi_z(h(z),z) = \Phi_r(h(z),z)$ has to be fulfilled, from which we can try to find the function $\varphi(z)$ approximately.

The uniform convergence of the consistent series, proved in Theorem 2, allows to solve the problems of the internal flow in nozzles [20] due to the specific of the series (13).

4. Conclusion

A method of special series is suggested for representation of solutions of nonlinear partial differential equations. The method consists in the expansion of the solution of the equation in the form of series with recurrently computed coefficients, which may be found from a sequence of linear ordinary differential equations. The recurrency of finding the coefficients is achieved by a special choice of the basic functions, which powers makes the series to be a solution. The basic functions can be functions of two types: universal basic functions and consistent basic functions, and can contain a functional arbitrariness. Universal basic functions are useful for representation of solutions for a wide class of nonlinear partial differential equations. Consistent basic functions take into account the specificity of the original non-linear equations and applicable to a wider class of nonlinear equations than universal basic functions.

Application of consistent basic functions is shown for the equation that describes an axisymmetric stationary flows of gas. We obtain a sequence of linear ordinary differential equations, which are equations for the series coefficients. The conditions under which this series converges to the solution of the equation are formulated. It turns out that a choice the consistent basic function allows to exactly satisfy the condition on the axis of symmetry, and by using the arbitrary function, and we can approximate and solve the problem of the internal gas flow in a nozzle.

Acknowledgments
The work was supported by by Russian Foundation for Basic Research 16–01–00401.

References

[1] Filimonov M Y 1991 *Diff. Equations* **27** 1158–1163
[2] Filimonov M Y, Korzunin L G and Sidorov A F 1993 *Russ. J. Numer. Anal. Math. Model.* **8** 101–125
[3] Filimonov M 2014 *AIP Conf. Proc.* **1631** 218–224
[4] Filimonov M and Masih A 2016 *Journal of Physics: Conference Series* **722**
[5] Filimonov M and Masih A 2016 *CEUR Workshop Proceedings* **1662** 245–252
[6] Filimonov M Y 2004 *Proc. of the Steklov Inst. of Math.* **27**(suppl. 1) 1625–1632
[7] Filimonov M 2015 *AIP Conf. Proc.* **1690**
[8] Filimonov M Y 2003 *Diff. Equations* **39** 1159–1166
[9] Filimonov M Y 2003 *Diff. Equations* **39** 844–852
[10] Kazakov A L and Lempert A 2016 *Journal of Physics: Conference Series* **722**
[11] Filimonov M Y 2008 *Proc. of the Steklov Inst. of Math.* **261**(suppl. 1) S55–S76
[12] Kurmaeva V V and Titov S S 2005 *Journal of Applied Mechanics and Technical Physics* **46** 780–790
[13] Kurmaeva V V and Titov S S 2007 *Journal of Applied and Industrial Mathematics* **1** 343–350
[14] Filimonov M Y 2000 *Diff. Equations* **36** 1685–1691
[15] Bashurov V V, Kropotov A I, Vaganova N A, Filimonov M Y, Pchelintsev M V and Skorkin N A 2012 *J. of Appl. Mech. and Techn. Phys.* **53**(1) 43–48
[16] Filimonov M Y and Vaganova N A 2012 *Proceedings of ASME 45158* **4** 133–141
[17] Vaganova N A and Filimonov M Y 2015 *AIP Conf. Proc.* **1690**
[18] Filimonov M Y and Vaganova N A 2016 *Journal of Physics: Conference Series* **754**
[19] Vaganova N A and Filimonov M Y 2015 *Lect. Notes in Comp. Sci.* **9045** 393–399
[20] Pirumov U and GSRoslyakov 1986 *Gas Flows in Nozzles* (Berlin: Springer)

A General Model of an Open Geothermal System

N A Vaganova[1,2] **and M Yu Filimonov**[1,2]

[1] IMM UB RAS, Yekaterinburg, Russia

[2] Ural Federal University, Yekaterinburg, Russia

E-mail: fmy@imm.uran.ru

Abstract. Geothermal aquifers may be considered as a renewable resource of heat and energy. Internal earth heat may be transported with underground water to the surface by wells, and to not overexploit the aquifer, this water should be returned back. As a rule such systems consist of two wells. Hot water from the producing well is used, as example, for greenhouse complex or other buildings needs, which cools the water, and the injection well returns the cold water into the aquifer. To simulate this open geothermal system a three-dimensional nonstationary mathematical model and numerical algorithms are developed taking into account the most important physical and technical parameters of the wells to describe the heat distribution and thermal water transportation in the aquifer. Results of numerical calculations are presented.

1. Introduction

Geothermal energy is a type of renewable energy that encourages conservation of natural resources [1]. The most effective way is direct producing electricity from steam and very hot water (more than 150°C) that can drive turbines. But such resources are in general located in areas with high volcanic activity. Water at intermediate temperatures, from about 50°C to 150°C, can be used directly for industrial processes or for heating buildings. Below 50°C, heat pumps are used to increase the temperature. Even water with a temperature of about 5°C can be used as an energy source, provided there is enough of it. Underground water with medium and low temperature is a widespread energy. In this paper we will consider one type of geothermal systems, which allow to trasfer the heat from deeper layers of ground: an open loop. A geothermal open loop system (fig. 1) is a system of two wells which drilled to tap an aquifer with hot water. The wells are injection (1) and producing (2), respectively. Temperature of water in the producing well is determined by temperature in the aquifer. This water then used for heating and then returns by pumps into the injection well with more cold temperature. During the time of exploitation of the system the cold water is distributed from injection to the producing well. The basic parameters of optimisation are the temperature of water in the producing well and the time when the water will be hot enough.

2. Mathematical model

The aquifer is a complex system and as a rule may include underground lakes, rivers, and kreeks, different porous and waterproofing layer (fig. 1). To simplify the consideration let conside a layer with water filtrating from injection to the producing well.

To simulate underground flows of water in porous media, Darcy's law and law of mass conservation (continuity equation) are used [2]. A convection-diffusion equation with dominant

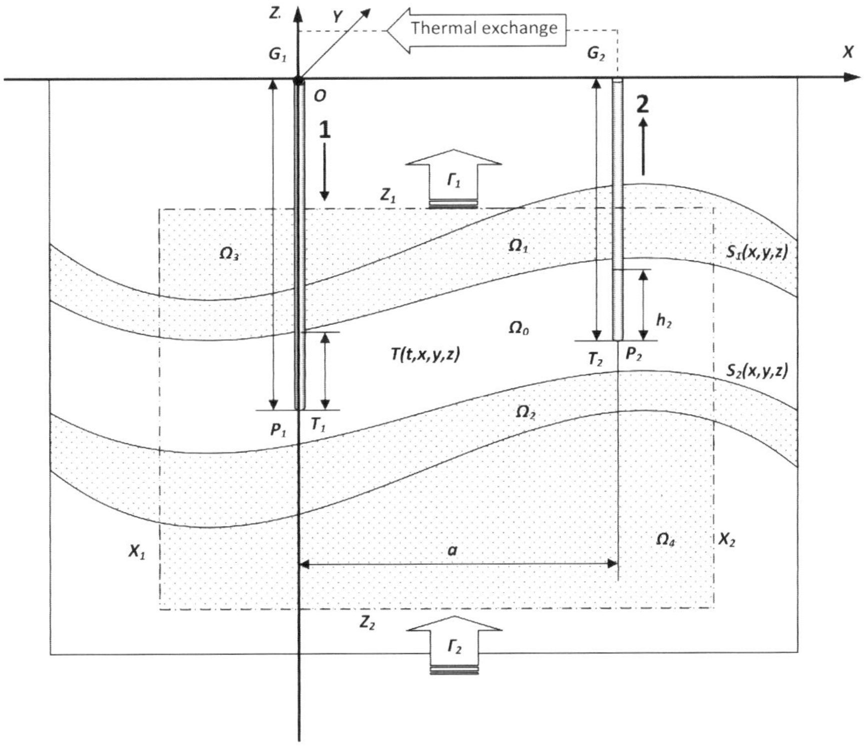

Figure 1. An aquifer with two wells. (1) and (2) are injection and producing well, respectively.

diffusion due to low velocity of filtration is considered. Let $T(t, x, y, z)$ be the temperature in the aquifer, $\mathbf{V} = (u, v, w)$ be a vector of filtration velocity. Thermal exchange is described by equation

$$\frac{\partial T}{\partial t} + b\left(\frac{\partial T}{\partial x}u + \frac{\partial T}{\partial y}v + \frac{\partial T}{\partial z}w\right) = \lambda_0 \Delta T, \tag{1}$$

here $b = \dfrac{\sigma\rho c_f}{\rho_0 c_0(1-\sigma) + \rho c_f \sigma}$, $\lambda_0 = \dfrac{\kappa_0}{\rho_0 c_0(1-\sigma) + \rho c_f \sigma}$, ρ_0 and ρ_f are density of aquifer soil and of water, c_0 and c_f are specific heats of aquifer soil and of water, κ_0 is thermal conductivity coeffitient of soil, σ is porosity, (u, v, w) is vector of velosity of water filtration in the soil. The aquifer has an initial temperature T_0 and in injection well temperature is set as "cold water" with temperature T_1, which returns from the producing well after using. Analytical approach to describe filtration processes meets difficulties and restrictions [3] so it is necessary to use numerical methods [4, 5].

To compute the velocity we use pseudoviscosity method with splitting by physical processes [6]. We solve an equation for pressure $p = p(t, x, y, z)$

$$\frac{\partial^2 p}{\partial x^2} + \frac{\partial^2 p}{\partial y^2} + \frac{\partial^2 p}{\partial z^2} = 0 \tag{2}$$

with the following boundary conditions for the surfaces of injection and producing wells

$$P(t, x, y, z)\Big|_{\Omega_1} = P_1 - \rho g z, \quad P(t, x, y, z)\Big|_{\Omega_2} = P_2 - \rho g z. \tag{3}$$

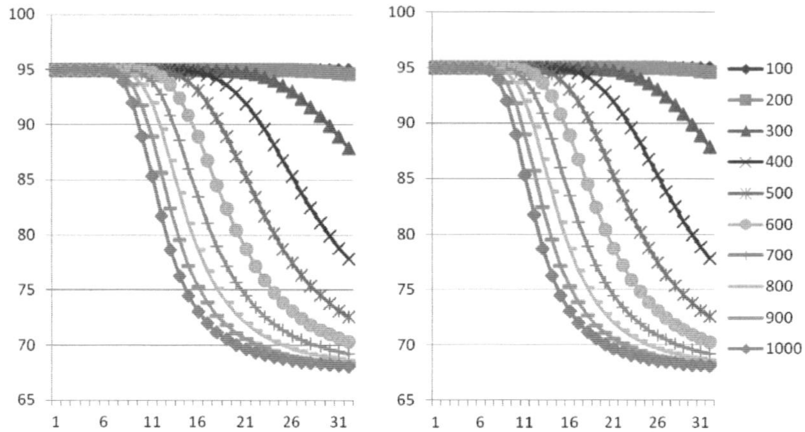

(a) (b)

Figure 2. Temperature in the producing well for different pressure values [kPa] in injection and producing wells in dependence with time of operation [years]. The distances between wells are 600m and 700m in fiqures (a) and (b), respectively

At the lateral boundaries of Ω for temperature and pressure we set the zero flux conditions. Size of the computational domain are large enough as to avoid the influence of the boundary conditions. Due to low velocity of filtration we can use a steady-state flow to describe convective transport terms in equation (1). To compute the components of velosity we may use the following equation for the previously obtained pressure field [6]:

$$\frac{\partial p}{\partial t} + \nabla \mathbf{V} = 0. \tag{4}$$

The equations (1)–(4) for temperature and pressure in aquifer is solved using a finite difference method based on an approach of works of A.A.Samarskii and P.N.Vabishevich [7]. A finite difference method with splitting by the spatial variables in three-dimensional domain is used to solve the problem. We construct an orthogonal grid, uniform, or condensing near the layer boundaries or to the surfaces of the wells. The original equation for each spatial direction is approximated by an implicit central-difference scheme and a three-point sweep method to solve the system of linear algebraic difference equations is used. This approach was successively used for the problems of describing a thermal trace from underground pipeline taking into account filtration and evaporation of fluid from soil surface [8, 9]. After finding the pressure field, a vector of velocity of filtered water is determined in the aquifer [10, 11]. It is also possible to take into account in the model a phaze transition [12, 13].

3. Numerical results

To solve the problems a finite-difference implicit central-difference upwind scheme and a three-point sweep method is used with splitting by the spatial variables in three-dimensional domain with orthogonal grid, uniform, or condensing near the surfaces of wells. For computing pressure p an iterative steady-state method is used with so called "weak" boundary conditions. Because of the processes are slow and quasi-stable we use the sequence pressure (steady state)$\rightarrow$velosity field (steady state)→temperature in dependence with time. Computations are carried out with 1 day time step for 31 years.

We present the results of calculations for the differential pressure between producing and injection wells in the range from 100 kPa ($P_1 = -P_2$=50kPa) and 1000 kPa ($P_1 = -P_2$=500kPa). Effective life of an open geothermal system depends not only from distance between the wellheads in aquifer (fig. 2), but mostly from the power of using, i.e. value of pressure difference. The pressure in the wells may be changed due to seasonal needs variation or others restrictions.

4. Conclusion

A model of geothermal aquifer is considered. This model includes both temperature distribution in underground layers and fluid filtration in porous media. This model is a base for computer simulation of an open geothermal loop consisted of two wells: injection and producing.

Computations allow simulate different regimes of exploitation and to estimate parameters of the open geothermal system, in particular to determine an appropriate distance between injection and producing wells and pump pressure depending on the operating conditions of the geothermal system. Investigations of different types and forms of injection and producing wells will follow.

4.1. Acknowledgments

This work was supported by RFBR projects (16-01-00401).

References

[1] Kreith F and Goswami D Y 2008 *Heat Pumps, Energy Management and Conservation Handbook* (CRC Press)
[2] Polubarinova-Cochina P Y 1977 *Theory of motion of ground water* (Moscow: Nauka)
[3] Vaganova N A 2003 *Proceedings of the Steklov Institute of Mathematics* **S2** S182
[4] Vaganova N A and Filimonov M Y 2015 *Lect. Notes Comput. Sci.* **9045** 385
[5] Filimonov M Y and Vaganova N A 2014 Proc.of IM UrB RAS Problems of Mechanics and Materials Science (Izhevsk: IM UrB RAS)
[6] Chorin A J 1967 *Journal of Computational Physics* **2** 12
[7] Samarsky A A and Vabishchevich P N 1995 *Computational Heat Transfer, Volume 2, The Finite Difference Methodology* (Wiley)
[8] Bashurov V V, Vaganova N A and Filimonov M Y 2011 *Computational technologies* **16(4)** 3
[9] Vaganova N 2014 (*AIP Conference Proceedings* vol 1631) (New York: AIP Publishing)
[10] Vaganova N A and Filimonov M Y 2015 *Lect. Notes in Comp. Sci.* **9045** 393–399
[11] Vaganova N A and Filimonov M Y *Journal of Physics: Conference Series* **754**
[12] Vaganova N A and Filimonov M Y 2015 *AIP Conf. Proc.* **1690**
[13] Filimonov M Y and Vaganova N A 2016 *Journal of Physics: Conference Series* **754**

Astaxanthin production from sewage of traditional Thai rice vermicelli

Chutinut Sujarit[1]*, Waigoon Rittirut[2], Doungporn Amornlerdpison[3] and Chairat Siripatana [4]

[1] Faculty of Science and Fisheries Technology, Rajamangala Univesity of Technology Srivijaya, 92150, Thailand

[2] Used to work at Phyayao University, Thailand

[3] Faculty of Fisheries Technology and Aquatic Resources Maejo University, 50290, Thailand

[4] School of Engineering and Resources, Walailak University, Nakhon Si Thammarat 80160, Thailand

E-mail: s.chutinut58@gmail.com

Abstract. This research aimed to investigate an optimal condition for astaxanthin production by *Phaffia rhodozyma* TISTR 5730 in two different media: synthetic YM medium and the medium added with coconut water and diluted with sewage from Thai traditional rice vermicelli plant (coconut water: sewage of 1:0, 1:1, 1:3 and 1:5 ration respectively). The basic medium formulation was composed of 10 g/L glucose, 3 g/L yeast extract, 0.1 g/L K_2HPO_4, 0.01 g/L NaCl, 0.01 g/L $MgSO_4$ and 0.01 g/L $CaCl_2$ with initial pH 5.5. The cultures were cultivated on 200 rpm shaking bath at 50 °C for 120 hr. It was found that *P. rhodozyma* TISTR 5370 grew optimally when cultivated in a mixture of coconut water and Thai rice vermicelli sewage (ratio of 1:3), with growth of 3.23 g dry biomass/L and specific astaxanthin production of 680 μg/g dry cell respectively. When fan palm sugar was added to increase reducing sugar from 10 to 15, 20 and 25 g/L, it was demonstrated that the 15 g/L formulation produced highest both dry cell weight (9.66 g/L) and astaxanthin (810 μg/g dry cell weight). Furthermore, when 0.5, 1.0 and 1.5 g/L citric acid was added as supplement, it was found that 1.0-g/L citric acid formulation gave the best result: 10.30 g/L dried cell weight and 930 μg/g dry cell weight astaxanthin. This study provides a promising alternative method of sewage reduction and valorization of wastewater from Thai traditional rice vermicelli plant.

1. Introduction

Rice Vermicelli (a traditional rice noddle) is very famous in Thailand, Malaysia and Indonesia. Its demand has been met by established factories which exist in quite different scales ranging from house-hold, small, medium and big scales. In Thailand, the noddles are produced from rice flour and the process uses large amount of water, thus creating a large volume of wastewater which must be treated in some ways. In normal processing condition, wastewater from this process has BOD of 3,000-5,000 mg/L and COD of 4,000-5000 mg/L approximately. Large amount of wastewater with this level of COD/BOD and low pH (approximately 5.5). gives rise to a considerably high cost in compulsive wastewater treatment. To mitigate the problem, some rice vermicelli factories turn this wastewater into biogas. However, this option is not very attractive because the COD of the wastewater is not high enough to be cost effective considering high investment in infrastructure and equipment.

A few researchers have explored other alternatives to better utilize the wastewater including that for producing bioethanol [1,2,3]. Probably one of the better idea is to turn this wastewater into high-value products by fermentation. Many researchers have explored this approach for various type of wastewater and astaxanthin is one product that has fulfilled this potential. Astaxanthin is a pigment in the xanthophyll family of carotenoidsis which is considered to be one of the most valuable compounds with a wide range of applications in the food, feed, cosmetics and pharmaceutical industry [4,5,6]. Its main characteristic is anti-free radical. It helps to prevent cancer in large intestine, digestion system, urine excretion system [7]. It is very popular as a feed supplement in salmon, shrimp and crab farming to induce red color in skin and their meat. High-grade asthaxanthin is a desirable, high-value ingredient for cosmetics. Currently all astaxanthin used in Thailand is imported. Global market for astaxanthin in aqua feed, nutraceuticals, cosmetics and food & beverages is estimated at 280 metric tons valued at US$447 million in 2014. This figure is climbing up gradually and it is predicted that in 2020 the total value of astaxanthin will exceed US$ 1.1 billion [8]. This work investigates the potential of astaxanthin production using wastewater (or sewage) from Thai rice vermicelli factories (called TRVF wastewater). Our long-term aim is to convert considerable wastewater from this source into some high-value product such as astaxanthin in such a way that the business is more profitable while creating more environmental-friendly communities.

2. Material and methods

3.1 Inoculum preparation

Phaffia rhodozyma TISTR 5370 was obtained from Thai Institute of Science and Technology, Phatumthani, Thailand. It was maintained in agar slant containing YM in 40 % glycerol. Before carrying experiments, the yeast was activated by growing in 50 ml of liquid medium (in 250-ml flask) containing YM broth which consisted of 1% glucose, 0.5% peptone, 0.3% malt extract and 0.3% yeast extract. The culture was incubated in a shaking incubator controlled at 25 °C, 200 rpm, the cool-white-fluorescent light having intensity of 500 lux, for period of 24-48 hr. The initial pH of medium was adjusted to 5.5 by 1N HCl before being sterilized at 121 °C for 15 min. The starting inoculum was 10% of the medium content which is equivalent to 10^7 – 10^8 cfu/ml (measured as absorbance of 0.5 nm) [9]

3.2 Wastewater handling and pre-treatment

The wastewater from traditional Thai rice vermicelli processing was collected from a factory in Kanthang district, Trang province. To ensure the uniformity as well as minimum contamination, the cooled wastewater left from the noddle-boiling step inside the boiling pan was used in all experiments. After being cooled, it was collected in plastic containers (500 L total volume) which were covered with ice during transportation from the factory to our laboratory. At the laboratory, all transported wastewater was mixed together and heated at 60 °C for 30 min. Then it was left to cool down and frozen until used in the experiments.

3.3 Media preparation

In this work, YM broth was used as the controlled medium as well as for inoculum preparation. The media used in the experiments were adapted from [9] and [10]. They (called CN media) have coconut-water-to-wastewater ratios of 1:1, 1:3 and 1:5 respectively. All formulas were supplemented with nutrients to contain 10 g/L glucose, 3 g/L yeast extract, 0.1 g/L K_2HPO_4, 0.01 g/L NaCl, 0.01 g/L $MgSO_4$ and 0.01 g/L $CaCl_2$. The initial pH was adjusted to 5.5. This is based on the work of Hu *et al.* [15] as they found that the best optimal initial pH for cell growth was 6.0 whereas the optimal pH was 5.0 for astaxanthin accumulation. Thus the compromised initial pH would be at the neighbourhood of

5.5. Two sources of carbon were investigated, namely: fan palm sugar (the media were adjusted to contain 15, 20, and 25 g/L glucose) and citric acid (citric acid was added and adjusted to contain 0.5, 1.0, and 1.5 g/L citrate respectively. In case of citric acid addition, it is considered as a precursor for astaxanthin production rather than the main carbon source.

3.4 Fermentation

The culture was grown in 250-ml flask having 50-ml broth which was prepared as in previous section. The flasks were incubated in a 200-rpm shaking incubator equipped with 500-Lux from cool white fluorescent lamp, 25 °C for 6 days (144 hr).

3.5 Dry cell and chemical analysis

Microbial growth was measured by monitoring the turbidity (by measuring light absorbance at 660 nm wavelength) as well as by measuring dry-cell content according to Kurane *et al.* [11]. Reducing sugar analysis was in accordance with Somogyi-Nelson method [12]. Astaxanthin content was analysed by a method adapted from [13] and [14]. From previous experiences, which is confirmed by the results in this study, insignificant astaxanthin is produced and accumulated before 48 hr after starting fermentation, thus we started analysing astaxanthin after 48 hr of fermentation until day 6 (120 hr). COD, BOD, total solid (TS), total soluble solid (TSS), oil, volatile soluble solid (VSS), total Kjeldahl nitrogen (TKN) were analysed according to APHA, WCPF standard analysis for wastewater [17].

3.6 Calculation of kinetic parameters

Maximum specific growth rate μ_{max} was calculated from the initial slope of each growth curve (g dry cell mass versus time) of the experiment divided by the instant biomass. Biomass yield coefficient $Y_{x/s}$, astaxanthin yield coefficient $Y_{p/s}$, and specific astaxanthin yield $Y_{p/x}$ were calculated by the following formulas.

$$Y_{x/s}=\frac{\text{dry biomass produced in a time period}}{\text{glucose consumed in that time period}}$$

$$Y_{p/s}=\frac{\text{astaxanthin produced in a time period}}{\text{glucose consumed in that time period}}$$

$$Y_{p/x}=\frac{\text{astaxanthin produced in a time period}}{\text{dry biomass produced in that time period}}$$

All time-course data presented in this work were obtained by averaging results from triplicate experiments.

4. Result and discussion

4.1 Characteristics of wastewater from processing of traditional Thai rice vermicelli

Traditionally, the process of Thai rice vermicelli started from fermenting flour-cake anerobically for 1-2 days, then it was boiled to turn into noodle-like vermicelli while the used water became wastewater to be treated before being released into the natural waterways. The wastewater, having white color, was filtered with thin cloth to remove particulates, kept in the fridge and analysed for COD, BOD, total solid (TS), total soluble solid (TSS), oil, volatile soluble solid (VSS), total Kjeldahl nitrogen (TKN) and initial pH. The results are shown in Table 1. This results suggested that the wastewater fell within the standard for industrial wastewater which can be disposed after treatment. Furthermore, the initial pH of the wastewater is suitable for astaxanthin production [15] given all nutrients are sufficient and balanced. However, the protein and fat concentration were too low to be used directly for astaxanthin production. Thus nutrient supplement was added as elaborated in previous section.

4.2 Growth of Phaffia rhodozyma TISTR 5730 in coconut-water mixed with TRVF wastewater

IOP Conf. Series: Journal of Physics: Conf. Series **820** (2017) 012011 doi:10.1088/1742-6596/820/1/012011

Table 1. Characteristics of sewage from traditional Thai rice vermicelli plant

Parameters	Values
BOD (mg/L)	4,000
COD (mg/L)	5,288
TS (mg/L)	3,250
TSS (mg/L)	1,500
Oil (mg/L)	45
VSS (mg/L)	500
TKN (mg/L)	150
pH	5.5

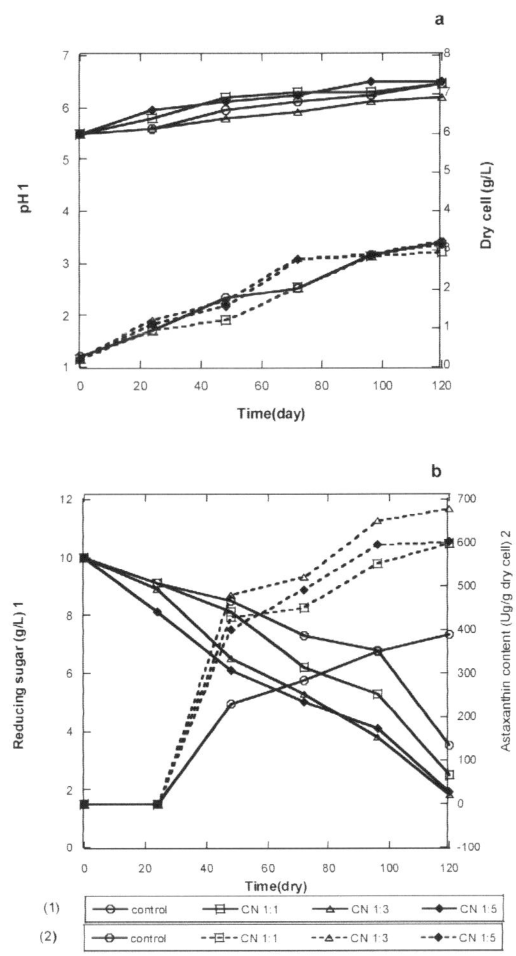

Figure 1 Growth and astaxanthin production of *P. rhodozyma* TISTR 5730 cultured on three different coconut-water-to-wastewater ratios of 1:1, 1:3 and 1:5 respectively: a) pH and dry cell and, b) Reducing sugar (or approximately glucose) and astaxanthin content

The experimental results from batch culturing of *Phaffia rhodozyma* TISTR 5730 in coconut water (CN) diluted with different ratio of coconut-water-to-TRVF-wastewater (1:0, 1:1, 1:3 and 1:5 respectively) are shown in Figure 1. The corresponding parameters are presented in Table 2.

Table 2. Effect of different coconut-water-to-wastewater ratios (1:1, 1:3 and 1:5) on the pH, biomass, astaxanthin content, maximum specific growth rates and yield coefficients for *P. rhdozyma* TISTR 5730 grown at 25 °C and 200 rpm under a light intensity of 500 lux for 120 h.

Coconut-water-to-wastewater ratio →	Control	CN 1:1	CN 1:3	CN 1:5
pH	6.45	6.50	6.20	6.50
Biomass (g dry cell/L)	3.19	2.95	3.23	3.15
Maximum growth rate, μ_{max} h^{-1}	0.15	0.14	0.11	0.13
Glucose consumed after 120 h, (g glucose/L)	6.3	7.2	8.3	8.1
Biomass yield coefficient, $Y_{x/s}$ (g dry cell/g glucose)	0.51	0.41	0.39	0.39
Astaxanthin conc. (µg/L)	1,244	1,741	***2,210***	1,922
Specific astaxanthin yield $Y_{p/x}$ (µg/g dry cell)	390	590	***680***	610
Astaxanthin yield coeeficient, $Y_{p/s}$ (µg/g glucose)	197	242	***265***	238

Starting from pH 5.5, the final pH after 120 hr was in the range of 6.2 to 6.5 and CN 1:3 had the lowest final pH of 6.2 which is closed to an optimal pH (5.0 to 6.0) for *P. rhodozyma*, particularly optimal pH for cell growth [15]. It was evident that the best coconut-water-to-TRVF-wastewater ratio was 1:3 after cultured in 250 ml flask, 200-rpm on shaker incubated at 25 °C for 120 hr. At this condition, 3.23 g/L of dry cell and 680 µg astaxanthin/g dry cell weight was achieved with small pH change (pH 5.5-6.2). It was also found that at the ratio of 1:1 and 1:5 the total cell growth was less than control experiment (basic medium without TRVF wastewater addition). However, the amount of astaxanthin concentration produced was not directly related to the amount of cell dry weight. Although the ratio of 1:1 and 1:2 gave lower growth (2.95 and 3.15 g dry cell/L respectively) than the control one (3.19 g dry cell/L), the astaxanthin produced was about double of the control. This indicated that adding TRVF wastewater into coconut-water medium induced astraxanthin production considerably. It was also noticed that the amount of astaxanthin produced followed closely the final pH, of which, in the experimental range, lower pH was associated with higher astaxanthin concentration. This is consistent with results of Sujarit et al [9,10] and Hu et al [15] who concluded that, for *P. rhodozyma*, pH 6.0 is optimal pH for cell growth whereas the optimal pH for astaxanthin production is 5.0.

4.3 Effect of sugar concentrations

To understand the effect of sugar concentration, palm sugar were added into the best medium chosen from the previous experiments (coconut juice medium mixed with TRVF wastewater at the ratio of 1:3) to adjust the initial reducing sugar concentration such that additional three levels of the concentration (15, 20 and 25 g/L respectively) were obtained. Note that one liter of this palm sugar syrup contains 28.8 g glucose, 26.0 g sucrose, 28.3 g fructose, 0.2-0.3 g protein and 0.98 g citric acid [16]. It's initial pH was 4.69. However, it should noticed that we added crystallized palm sugar not the syrup one.

Figure 2 shows the experimental results from batch culturing of *Phaffia rhodozyma* TISTR 5730 in coconut water (CN) adjusted to three glucose concentrations (15, 20 and 25 g/L respectively) by adding fan palm sugar. The corresponding parameters are presented in Table 3.

Again, the batch fermentation was carried out in 250 ml flask, 200-rpm on shaker incubated at 25 °C, white fluorescent 500-Lux light intensity, for 120 hr. It was evident that, with the sugar range of 0-25 g/L, *P. rhodozyma* TISTR 5370 grew well and consumed most of carbon source for all levels of initial sugar concentrations. As the reducing sugar increased from 10 g/L to 15 g/L, cell growth was higher and biomass yield coefficient $Y_{x/s}$ increased from 0.51 to 0.77 g dry cell weight /g sugar consumed. However, although the sugar was consumed and brought to a low level (< 3.5 g/L), biomass yield coefficient became lower as sugar concentration got higher if the initial sugar concentration was higher than 15 g/L. This indicated that, in the current physico-chemical conditions, the yeast appeared to have a growth limit (which is in the neighbourhood of 10 g/L) and the increase in sugar concentration did not help to promote growth to exceed this limit. This limit could be lifted up by increase oxygen transfer rate which can be achieved by stronger mixing [19].

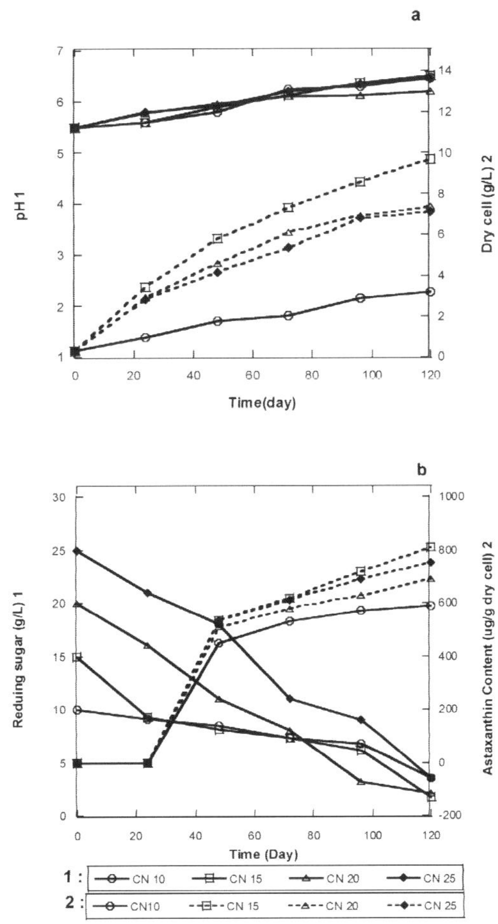

Figure 2 Growth and astaxanthin production of *P. rhodozyma* TISTR 5730 cultured on three levels of the glucose concentration - 15, 20 and 25 g/L respectively (fixed coconut-water-to-wastewater ratios of 1:3) a) pH and dry cell and, b) Reducing sugar (or approximately glucose) and astaxanthin content

Higher sugar concentration than 15 g/L tended to decrease growth rate as well as the maximum biomass attained at the stationary phase. As it is well known that astaxanthin is an intra-cellular pigment, less cell/biomass is generally associated with less astaxanthin yield. However, astaxanthin is a secondary metabolite which is produced during the fermentation by *P. rhodozyma* in both growth-associated and non-growth associated manners. Thus in general, we can infer that, with a certain physiological state of *P. rhodozyma*, the astaxanthin concentration in the cell should be approximately constant so more biomass is associated with more astaxanthin content. However, since the physiological state of *P. rhodozyma* is varied according to physico-chemical condition of the environment as well as its internal factors, we also expected the non-growth associated astaxanthin accumulation played important role in the fermentation period.

Table 3. Effect of glucose concentration (at fixed coconut-water-to-wastewater of 1:3) on the pH, biomass, astaxanthin content, maximum specific growth rates and yield coefficients for *P. rhdozyma* TISTR 5730 grown at 25 °C and 200 rpm under a light intensity of 500 lux for 120 h.

Media with different glucose conc. →	Control 10 g/L	CN 15 g/L	CN 20 g/L	CN 25 g/L
pH	6.45	6.50	6.20	6.43
Biomass (g dry cell/L)	3.19	9.66	7.35	6.78
Maximum growth rate, μ_{max} h^{-1}	0.15	0.17	0.11	0.18
Glucose consumed after 120 h, (g glucose/L)	6.3	12.6	17.3	21.3
Biomass yield coefficient, $Y_{x/s}$ (g dry cell/g glucose)	0.51	0.77	0.42	0.32
Astaxanthin conc. (µg/L)	1,244	***7,824***	3,807	4,407
Specific astaxanthin yield $Y_{p/x}$ (µg/g dry cell)	390	***810***	518	650
Astaxanthin yield coeeficient, $Y_{p/s}$ (µg/g glucose)	197	***620***	220	207

Higher sugar concentration corresponds to high C/N ratio. At first look, our results may not agree with those of Liu and Wu [18] who found that carotenoid biosynthesis is promoted by high C/N ratio, low carbon and nitrogen concentrations, and slightly acidic condition when cell growth is suppressed. However, while high sugar concentration means high C/N ratio (promote astaxanthin synthesis), it also corresponding high carbon concentration (demote the synthesis). Thus, our results are actually supporting the finding of the previous work [18,20] and indicate that the glucose concentration of 15 g/L met the balance between the two opposite effects of high sugar concentration and give the best result.

4.4 Effect of citric acid as an astaxanthin-enhancing supplement

There were considerable attempts to induce the production and accumulation of astaxanthin in the culture of both *Haematococcus pluvialis* and *P. rhodozyma*. Many researchers tried to optimize astaxanthin biosynthesis by pH control strategy [15,22], fed-batch strategy [23,25,27], light type and intensity [24], nitrogen source [21] and citrate supplement [26].

Figure 3 shows the experimental results from batch culturing of *Phaffia rhodozyma* TISTR 5730 in coconut water (CN) adjusted citrate concentration to 0.5, 1.0 and 1.5 g/L (fixed glucose concentrations at 15 g/L by adding fan palm sugar). The corresponding parameters are presented in Table 4.

In our study, when 0.5, 1.0 and 1.5 %wt/volume was added to the mixture of coconut juice and TRVF wastewater (1:3 by volume), initial pH 5.5, after cultured in 250 ml flask, 200-rpm on shaker incubated at 25 °C, white fluorescent 500-Lux light intensity, for 120 hr. It was found that *P. rhodozyma* TISTR 5370 grew much better than in control and CN 1:3 if without citric-acid addition (Table 5 and Figure 4). Addition of citric acid to 1.0 g/L gave the best growth and enhanced astaxathin production considerably. In this case, biomass concentration, in term of dry-cell mas, reached 10.33 g/L and astaxanthin amount of 930 microgram/ g dry cell was attained. This result followed the same trend as observed by Sujarit (2009) for the same microorganisms, but grown in artificial MB medium, with 1.5 g/L citric acid addition. In her work, the referred condition gave the cell and astaxanthin concentration of 13.25 g dry cell/L and 1,051 μg/g dry cell respectively.

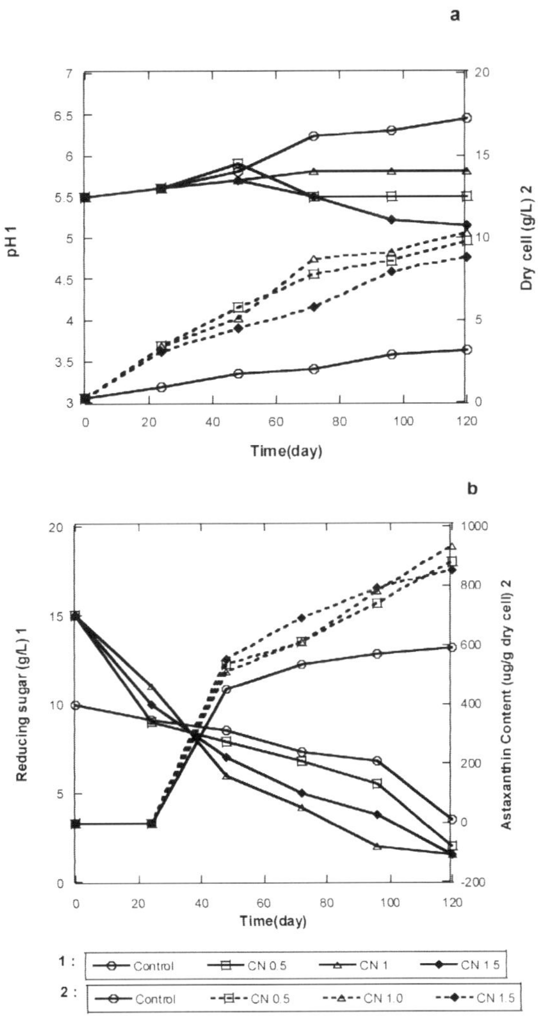

Figure 3 Growth and astaxanthin production of *P. rhodozyma* TISTR 5730 cultured on three levels of citrate concentration – 0.5, 1.0 and 1.5 g/L respectively (fixed coconut-water-to-wastewater ratios of 1:3) a) pH and dry cell and, b) Reducing sugar (or approximately glucose) and astaxanthin content

4.5 Comparison with other works

So far the best result in our work is 9,246 μg/L, 920 μg/g dry cell and 708 μg/g glucose for astaxanthin concentration, specific astaxanthin yield $Y_{p/x}$ and astaxanthin yield coefficient, $Y_{p/s}$ respectively. This is a typical good result for wild type strains grown in various media and physicochemical conditions [29,31,33,35]. However, our achieved quantities were far less than those obtained from commercial production from which well-developed microbial strains were used [28,30,32,34]. For example, Hu *et al.* [28] achieved the maximum astaxanthin concentration of 39.47 mg/L after 132 hours of fermentation, which is more than four-folds of the best result in this work. This is not surprised and our results could be improve much further if better feeding strategy and more optimized condition is used. However, suffice is to say that sewage from Thai traditional rice vermicelli plant mixed with coconut water can provide a good alternative for better waste utilization to produce astaxanthin in the future. Certainly, much more future work is needed.

Table 4. Effect of citrate concentration (at fixed coconut-water-to-wastewater of 1:3) on the pH, biomass, astaxanthin content, maximum specific growth rates and yield coefficients for *P. rhdozyma* TISTR 5730 grown at 25 °C and 200 rpm under a light intensity of 500 lux for 120 h.

Media with different citrate conc. →	Control 10 g glucose/L	CN 0.5 g/L 15 g glucose/L	CN 1.0 g/L 15 g glucose/L	CN 1.5 g/L 15 g glucose/L
pH	6.45	5.40	5.75	5.10
Biomass (g dry cell/L)	3.19	9.66	10.05	8.10
Maximum growth rate, μ_{max} h^{-1}	0.15	0.17	0.11	0.18
Glucose consumed after 120 h, (g glucose/L)	6.3	12.6	13.0	13.0
Biomass yield coefficient, $Y_{x/s}$ (g dry cell/g glucose)	0.51	0.77	0.77	0.62
Astaxanthin conc. (μg/L)	1,882	8,404	***9,246***	6,925
Specific astaxanthin yield $Y_{p/x}$ (μg/g dry cell)	590	870	***920***	855
Astaxanthin yield coeeficient, $Y_{p/s}$ (μg/g glucose)	301	670	***708***	530

3. Conclusion

Optimal condition for cultivating cell and producing astaxanthin in batch culture using wild-type *Phaffia rhodozyma* TISTR 5730, grown in a medium formulated by mixing coconut juices with wastewater from Rice Vermicelli factory, supplemented with 3 g/L yeast extract, 0.1 g/L K_2HPO_4, 0.01 g/L NaCl, 0.01 g/L $MgSO_4$, and 0.01 g/L $CaCl_2$, was found with respect to CW-to-TRVF ratio, sugar concentration, and concentration of citric acid supplement. For fixed initial pH of 5.5, cultivated in 200-rpm shaken flask, exposed to 500-lux fluorescent lamps, at 25 °C for 120 hr, the optimal CW-to-TRVF ratio, sugar and citrate concentration was 1:3, 15 g/L and 1.0 g/L respectively.

4. Acknowledgement

This research were supported by Walailak University (analytical equipment and devices as well as the laboratory space) and Rajamanklala Srivijaya (finacial support).

5. References

[1] Siripattanakul-Ratpukdi S 2012 *Applied Water Science* **2** 1 47-53
[2] Siripattanakul S Ratanapongleka K, Sangthean S, Yoottachana K and Pimwongnok K 2010 *Water Practice and Technology* **5** 3
[3] Ratanapongleka K, Siripattanakul S, Suvannapen W and Tummavong J 2010 *International Journal of Chemical and Environmental Engineering* **1** 1 13-17

[4] Borowitzka M 2013 *Journal of Applied Phycology* **25** 743-757
[5] Koller M, Muhr A and Braunegg G 2014 *Algal Research* **6** 52-63
[6] Pérez-López P, González-García S, Jeffryes C, Agathos S N, McHugh E, Walsh D, Murray P, Moane S, Feijoo G and Moreira M 2014 *Journal of Cleaner Production* **64** 332–344.
[7] Jyonouchi H, Sun S, Iijima K, Gross M D 2000 *Nutrition and Cancer* **36** 59–65
[8] Research and Market 2015 *Global Astaxanthin Market - Sources, Technologies and Application*
[9] Sujarit C 2009 *Optimization and Modeling of Astaxanthin Production by Phaffia rhodozyma TISTR 5730*. PhD Thesis Program in Biotechnology, Walailak University, Thailand
[10] Sujarit C, Siripatana C and Rittirut W 2012 Production of Astaxanthin from Sago Flour by *Phaffia rhodozyma* TISTR 5730 *1st Mae Fah Luang University International Conference* 1-8
[11] Kurane R, Hatamochi K, Kakuno T, Kiyoshara M, Hirano M, Taniguchi Y 1994 *Bioscience Biotechnology and Biochemistry* **58** 428–429.
[12] Nelson N 1944 *J. Biol. Chem.* **153** 375-380
[13] Britton G 1976 *Journal of the Chemical Society, Chemical Communications* **3** 49
[14] Jian-Ping Y, Xiian, D I and Feng C 1997 *Journal of Agricultural and Food Chemistry* **45** 1952-1956.
[15] Hu Z-C, ZhengY-G, Wang Z and Shen Y-C 2006 *Enzyme and Microbial Technology* **39** 4 586-590
[16] Sujarit C and Rittirut W 2014 *Journal of Fisheries Technology Research* **8** 2
[17] APHA, WCPF 1992 Standard methods for the examination of waters and wastewaters, 17th edn. Washington, DC
[18] Liu Y-S, Wu J-Y 2007 *Biochemical Engineering Journal* **36** 182-189
[19] Liu Y-S, Wu J-Y and Ho K-p 2006 *Biochemical Engineering Journal* **27** 331-335
[20] Yamane Y, Higashida K, Nakashimada Y, Kakizono T and Nishio N 1997 *Biotechnology Letters* **19** 1109-11
[21] Fang T J and Wang J-M 2002 *Process Biochemistry* **37** 1235-1245
[22] Luo H, Niu Y, Duan C, Su H and Yan G 2013 *Process Biochemistry* **48** 195-200
[23] Moriel D G, Machdo I M P, Fontana J D and Bonfim T M B 2004 *Brazilian Journal of Pharmaceutical Sciences* **40** 3 421-424
[24] Stachowiak B and Czarnecki Z 2007 *Polish Journal of Food and Nutrition Sciences* **57** 3 129-131
[25] Ramírez J, Obledo N, Arellano M and Herrera 2006 *e-Gnosis* **4** Art. 5
[26] Flores-Cotera L B, Martín R and Sánchez S 2001 *Applied Microbiology and Biotechnology* **55** 341-347
[27] Moriel D G, Chociai M B, Machdo I M P, Fontana J D and Bonfim T M B 2004 *Brazilian Archives of Biology and Technology* **48** 3 397-401
[28] Hu Z-C, ZhengY-G, Wang Z and Shen Y-C 2007 *Food Technol. Biotechnol.* **45** 2 209–212
[29] Bhatt P C, Ahmad M and Panda B P 2013 *Biocatalysis and Agricultural Biotechnology* **2** 58–63
[30] Fang T J and Chiou T-Y 1996 *Journal of Industrial Microbiology* 16 p 175-181
[31] Luna-Flores C H, Ramírez-Cordova J J, Pelayo-Ortiz C, Femat R and E.J. Herrera-López E J 2010 *Biochemical Engineering Journal* **53** 131–136
[32] Park K-M, Song M-W, Kang S-J and Lee J-H 2007 *J. Microbiol. Biotechnol.* **17** 7 1221-1225
[33] Jirasripongpun K, Pewlong W, Kitraksa P and Krudngern C 2008 *Letters in Applied Microbiology* **47** 112–116
[34] Xiao A, Jiang X, Ni H, Yang Q and Cai H 2015 *Electronic Journal of Biotechnology* **18** 148-153
[35] Domínguez-Bocanegra A R and Torres-Muñoz J A 2004 *Applied Microbiology and Biotechnology* **66** 3 249-252

IOP Conf. Series: Journal of Physics: Conf. Series **820** (2017) 012012 doi:10.1088/1742-6596/820/1/012012

Experimental Demonstration of the Dependence of Temperature Decrease on the Hydraulic Radius of Regenerator in a Traveling-Wave Thermoacoustic Refrigerator

Ikhsan Setiawan, Muhammad N M Fadly, Agung B S Utomo

Departement of Physics, Faculty of Mathematics and Natural Sciences, Universitas Gadjah Mada, Sekip Utara BLS 21 Yogyakarta 55281 Indonesia

E-mail: ikhsan_s@ugm.ac.id

Abstract. Thermoacoustic refrigerators are cooling devices that employ sound wave in heat removal process. They mostly use air or other inert gas as working fluid and hence they are environmentally benign. The cooling process takes place in a porous medium (regenerator or stack). This paper presents an experiment on a loudspeaker-driven traveling-wave thermoacoustic refrigerator to demonstrate the dependence of temperature decrease on the hydraulic radius (r_h) of regenerator. The refrigerator has a looped tube and a resonator tube, with the length of 260 cm and 150 cm, respectively. Six different regenerators are used, each of which is made of a tight stack of stainless-steel wire-mesh screens with different mesh sizes, providing r_h from 0.07 mm until 0.28 mm. The regenerator length is 60 mm. Atmospheric air is used as the working gas, driven at operating frequency of 30 Hz, giving thermal penetration depth (δ_κ) of 0.49 mm. The experimental results show that there is an optimum r_h which gives the largest temperature decrease at the cooling point. It is found, in this case, that the optimum r_h is 0.10 mm (that is, $r_h/\delta_\kappa = 0.20$) and the corresponding temperature decrease is 18.7 °C (i.e. from 28.0 °C down to 9.3 °C).

1. Introduction

Thermoacoustic refrigerators are cooling devices that employ sound wave to absorb heat from a cold region and release more heat to the environment. They mostly use air or other inert gas as working fluid and hence they are environmentally benign.[1, 2] The cooling process is induced by thermoacoustic effects which are thermal interactions between the oscillating gas within the flow channels of a porous medium and the solid walls of the channels. Detail description on the thermal interaction and thermodynamic cycle executed by the oscillating gas which interacts with the solid wall can be found elsewhere.[1, 3, 4]

The important parameter of the porous medium is the hydraulic radius (r_h) of the flow channels. In thermoacoustic devices, the hydraulic radius is usually compared to the thermal penetration depth (δ_κ) of the working gas to specify the thermal contact condition between the gas and channel walls.[4, 5] If $r_h \gg \delta_\kappa$, there is no thermal contact between the gas and the walls. The gas parcels will expand and contract adiabatically and reversibly, no heat transfer will take place, and hence the thermoacoustic effects will not occur. On the other hand, if

$r_h \ll \delta_\kappa$, a perfect thermal contact condition between the gas and the walls is obtained, and the porous medium is called a "regenerator", whereas if $r_h \sim \delta_\kappa$, the situation is referred to an imperfect thermal contact because there is a time delay between gas motion and heat transfer, and the porous medium is called a "stack". Standing wave thermoacoustic devices require the imperfect thermal contact condition to work, while the traveling wave thermoacoustic devices can operate under the perfect thermal contact circumstance.[3, 4] The former applies Brayton cycle on the working gas, consists of two reversible adiabatic steps and two irreversible isobaric steps, whereas the latter is based on Stirling cycle which is inherently reversible, and therefore their thermal efficiencies can easily surpass that of standing wave devices.[6]

Because the thermal performance (efficiency) of the traveling-wave refrigerators is higher than those of standing-wave refrigerators, and due to the importance of the role of regenerator hydraulic radius in thermoacoustic devices as described above, we conduct an experiment on a traveling-wave thermoacoustic refrigerator with various hydraulic radius of regenerator. The aim is to experimentally demonstrate the influence of the hydraulic radius on the maximum temperature decrease that can be achieved by the refrigerator. The detail of the experimental setup and method is described in the following section.

2. Experimental Setup and Method

The traveling-wave thermoacoustic refrigerator used in this experiment is driven by a loudspeaker and mainly consists of a looped tube, a resonator tube, a regenerator, and a loudspeaker, as shown schematically in Fig. 1. The looped tube and resonator are made of polyvinyl chloride (PVC) pipes with inner diameter of around 55 mm. The total length of the looped tube is 220 cm, whereas the resonator tube has the length of 150 cm. The working gas inside the looped-tube and resonator is free air at atmospheric pressure and room temperature ($\approx 28\,^\circ$C). A 8-inch loudspeaker is used to provide acoustic wave into the tubes. The sound frequency is controlled by using a digital function generator, while the electrical input power is adjusted by an audio amplifier and calculated from the electric current and voltage measured by a digital ampere-meter A and a digital volt-meter V, respectively. A regenerator, with length of 60 mm, is placed inside the looped tube at a distance of 25 cm from the top left corner of the looped-tube illustrated in Fig. 1 (measured to the center of regenerator). The temperatures at both ends of regenerator, denoted as T_h and T_c, are measured by using LM35 sensors which are connected to a data logger and a computer.

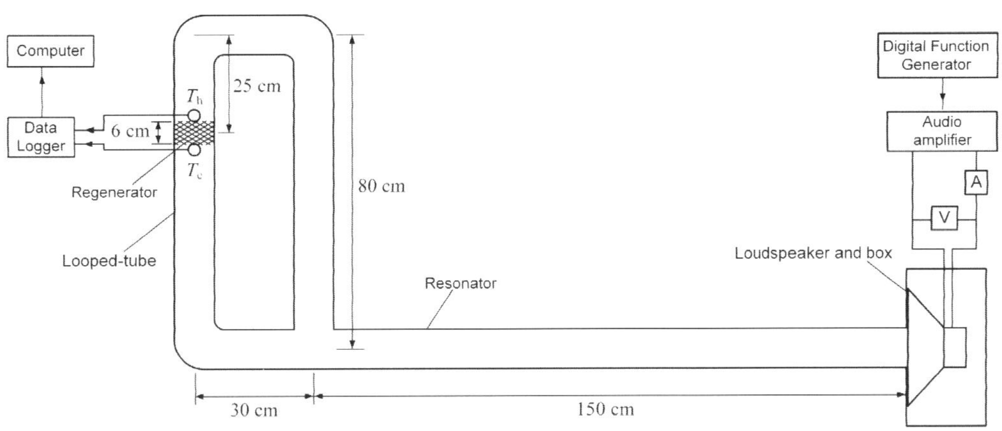

Figure 1. The schematic diagram of the experimental setup.

Table 1.
Six regenerators with different mesh sizes (n) and wire diameter (D) used in this experiment. The estimated hydraulic radius of the regenerator is r_h.

Regenerator	n (inch^{-1})	D (mm)	r_h (mm)
1	24	0.21	0.28
2	30	0.18	0.22
3	40	0.16	0.16
4	50	0.14	0.13
5	60	0.14	0.10
6	80	0.12	0.07

In order to demonstrate the influence of the hydraulic radius of regenerator on the temperature decrease, six regenerators with different hydraulic radius are used. They are made of stainless-steel wire-mesh screens which are stacked tightly inside the PVC tube. Various hydraulic radius of the regenerators are obtained by using different mesh sizes (n) of the wire-mesh screen for each regenerator. (The mesh size represents the number of holes per unit length, usually expressed in inch^{-1}.) Table 1 presents various mesh sizes which are used in this experiment, together with the wire diameter D, and the hydraulic radius (r_h) of the corresponding regenerators. The r_h is estimated by using equation [7]

$$r_h = D\frac{\phi}{4(1-\phi)}, \tag{1}$$

where $\phi = 1 - \pi n D/4$ is the porosity of regenerator. From Table 1 we see that the hydraulic radius of the regenerators vary from 0.07 mm until 0.28 mm.

The experimental procedure consists of two steps as followed. The first step is to determine the operating frequency of the refrigerator. The frequency of sound produced by the loudspeaker is varied by using the digital function generator. For each frequency variation, the temperatures T_h and T_c at both ends of regenerator are measured as the refrigerator operated, i.e. as functions of time, and the maximum temperature decrease $\Delta T_{c,\max}$ that can be achieved by the refrigerator for each frequency is recorded. The operating frequency is then determined as the frequency that gives the largest $\Delta T_{c,\max}$. The second step is to see the influence of different hydraulic radius r_h of regenerator on the $\Delta T_{c,\max}$. The hydraulic radius is varied by using six regenerators which have various hydraulic radius as listed in Table 1. For each r_h variation, the temperatures T_h and T_c are measured as functions of time and the $\Delta T_{c,\max}$ is noted. In this case, the refrigerator is driven at its operating frequency which has been found in the first step described above.

3. Results and Discussion

For a given regenerator (that is, for a given n or r_h), the sound frequency is varied from lowest value of 30 Hz, which is the lower limit of frequency range of the loudspeaker that we used, with the increment of 5 Hz. The typical measurement results of the temperatures T_h and T_c as functions of time are shown in Fig. 2 which presents the result for the regenerator 5 (i.e. with r_h = 0.10 mm and n = 60 inch^{-1}) and frequency of 30 Hz. It reveals, in this case, that at steady state (after operates for about 500 s) the temperature at the cooling point (T_c) is 9.3 °C. It is the lowest temperature that can be achieved by the refrigerator which operates at the condition mentioned above. It means that the maximum temperature decrease ($\Delta T_{c,\max}$) is 18.7 °C recalling that the initial temperature is 28.0 °C.

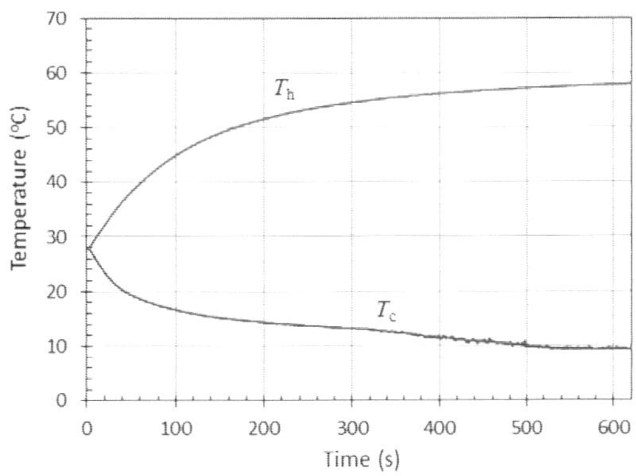

Figure 2. A typical temperatures history at the both ends of regenerator (T_h and T_c). In this case, the regenerator 5 is used with sound frequency of 30 Hz.

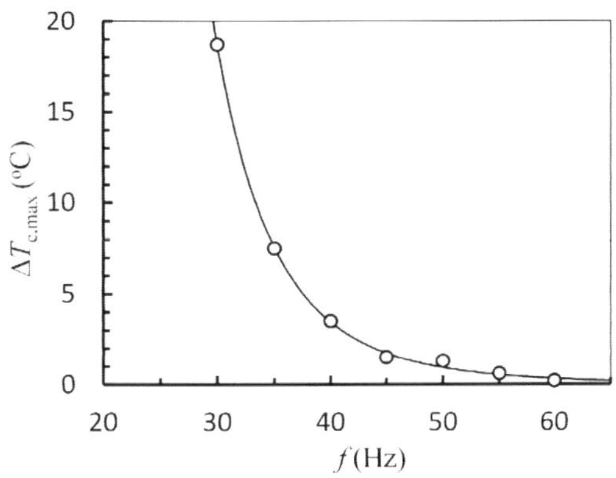

Figure 3. The influence of the sound frequency f on the maximum temperature decrease $\Delta T_{c,\max}$. The solid line is a guide to the eye.

Next, Fig. 3 shows the influence of sound frequency on $\Delta T_{c,\max}$ at the cooling point in the thermoacoustic refrigerator with regenerator 5. This figure shows the fact that $\Delta T_{c,\max}$ is significantly depending on the driving frequency, and the largest $\Delta T_{c,\max}$ is obtained by using 30 Hz sound frequency. Because 30 Hz is the lower limit of frequency range of the loudspeaker that we used, it is deduced that the operating frequency of the thermoacoustic refrigerator is 30 Hz. This frequency produces the largest $\Delta T_{c,\max}$ probably because it is close to the resonance frequency of the refrigerator.

The thermal penetration depth δ_κ of gas is dependent on frequency according to the equation[7]

$$\delta_\kappa = \sqrt{\frac{2\kappa}{\omega}} \tag{2}$$

where κ is the thermal diffusivity of gas and ω is the angular frequency of sound wave. Therefore, our thermoacoustic refrigerator with atmospheric air ($\kappa = 2.26 \times 10^{-5}$ m^2/s) as working gas and

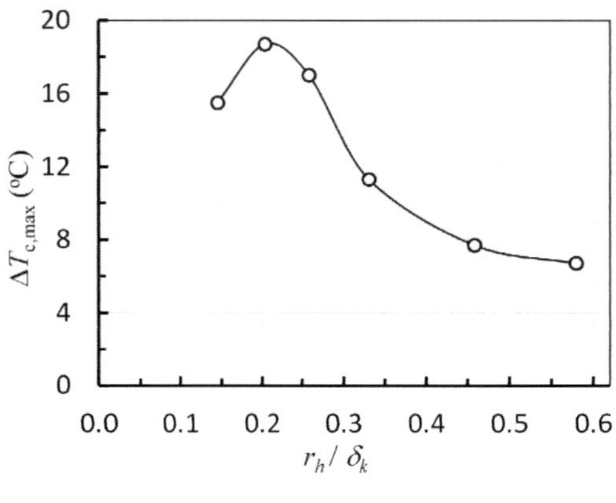

Figure 4. The influence of normalized hydraulic radius (r_h/δ_κ) on the maximum temperature decrease ($\Delta T_{c,\max}$). The solid line is a guide to the eye.

driven at the operating frequency of 30 Hz will have $\delta_\kappa = 0.49$ mm. Furthermore, if we compare this thermal penetration depth with the hydraulic radius presented in Table 1, we find that the condition of $r_h \ll \delta_\kappa$ occurs in our traveling-wave thermoacoustic refrigerator.

Figure 4 depicts the dependence of $\Delta T_{c,\max}$ on the hydraulic radius of regenerator which is normalized by the thermal penetration depth, r_h/δ_κ. It can be seen that there is an optimum value of r_h/δ_κ ratio which gives the largest $\Delta T_{c,\max}$. In this case, the optimum r_h/δ_κ is 0.20 (i.e. $r_h = 0.10$ mm) and the related temperature decrease $\Delta T_{c,\max}$ is 18.7 °C. This fact indicates that the use of regenerator with r_h about $\frac{1}{5}\delta_\kappa$ yields the most effective heat pumping in this thermoacoustic refrigerator. When $r_h < \frac{1}{5}\delta_\kappa$, the channel size is relatively too small ($r_h < 0.10$ mm), resulting in a large viscous dissipation in the regenerator and hence a less effective heat pumping process along the regenerator. On the other hand, when $r_h > \frac{1}{5}\delta_\kappa$ (that is, for the right branch of the curve in Fig. 4), the channel is relatively wide ($r_h > 0.10$ mm), the distance of gas parcels to channel wall is too far, thus the heat exchange between the gas and channel wall does not effectively occur, and the cooling process is not optimal.

4. Conclusion and Future Work

The dependence of temperature decrease on the hydraulic radius of regenerator in a loudspeaker-driven traveling-wave thermoacoustic refrigerator has been experimentally demonstrated. The experimental results show that there is an optimum hydraulic radius which gives the largest temperature decrease. It is found that the optimum hydraulic radius is 0.20 times thermal penetration depth of the working gas. In this thermoacoustic refrigerator, in which the atmospheric air is used as the working gas, the optimum hydraulic radius is 0.10 mm, and the related temperature decrease is 18.7 °C (i.e. from 28.0 °C down to 9.3 °C). This result implies that we need to design a regenerator with hydraulic radius of around 0.20 thermal penetration depth in order to get the lowest temperature decrease in a traveling wave thermoacoustic refrigerator.

Improving the performance of the thermoacoustic refrigerator is very important, and having an optimum hydraulic radius of the regenerator is one of the keys. The next effort that could be taken in order to upgrade the performance is by employing high pressure working gas. In this case, in designing a proper regenerator (with optimum hydraulic radius), we should be aware that the gas pressure will affect the operating (resonance) frequency and thermal penetration depth of the working gas [8].

Acknowledgement
This research is financially supported by the Annual Research Grant in Department of Physics, Faculty of Mathematics and Natural Sciences, Universitas Gadjah Mada, Yogyakarta, Indonesia.

References

[1] Jin T *et al* 2015 Thermoacoustic prime movers and refrigerators: Thermally powered engines without moving components *Energy* **93** 828–53
[2] Ghorbanian K and Karimi M 2014 Design and optimization of a heat driven thermoacoustic refrigerator, *Appl. Therm. Eng.* **62**, 653–61
[3] Garret S L 2016 Thermoacoustic engines and refrigerators *The 13th French Acoustics Congress* April 11–15 Le Mans, France
[4] In't Panhius P H M W 2009 Mathematical aspects of thermoacoustics *Ph.D. Thesis* Universiteit Eindhoven Netherlands ISBN 978-90-386-1862-3
[5] Garret S L 2004 Resource letter: TA-1: Thermoacoustic engines and refrigerator *Am. J. Phys.* **72**, 11–7
[6] Backhaus S and Swift G W 2000 A thermoacoustic-Stirling heat engine: Detailed study *J. Acoust. Soc. Am.* **107** 3148–66
[7] Swift G W 2002 *Thermoacoustics: A unifying perspective for some engines and refrigerators* (New York: Acoust. Soc. Am.)
[8] Ikhsan Setiawan *et al* 2016 Experimental study on a standing wave thermoacoustic prime mover with air working gas at various pressures *J. Phys.: Conf. Ser.* **710** 012031.

IOP Conf. Series: Journal of Physics: Conf. Series **820** (2017) 012013 doi:10.1088/1742-6596/820/1/012013

Solution of electromagnetic wave equation in graded interface for exponential permeability and permittivity profile

A Suparmi, Beta Nur Pratiwi , C Cari and Dewanta Arya Nugraha

Physics Department, Graduate Program, Sebelas Maret University, INDONESIA

Email: soeparmi@staff.uns.ac.id, pratiwi.bn@gmail.com

Abstract. The electromagnetic wave equation in graded interface for exponential permeability and exponential permittivity profile was solved using asymptotic iteration method (AIM). In exponential function, we obtained the graded graphs of permittivity and permeability. By using exponential permeability and permittivity profile in Maxwell equation, we reduced Maxwell equation into the second orde differential equation. The second orde differential equation was reduced to AIM-type differential equation using separation variable and suitable parameter. By AIM, the wave vector and field equations were obtained. The field equations were expressed in hypergeometric function.

1. Introduction

During the last decade, a new class of artificial composite materials has attracted many researchers. The artificial composite materials which unique properties from those normally found in nature. One of best-known classes of these materials is left-handed material (LHM) with negative permeability and negative permittivity.The negative permeability was obtained from the double split-ring resonator and negative permittivity from nano-wire.[1] Materials which normally found in nature is classified in right-handed material (RHM).

LHM has unusual properties, the direction of the Pointing vector in these materials is opposite to that of the wave vector. The first theoretical consideration on the topic, written by Vaselago in 1968 [2], and then in 1996, Pendry brought the concept to practical [3]. LHM could be useful in many various applications, for instance in lensing, filtering and antireflection coatings. [4]

Permeability and permittivity are the fundamental characteristics to control the propagation of electromagnetic waves. The graded graph for permeability and permittivity profile was solved in some earlier research. LHM with the graded refractive index interface has been given by Dalarsson and Tassin [5]. The function of hyperbolic tangent is the mostly used for graded profile. [6-7]. In recently, Husein et al. give the approximate solution of wave propagation in through a graded interface positive-negative [8]. Some methods to solve the second order differential equation has been used by many researchers, such Nikiforov-Uvarov [9-10], the Supersymmetry Quantum Mechanics (SUSY QM) [11], the polynomial Romanovski[12] and asymptotic iteration method (AIM) [13-15].

In this paper, we solved the electromagnetic wave equation in transverse magnetic mode for a graded interface using exponential permeability and permittivity profile. By using AIM, we obtained the fields equation and wave vector. The basic theory is presented in section 2. The results are presented in section 3, in here we restrict just on the calculations and a conclusion in section 4.

2. The basic theory

2.1. The fields equation

Maxwell equation for dielectric, are given as

$$-\nabla\times\vec{E}=i\omega\mu(\omega,y)\vec{H},\quad \nabla\times\vec{H}=i\omega\varepsilon(\omega,y)\vec{E} \tag{1}$$

where $\mu(\omega,y)$ and $\varepsilon(\omega,y)$ are frequency and spatial dependent magnetic permeability and dielectric permittivity, respectively. [8] To simplify the calculations, we set the *yz* plane as the plane of ray incident, so $(d/dx)=0$.In addition, the optical characteristics of $\varepsilon(y)$ and $\mu(y)$ are considered as functions of material thickness, *y*. Using Maxwell equation and considering the geometry of the material, we obtain a second order differential equation of electromagnetic waves in inhomogeneous materials on mode transverse magnetic (TM). Actually, in two-dimensional study, there are two independent modes i.e. TM mode, the group of fields: $\{E_x,E_y,H_z\}$, $\{E_y,E_z,H_x\}$,$\{E_z,E_x,H_y\}$ and transverse electric (TE) mode, the group of fields: $\{H_x,H_y,E_z\}$, $\{H_y,H_z,E_x\}$,$\{H_z,H_x,E_y\}$.In principle, these two modes (TM and TE) are similar to each other, so we only need to do one calculation, e.g. the TM mode. By using the plane wave solution which is harmonic in time, is given

$$H_x=H(y,z)e^{i\omega t} \tag{2}$$

For the group fields in TM mode which we use i.e.$\left[E_y,E_z,H_x\right]$. From equation (1) and equation (2) we obtain

$$\frac{\partial^2 H_x}{\partial y^2}-\frac{\varepsilon'}{\varepsilon}\frac{\partial H_x}{\partial y}+\omega^2\mu\varepsilon H_x+\frac{\partial^2 H_x}{\partial z^2}=0 \tag{3}$$

and by using $H_x=H(y,z)=H(y)H(z)$ in equation (3), we obtained

$$\frac{\partial^2 H(z)}{\partial z^2}+k_z^{\,2}H(z)=0 \tag{4}$$

$$\frac{\partial^2 H(y)}{\partial y^2}-\frac{\varepsilon'}{\varepsilon}\frac{\partial H(y)}{\partial y}+\left(\omega^2\mu\varepsilon-k_z^{\,2}\right)H(y)=0 \tag{5}$$

where k_z is the wave vector in *z* direction. Equation (4) is the simple differential equation which it has the solution,

$$H(z)=e^{\pm ik_z z} \tag{6}$$

while, for equation (5) with the graded interface permittivity and permeability is solved using AIM.

2.2. Asymptotic Iteration Method

AIM-type differential equation in terms: [13-15]

$$f_n''(d)-\lambda_0(d)f_n'(d)-s_o(d)f_n(d)=0 \tag{7}$$

where $\lambda_0(d)\neq 0$ and $s_o(d)$ are coefficients of the differential equation and are well-defined functions as well as sufficiently differentiable. The solution of equation (7) can be obtained by using iteration of λ_j and s_j :

$$\lambda_j(d)=\lambda_{j-1}'+\lambda_{j-1}\lambda_0+s_{j-1};s_j(d)=s_{j-1}'+s_0\lambda_{j-1,}j=1,2,3,... \tag{8}$$

Eigenvalues can be obtained using equation (9) below,

$$\lambda_j(d)s_{j-1}(d)-\lambda_{j-1}(d)s_j(d)=0=\Delta_j,j=1,2,3... \tag{9}$$

On the other hand, equation (7) can be written in term

$$f^{''}(d)=2\left(\frac{ad^{N+1}}{1-bd^{N+2}}-\frac{c+1}{d}\right)f^{'}(d)-\frac{Wl^{N}}{1-bd^{N+2}}f(d) \tag{10}$$

Equation (10) is AIM-type differential equation which is solved by using equation (11)

$$f_n(d)=(-1)^n C(N+2)^n(\sigma)_n\,{}_2F_1\left(-n,q+n,\sigma,bd^{N+2}\right);n=0,1,2,3... \tag{11}$$

where

$$\sigma=\frac{2c+N+3}{N+2};q=\frac{(2c+1)b+2a}{(N+2)b} \tag{12}$$

Here C is normalization constant, $(\sigma)_n$ is Pochamer symbol and ${}_2F_1$ is hypergeometric function. Equation (10) is used to obtain the fields equation.

3. Solution of the electromagnetic wave equations

The exponential permeability and permittivity are expressed as

$$\mu(\omega,y)=\frac{\mu_0\mu_r}{1+e^{(\rho y)}} \tag{13}$$

$$\varepsilon(\omega,y)=\frac{\varepsilon_0\varepsilon_r}{1+e^{(\rho y)}} \tag{14}$$

with ρ is abruptness's parameter describing the transition of wave propagation from RHM to LHM.. The graded graph of permeability with relative permeabiliy is negative is expressed in Figure 1.a and the graded graph of permittivity with relative permittivity is positive in Figure 1.b.

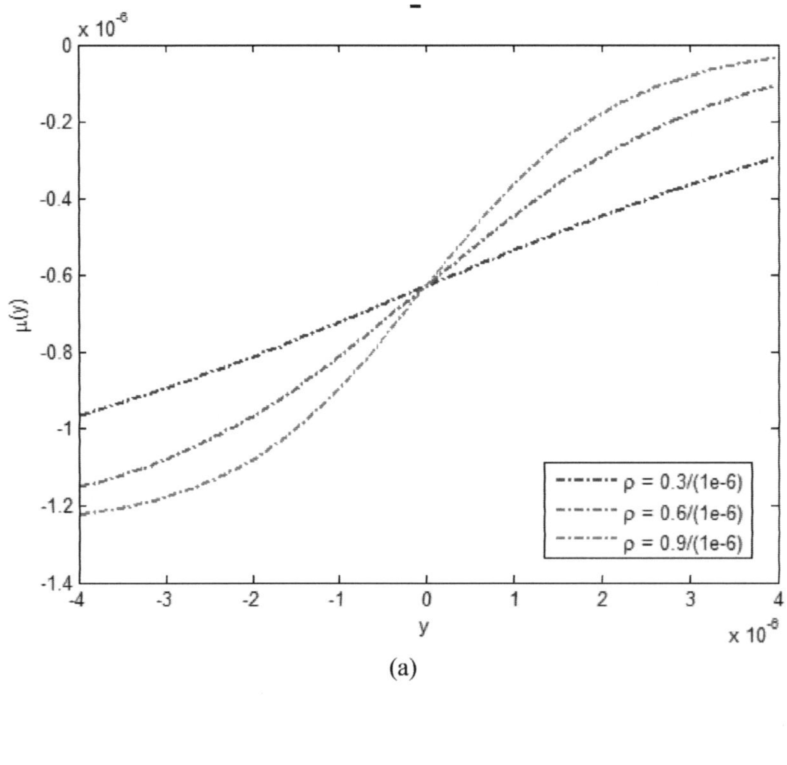

(a)

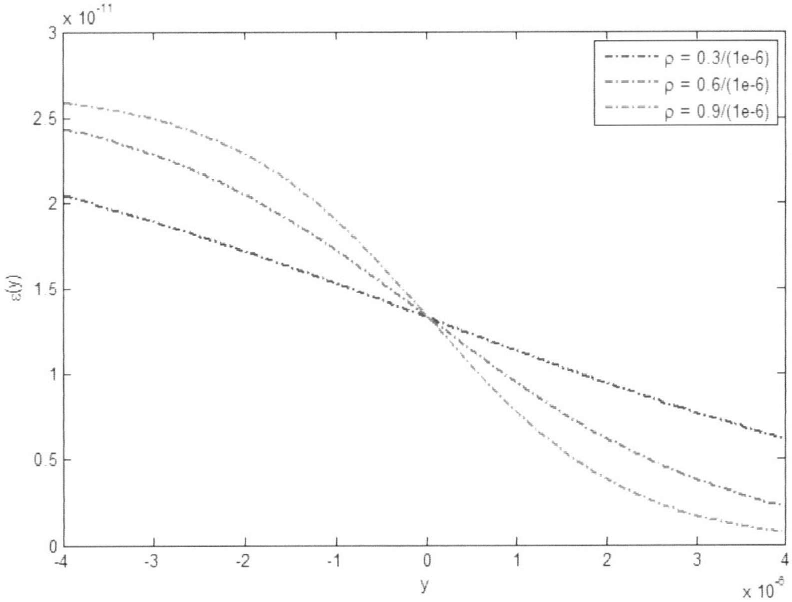

(b)

Figure 1. Graph of (a) permeability (b) permittivity

By inserting equations (13-14) into equation (5) we obtained the electromagnetic wave equation as follows

$$\frac{\partial^2 H(y)}{\partial y^2}+\frac{\rho e^{(\rho y)}}{1+e^{(\rho y)}}\frac{\partial H(y)}{\partial y}+\left(\omega^2\frac{\varepsilon_0\varepsilon_r\mu_0\mu_r}{\left(1+e^{(\rho y)}\right)^2}-k_z^{\ 2}\right)H(y)=0 \tag{15}$$

using the definition that $c=\left(1/\sqrt{\varepsilon_0\mu_0}\right)$; $\mu_r\varepsilon_r=n_\omega^2$; $k_0=(\omega/c)$ where c is light velocity in vacuum, n_ω is the refraction index and k_0 is wave vector in vacuum in equation (15), so equation (15) can be rewritten,

$$\frac{\partial^2 H(y)}{\partial y^2}+\frac{\rho e^{(\rho y)}}{1+e^{(\rho y)}}\frac{\partial H(y)}{\partial y}+\left(\frac{k_0^{\ 2}n_\omega^2}{\left(1+e^{(\rho y)}\right)^2}-k_z^{\ 2}\right)H(y)=0 \tag{16}$$

If we set $u=\rho y$, equation (16) become

$$\frac{d^2 H(y)}{du^2}+\frac{e^{(u)}}{1+e^{(u)}}\frac{dH(y)}{du}+\frac{1}{\rho^2}\left(\frac{k_0^{\ 2}n_\omega^2}{\left(1+e^{(u)}\right)^2}-k_z^{\ 2}\right)H(y)=0 \tag{17}$$

by changing the suitable variable

$$1+e^{(u)}=g \tag{18}$$

in equation (17), we obtained

$$\frac{d^2 H(y)}{dg^2}+\frac{1}{g}\frac{dH(y)}{dg}+\frac{1}{\rho^2}\frac{1}{(g-1)^2}\left(\frac{k_0^{\ 2}n_\omega^2}{g^2}-k_z^{\ 2}\right)H(y)=0 \tag{19}$$

Setting the parameter

$$H(y)=g^{\alpha}(1-g)^{\beta}f(g) \tag{20}$$

into equation (19), we found

$$f^{''}(g)+\left(\frac{(2\alpha+1)-(2\alpha+2\beta+1)g}{g(1-g)}\right)f^{'}(g)-\left(\frac{2\alpha\beta+\beta+\left(2k_0^{\,2}n_\omega^2/\rho^2\right)}{g(1-g)}\right)f(g)=0 \tag{21}$$

where

$$\alpha=i\frac{k_0 n_\omega}{\rho};\qquad \beta=\sqrt{\frac{k_z^{\,2}-k_0^{\,2}n_\omega^{\,2}}{\rho^2}+\frac{1}{4}}+\frac{1}{2} \tag{22}$$

Equation (21) is AIM-type differential equation. By compare equation (21) and equation (7), we found

$$\begin{aligned}\lambda_0(s)&=\left(\frac{(2\alpha+2\beta+1)g-(2\alpha+1)}{g(1-g)}\right)=\frac{2\beta}{1-g}-\frac{(2\alpha+1)}{g}\\ s_0(s)&=\left(\frac{2\alpha\beta+\beta+\left(2k_0^{\,2}n_\omega^2/\rho^2\right)}{g(1-g)}\right)\end{aligned} \tag{23}$$

using equations (8-9) and equation (23), we obtained the generalized eigenvalue of equation (22), is given as

$$\Delta_j=\sum_{j=1}^{j}\left(2\alpha\beta+(2j-1)\beta+(2j-2)\alpha+(j-1)^2\right)+2k_0^{\,2}n_\omega^2/\rho^2=0 \tag{24}$$

From equation (24) and equation (22), we obtained the wave vector $k_z^{\,2}$,

$$\Delta_1=2\alpha\beta+\beta+2k_0^{\,2}n_\omega^2/\rho^2=0\rightarrow k_{z_1}^{\,2}=\left(\rho^2\left(\frac{-2k_0^{\,2}n_\omega^2}{\rho(2ik_0n_\omega+\rho)}-\frac{1}{2}\right)^2+k_0^{\,2}n_\omega^2-\frac{\rho^2}{4}\right) \tag{25}$$

$$\Delta_2=2\alpha\beta+3\beta+2\alpha+1+2k_0^{\,2}n_\omega^2/\rho^2=0\rightarrow$$

$$k_{z_2}^{\,2}=\left(\rho^2\left(\frac{1}{(2ik_0n/\rho)+3}\left(\frac{-2k_0^{\,2}n_\omega^2}{\rho^2}-\frac{2ik_0n_\omega}{\rho}-1\right)-\frac{1}{2}\right)^2+k_0^{\,2}n_\omega^2-\frac{\rho^2}{4}\right) \tag{26}$$

Equation (25-26) are wave vector equation $k_z^{\,2}$ which depends on abruptness's parameter, refraction index and wave vector in vacuum k_0 .

To obtained the field equations, we compare equation (21) with equation (10), and by using equation (12), we obtained the parameters:

$$\begin{aligned}&N=-1;\ b=1;\ t=\beta;\ c=\alpha-\tfrac{1}{2};\ W=-(2\alpha\beta+\beta+2k_0^{\,2}n_\omega^2/\rho^2)\\ &\sigma=(2\alpha+1);\ q=(2\alpha+2\beta)\end{aligned} \tag{27}$$

and from equation (11) and equation (27), we obtained

$$f_n(g)=(-1)^n C(1)^n(2\alpha+1)_n\,{}_2F_1\left(-n,(2\alpha+2\beta+n),(2\alpha+1),g\right) \tag{28}$$

Then by using equations (18, 20) and equation (28), we obtained the generalized magnetic field equation, is given as

$$\begin{aligned}H_n(u)&=\left(1+e^u\right)^{\alpha}\left(-e^u\right)^{\beta}(-1)^n C(1)^n(2\alpha+1)_n\\ &\times{}_2F_1\left(-n,(2\alpha+2\beta+n),(2\alpha+1),\left(1+e^u\right)\right);\ u=\rho y\end{aligned} \tag{29}$$

for

$$n=1 \rightarrow H_1(y) = -C\left(1+e^{(\rho y)}\right)^{\alpha}\left(-e^{u}\right)^{\beta}(2\alpha+1)\left(1-\frac{(2\alpha+2\beta+1)}{(2\alpha+1)}\left(1+e^{(\rho y)}\right)\right) \tag{30}$$

$$\begin{aligned} n=2 \rightarrow H_2(y) &= C\left(1+e^{(\rho y)}\right)^{\alpha}\left(-e^{u}\right)^{\beta}(2\alpha+1)(2\alpha+2) \\ &\times\left(1-\frac{2(2\alpha+2\beta+1)}{(2\alpha+1)}\left(1+e^{(\rho y)}\right)+\frac{2(2\alpha+2\beta+1)(2\alpha+2\beta+2)}{(2\alpha+1)(2\alpha+2)}\left(1+e^{(\rho y)}\right)^2\right) \end{aligned} \tag{31}$$

which α, β is defined in equation (22).

By using equations (30-31) and equation (6), we obtained the total magnetic field equation are given as

$$n=1 \rightarrow H_{x_1} = -C\left(1+e^{(\rho y)}\right)^{\alpha}\left(-e^{u}\right)^{\beta}\left(e^{ik_z z}\right)(2\alpha+1)\left(1-\frac{(2\alpha+2\beta+1)}{(2\alpha+1)}\left(1+e^{(\rho y)}\right)\right) \tag{32}$$

$$\begin{aligned} n=2 \rightarrow H_{x_2} &= C\left(1+e^{(\rho y)}\right)^{\alpha}\left(-e^{\rho y}\right)^{\beta}\left(e^{ik_z z}\right)(2\alpha+1)(2\alpha+2) \\ &\times\left(1-\frac{2(2\alpha+2\beta+1)}{(2\alpha+1)}\left(1+e^{(\rho y)}\right)+\frac{2(2\alpha+2\beta+1)(2\alpha+2\beta+2)}{(2\alpha+1)(2\alpha+2)}\left(1+e^{(\rho y)}\right)^2\right) \end{aligned} \tag{33}$$

Equation (32-33) are magnetic field equations $H_x = H(y)H(z)$ which give information about the distribution of magnetic field in x-direction.

4. Conclusion

This paper has developed by using exponential permeability and permittivity profile in Maxwell equation to obtained the wave vector and field equations using asymptotic iteration method. The magnetic field equations were expressed in hypergeometric function.

Acknowledgement

This research was partly supported by Sebelas Maret University Higher Education Project Grant with contract no. 632/UN27.21/LT/2016

References

[1] Smith D R, Padilla W J, Vier D C, Nemat-Nasser S C and Schultz S 2000 *Physical Review Letters* **84** 18
[2] Vaselago V G 1968 *Soviet Physics Uspekhi* **10** 509-514
[3] Pendry J B, Holden A J, Stewart W J and Youngs I 1996 *Physical Review Letters* **76** 4773–4776
[4] Pendry J B 2000 *Physical Review Letters* **85** 4184-4187
[5] Dalarsson M and Tassin P 2009 *Optics Express* **17** 6747-6752
[6] Dalarsson M, Norgren M, Asenov T and Doncov N 2013 *Progress in electromagnetics research* **137** 527-538
[7] Dalarsoon N 2005 Proc.1st *International Workshop on Nanoscience & Nanotechnology IWON*(Belgrade, Serbia and Montenegro) p 194
[8] Husein A S,Cari C, Suparmi A and Hadi M 2016 *AIP Conference Proceedings*vol **1719**(The 4th International Conference on Theoretical and Applied Physics (ICTAP), Bali) p 030048
[9] Suparmi A, Cari C, Pratiwi B N and Deta U A 2016 *AIP Conference Proceeding*vol **1710** (6TH Nanoscience and Nanotechnology Symposium (NNS), Solo Indonesia) p 030010
[10] Cari C, Suparmi A, Yunianto M and Husein A S 2016 *AIP Conference Proceeding*vol **1710** (6TH Nanoscience and Nanotechnology Symposium (NNS), Solo Indonesia) p 030009
[11] Cari C and Suparmi A 2014 *AIP Conference Proceeding*vol **1615** (4th International Conference of Advances in Nuclear Science and Enginerring , Bandung) p 101

[12] Suparmi A, Cari C and Deta U A 2014 *Chin. Phys. B* **23** 090304-1
[13] Ciftci H, Hall R L and Saad N 2003 *Journal of Physics A: Mathematical and General* **36** 11807
[14] Soylu A, Bayrak O and Boztosun I 2008 *Journal of Physics A: Mathematical and Theoretical* **41** 065308
[15] Falaye B J, Hamzavi M and Ikhdair S M 2012 *arXiv*:1207.1218v [nucl-th]

Asymptotic iteration method for solution of the Kratzer potential in D-dimensional Klein-Gordon equation

Dewanta Arya Nugraha, A Suparmi*, C Cari and Beta Nur Pratiwi

Physics Department, Graduate Program, Sebelas Maret University, INDONESIA

E-mail: dewanta.an@gmail.com

Abstract. The solution of D-dimensional Klein-Gordon equation for Kratzer potential was presented using asymptotic iteration method. The D-dimensional Klein-Gordon equation was reduced to one-dimensional radial differential equation. The Kratzer central potential was used in radial part solution to obtain the energy eigenvalues and radial wave function. The bound-state energy eigenvalues were calculated for various values of quantum numbers and dimensions using Matlab software.

1. Introduction

In many fields of physics, particularly quantum physics, atomic physics, and nuclear physics, for the relativistic case, Klein Gordon equation is the one of fundamental equation to describe the behavior of spinless microscopic particles. Solutions of Klein Gordon equation have been studied in many researches, both analytically and approximately. In earlier research, solution of Klein Gordon equation give the eigenvalue and eigen function which depend on the certainly potentials and certainly methods, such as Hulthen potential using Nikiforov–Uvarov [1], Hulthén-Type Potential using supersymmetric quantum mechanics (SUSYQM) [2], Makarov potential using factorization method [3], q-Deformed Manning-Rosen Potential using Asymptotic Iteration Method(AIM) [4], [5] and other. Recently, AIM has received much attention as a method for solving the Klein-Gordon equation [4], [6]–[8]. It has been applied to a large number of physically interesting potentials and has often yielded highly-accurate results.

The solution of Klein Gordon equation was mostly studied in three dimensional. But in recently, the solution of quantum system was extended in higher dimensions. Three dimensional space is embedded in a higher dimensional space. It was amazing that the force which influences until infinity range is so small compared to the electromagnetic force for example. Maybe we experience only a small part of it in our three dimensions and the major part is distributed to other dimensions we can't perceive. So it is useful to deal with quantum mechanics in higher dimensions. The higher dimensions system in quantum physics has been studied in many cases, such as the solution of Klein-Gordon equation in D-dimensions for Hulthen potential [1], [2], Manning- Rosen potential [4], Kratzer potential [7], Morse potential [8], Poschl-Teller potential including centrifugal term [9], Hylleraas Potential [10], trigonometric Poschl-Teller potential [11], and other.

In this research, we investigated the Kratzer potential for D-dimensional Klein-Gordon using AIM. Kratzer potential is defined as [7], [12]

*Corresponding: soeparmi@staff.uns.ac.id

$$V(r)=-\frac{V_1}{r}+\frac{V_2}{r^2},\quad S(r)=-\frac{S_1}{r}+\frac{S_2}{r^2} \tag{1}$$

The Klein-Gordon equation for Kratzer potential was reduced to Schodinger like type equation, so it can be solved using AIM. AIM is one of alternative method. It has high efficiency and accuracy to determine eigen functions and eigen energies. [13]

2. Klein-Gordon Equation in D-dimensions

The time independent KG equation in D-dimensional with potentials of vector and scalar $V(r)$ and $S(r)$, respectively, with $r=|\mathbf{r}|$ describing a spinless particle. In the general form the equation can write as [14]–[16]

$$\left\{\nabla_D^2+\frac{1}{\hbar^2c^2}\left[\left[E-V(r)\right]^2-\left[Mc^2+S(r)\right]^2\right]\right\}\psi_{\ell_1,...,\ell_{D-2}}^{(\ell_{D-1}=\ell)}(\mathbf{x})=0, \tag{2}$$

With E and M are the KG energy and the mass of particle. The $\mathbf{x}$ in equation (2) is a D-dimensional position vector. The laplacian ∇_D^2 is

$$\nabla_D^2=\frac{1}{r^{D-1}}\frac{\partial}{\partial r}\left(r^{D-1}\frac{\partial}{\partial r}\right)+\frac{1}{r^2}\sum_{j=1}^{D-2}\frac{1}{\sin^2\theta_{j+1}\sin^2\theta_{j+2}...\sin^2\theta_{D-1}}\times\left\{\frac{1}{\sin^{j-1}\theta_j}\left(\frac{\partial}{\partial\theta_j}\sin^{j-1}\theta_j\frac{\partial}{\partial\theta_j}\right)\right\}+ \frac{1}{r^2}\left\{\frac{1}{\sin^{D-2}\theta_{D-1}}\left(\frac{\partial}{\partial\theta_{D-1}}\sin^{D-2}\theta_{D-1}\frac{\partial}{\partial\theta_{D-1}}\right)\right\}, \tag{3}$$

And

$$\psi_{\ell_1,...,\ell_{D-2}}^{(\ell)}(\mathrm{x})=R_\ell(r)Y_{\ell_1,...,\ell_{D-2}}^{(\ell)}(\theta_1,\theta_2,\cdots,\theta_{D-1}) \tag{4}$$

The wave function for radial part $R_\ell(r)$ is decomposed as

$$R_\ell(r)=r^{-(D-1)/2}F(r), \tag{5}$$

If $S=V$, it will produce a non-trivial nonrelativistic limit with a potential function $2V$ and not V. Equation (2) is reduced to KG equation in D-dimension for radial part with the potential function V is scaled as $2V$. (in the relativistic atomic units $\hbar=c=1$) [17]

$$\nabla_D^2\psi_{\ell_1,...,\ell_{D-2}}^{(\ell_{D-1}=\ell)}(\mathrm{x})+\left[E^2-M^2-(E+M)V(r)\right]\psi_{\ell_1,...,\ell_{D-2}}^{(\ell_{D-1}=\ell)}(\mathrm{x})=0, \tag{6}$$

3. Asymptotic Iteration Method (AIM)

AIM is used to solve the second order homogeneous equation of form [4], [6], [18], [19].

$$y_n''(x)=\lambda_0(x)y_n'(x)+s_0(x)y_n(x) \tag{7}$$

with $\lambda_0(x)\neq 0$ and prime symbol denotes the derivative with respect to x. The others parameter n is interpreted as the radial quantum number. The other variables, $s_0(x)$ and $\lambda_0(x)$ are differentiable. To get the solution, we have to differentiate Eq. (7) along x, and find

$$y_n'''(x)=\lambda_1(x)y_n''(x)+s_1(x)y_n(x) \tag{8}$$

Where

$$\lambda_1(x)=\lambda_0'(x)+s_0(x)+\lambda_0^2(x) \tag{9}$$

$$s_1(x)=s_0'(x)+s_0(x)\lambda_0(x) \tag{10}$$

where $\lambda_0(x)\neq 0$ and $s_0(x)$ is a function of $C\infty$ (coefficient of the differential equation) Asymptotic Iteration Method and can be applicated exactly in the different problem if the wave function has been known and fulfill boundary condition zero (0) and infinity (∞).

Equation (7) can be simple itterated until (k+1) and (k+2), k = 1, 2, 3, ... and then we get

$$y_n^{k+1}(x) = \lambda_{k-1}(x)y_n'(x) + s_{k-1}(x)y_n(x) \tag{11}$$

$$y_n^{k+2}(x) = \lambda_k(x)y_n'(x) + s_k(x)y_n(x) \tag{12}$$

where

$$\lambda_k(x) = \lambda_{k-1}'(x) + s_{k-1}(x) + \lambda_0(x)\lambda_{k-1}(x) \tag{13}$$

$$s_k(x) = s_{k-1}'(x) + s_0(x)\lambda_{k-1}(x) \tag{14}$$

Which is called recurrence relation. To determine the energy eigenvalues, we use quantization condition given by following equation

$$\Delta_k(x) = \lambda_k(x)s_{k-1}(x) - \lambda_{k-1}(x)s_k(x) = 0 \tag{15}$$

With the iteration number k is define as $k = 1, 2, 3, \ldots$. The wave function can be determined using the wave function equation as follows

$$y_k(x) = C_2 \exp\left(-\int^r \frac{s_k}{\lambda_k} dx'\right) \tag{16}$$

4. Results and Discussion

Substituting equations (1, 2 – 5) into equation (6) we obtain

$$\frac{1}{r^{D-1}}\frac{\partial}{\partial r}\left(r^{D-1}\frac{\partial}{\partial r}\frac{F(r)}{r^{(D-1)/2}}\right) - \left[(E+M)\left(-\frac{V_1}{r} + \frac{V_2}{r^2}\right)\right]\frac{F(r)}{r^{(D-1)/2}} + \left(E^2 - M^2\right)\frac{F(r)}{r^{(D-1)/2}} = \lambda_{D-1}\frac{F(r)}{r^2 r^{(D-1)/2}} \tag{17}$$

Where λ_{D-1} is the separation constant with $\lambda_{D-1} = l(l+D-2)$ and D is the dimension. Equation (17) can be simplify as

$$\left[\frac{d^2}{dr^2} - \frac{1}{r^2}\left\{\left(\frac{D-1}{2}\right)\left(\frac{D-3}{2}\right) - \lambda_{D-1} - (E+M)V_2\right\} + \frac{1}{r}(E+M)V_1 + \left(E^2 - M^2\right)\right]F(r) = 0 \tag{18}$$

By setting the part of equation (18),

$$A(A+1) = \left(\frac{D-1}{2}\right)\left(\frac{D-3}{2}\right) - \lambda_{D-1} - (E+M)V_2 \tag{19}$$

$$B = (E+M)V_1 \tag{20}$$

$$E_s^2 = \left(M^2 - E^2\right) \tag{21}$$

Then equation (18) become

$$\frac{d^2F(r)}{dr^2} - \frac{A(A+1)}{r^2}F(r) + \frac{B}{r}F(r) - E_s^2F(r) = 0 \tag{22}$$

By using Frobenius method, we let $F(r) = r^s \sum C_n r^n$. The equation (22) has an irregular singularity $r \to \infty$, where the normalized solution is $F(r) \sim e^{-E_s r}$. When it has the singularity $r \to 0$, the $F(r) \sim r^{A+1}$. To make equation (22) into form of equation (7), the wave function is set as follows

$$F(r) = r^{A+1}e^{-E_s r}g(r) \tag{23}$$

Substitute equation (23) into equation (22), we get

$$g''(r) = \left\{2E_s - \frac{2(A+1)}{r}r^A\right\}g'(r) + \left\{\frac{2(A+1)E_s - B}{r}\right\}g(r) \tag{24}$$

Equation (24) is the homogeneous linear differential equation in second-order form. It can be solve by using AIM. By comparing equation (24) with equation (7), then we can get $\lambda_0(r)$, and $s_0(r)$. The $\lambda_k(r)$, and $s_k(r)$ can be calculated as follows

$$\lambda_0 = \left\{2E_s - \frac{2(A+1)}{r}\right\}$$

$$s_0 = \left\{\frac{2(A+1)E_s - B}{r}\right\},$$

$$\lambda_1 = \frac{2(A+1)}{r^2} + \frac{2(A+1)E_s - B}{r} + \left(2E_s - \frac{2(A+1)}{r}\right)^2$$

$$s_1 = -\frac{2(A+1)E_s - B}{r} + \frac{2(A+1)E_s - B}{r}\left\{2E_s - \frac{2(A+1)}{r}\right\}, \text{ etc.} \tag{25}$$

From equations (12 – 14, 25), we found the general eigenvalue of equation (24) in general term as follows

$$E_{sn} = \frac{B}{2(A+1+n)} \tag{26}$$

Where n is quantum number with $n = 0,1,2,\ldots$. By substituting equation (19-21) to equation (26), the relativistic energy equation become

$$M^2 - E_n^2 = \left[\frac{(E_n + M)V_1}{2\left\{\sqrt{\left(\frac{D-1}{2}\right)\left(\frac{D-3}{2}\right) - \lambda_{D-1} - (E_n + M)V_2 + \frac{1}{4}} + \frac{1}{2} + n\right\}}\right]^2 \tag{27}$$

With V_1 and V_2 are the potential parameters where $V_1 = 2D_e a$ and $V_2 = D_e a^2$. Where D_e and a are the dissociation energy and the equilibrium internuclear separation. For the Kratzer potential, when $r \to 0$ then the potential $V(r) \to \infty$ because there is internuclear repulsion. When $r \to \infty$ then the potential $V(r) \to 0$, i.e., the molecule decomposes [20].

Table 1. Various diatomic molecules reduced mass and spectroscopic properties in ground state.[20],[21]

Parameter	CO	NO	O_2	I_2
D_e (eV)	10.84514471	8.043782568	5.156658828	1.581791863
a (Å)	1.1282	1.1508	1.208	2.662
M (amu)	6.860586000	7.468441000	7.997457504	63.45223502

Table 1. presents the parameters for the potential. The potential parameter must be convert to natural unit where $\hbar = c = 1$. The a and M units are converted to $1/\text{eV}$ and eV due to the natural units. The relativistic energies value were calculated numerically from equation (27) by using Matlab. The results can be seen in Table 2.

Table 2. The relativistic energy E_n in (GeV) with $l=1$

D	n	CO	NO	O_2	I_2
3	0	-6.391161	-6.957542	-7.450621	-59.106092
	1	-6.391725	-6.958273	-7.451656	-59.106787
	2	-6.392854	-6.959735	-7.453726	-59.108176
	3	-6.394546	-6.961928	-7.456831	-59.110261
	4	-6.396803	-6.964852	-7.460970	-59.113040
	5	-6.399623	-6.968507	-7.466145	-59.116513
4	0	-6.391020	-6.957359	-7.450363	-59.105918
	1	-6.391584	-6.958090	-7.451397	-59.106613
	2	-6.392713	-6.959552	-7.453467	-59.108003
	3	-6.394405	-6.961745	-7.456572	-59.110087
	4	-6.396662	-6.964669	-7.460712	-59.112866
	5	-6.399482	-6.968324	-7.465886	-59.116340
5	0	-6.390597	-6.956810	-7.449586	-59.105397
	1	-6.391161	-6.957542	-7.450621	-59.106092
	2	-6.392289	-6.959004	-7.455796	-59.107482
	3	-6.393982	-6.961197	-7.459935	-59.109566
	4	-6.396238	-6.964121	-7.465110	-59.112345
	5	-6.399059	-6.967776	-7.471319	-59.115819

From Table 2., it is shown that the relativistic energy value is decreasing due to the increase of quantum number n. For higher dimension, the energy value is decreasing too. The relativistic energies for diatomic molecules of CO, NO, O_2, and I_2 are negative. The wave functions were determined using equation (16). And the graph of the wave function for O_2 particle are shown in Figure 1.

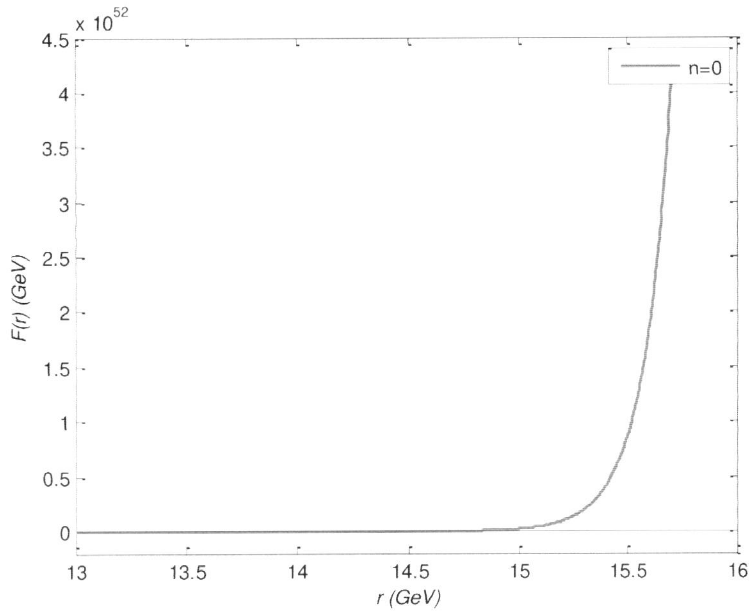

(a)

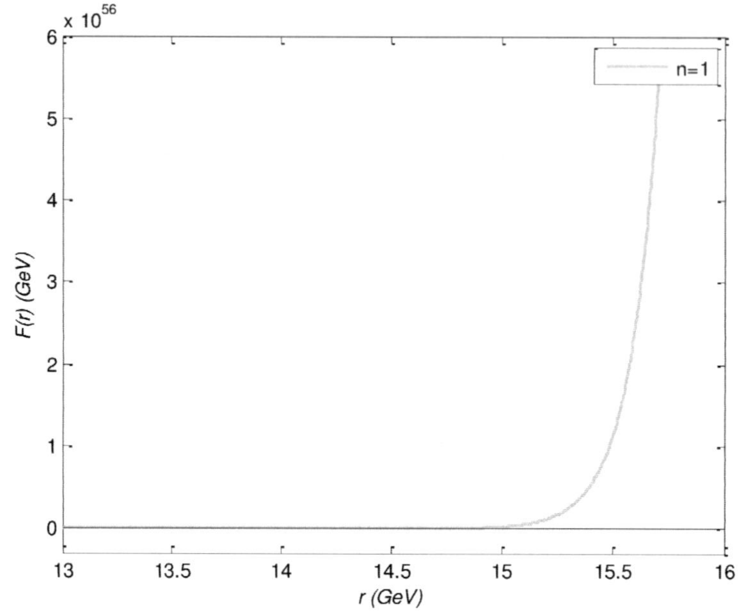

(b)

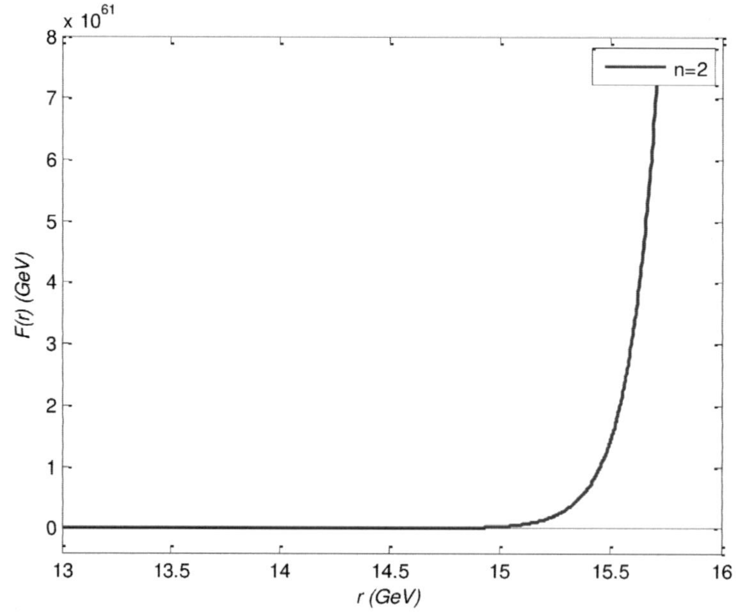

(c)

Figure 1. Four dimensional radial wave functions of O_2 with $l=1$ (a) $n=0$, (b) $n=1$, and (c) $n=2$

5. Conclusion

The bound-state energy eigenvalues of Klein-Gordon equation for Kratzer potential in D-dimension were obtained via asymptotic iteration method. From the results can be concluded that the relativistic energy value is decreasing due to the increase of quantum number n. For higher dimension, the energy value is decreasing. The energy of diatomic molecules of CO, NO, O_2, and I_2 are negative.

Acknowledgement

This research was partly supported by Sebelas Maret University Higher Education Project Grant with contract No. 632/UN27.21/LT/2016

References

[1] A. N. Ikot, L. E. Akpabio, and E. J. Uwah, "Bound State Solutions of the Klein Gordon Equation with the Hulth ´ en Potential," *Electron. J. Theor. Phys.*, vol. 25, no. 25, pp. 225–232, 2011.

[2] H. Hassanabadi, S. Zarrinkamar, and H. Rahimov, "Approximate Solution of D-Dimensional Klein – Gordon Equation with Hulth ´ Potential via SUSYQM," *Commun. Theor. Phys.*, vol. 56, no. 3, pp. 423–428, 2011.

[3] Z. Min-cang and W. Zhen-Bang, "Exact solutions of the Klein – Gordon equation with Makarov potential and a recurrence relation," *Chinese Phys.*, vol. 16, no. 7, pp. 1863–1867, 2007.

[4] T. Das, "Exact Solutions of the Klein-Gordon Equation for q-Deformed Manning-Rosen Potential via Asymptotic Iteration Method," *arXiv:1409.1457v1 [quant-ph]*, pp. 1–11, 2014.

[5] B. N. Pratiwi, A. Suparmi, C. Cari, and D. A. Nugraha, "The iterative solution of wave propagation in transverse magnetic mode for graded positive- negative," *J. Phys. Conf. Ser.*, vol. 795, p. 12024, 2017.

[6] T. Barakat, "The asymptotic iteration method for dirac and klein gordon equations with a linear scalar potential," *Int. J. Mod. Phys. A*, vol. 21, no. 19, pp. 4127–4135, 2006.

[7] N. Saad, R. L. Hall, and H. Ciftci, "The Klein-Gordon equation with the Kratzer potential in d dimensions," *Cent. Eur. J. Phys.*, vol. 6, no. 3, pp. 717–729, 2008.

[8] A. Soylu, O. Bayrak, and I. Boztosun, "Exact Solutions of Klein – Gordon Equation with Scalar and Vector," *Chinese Phys. Lett.*, vol. 25, no. 8, pp. 2754–2757, 2008.

[9] Y. Xu, S. He, and C.-S. Jia, "Approximate analytical solutions of the Klein–Gordon equation with the Pöschl–Teller potential including the centrifugal term," *Phys. Scr.*, vol. 81, no. 4, p. 45001, 2010.

[10] A. N. Ikot, O. A. Awoga, and B. I. Ita, "Exact Solutions of the Klein-Gordon Equation with Hylleraas Potential," *Few-Body Syst.*, vol. 53, no. 3–4, pp. 539–548, 2012.

[11] D. A. Nugraha, A. Suparmi, C. Cari, and B. N. Pratiwi, "Asymptotic iteration method for analytical solution of Klein-Gordon equation for trigonometric Pöschl-Teller potential in D-dimensions," *J. Phys. Conf. Ser.*, vol. 795, p. 12025, 2017.

[12] H. Hassanabadi, H. Rahimov, and S. Zarrinkamar, "Approximate solutions of Klein-Gordon equation with Kratzer potential," *Adv. High Energy Phys.*, vol. 2011, 2011.

[13] A. Suparmi, C. Cari, U. A. Deta, A. S. Husein, and H. Yuliani, "Exact Solution of Dirac Equation for q-Deformed Trigonometric Scarf potential with q-Deformed Trigonometric Tensor Coupling Potential for Spin and Pseudospin Symmetries Using Romanovski Polynomial," *J. Phys. Conf. Ser.*, vol. 539, p. 12004, 2014.

[14] O. Bayrak and I. Boztosun, "Arbitrary ℓ-state solutions of the rotating Morse potential by the asymptotic iteration method," *J. Phys. A. Math. Gen.*, vol. 39, no. 22, p. 6955, 2006.

[15] W. C. Qiang and S. H. Dong, "Arbitrary l-state solutions of the rotating Morse potential through the exact quantization rule method," *Phys. Lett. Sect. A Gen. At. Solid State Phys.*, vol. 363, no. 3, pp. 169–176, 2007.

[16] H. Ciftci, R. L. Hall, and N. Saad, "Asymptotic iteration method for eigenvalue problems," *J. Phys. A. Math. Gen.*, vol. 36, no. 47, pp. 11807–11816, 2003.

[17] A. D. Alhaidari, H. Bahlouli, and A. Al-Hasan, "Dirac and Klein-Gordon equations with equal scalar and vector potentials," *Phys. Lett. Sect. A Gen. At. Solid State Phys.*, vol. 349, no. 1–4, pp. 87–97, 2006.

[18] R. A. Sari, A. Suparmi, and C. Cari, "Solution of Dirac equation for Eckart potential and trigonometric Manning Rosen potential using asymptotic iteration method," *Chinese Phys. B*, vol. 25, no. 1, 2015.

[19] S. Pramono, A. Suparmi, and C. Cari, "Relativistic Energy Analysis of Five-Dimensional q - Deformed Radial Rosen-Morse Potential Combined with q -Deformed Trigonometric Scarf Noncentral Potential Using Asymptotic Iteration Method," *Adv. High Energy Phys.*, vol. 2016, 2016.

[20] O. BAYRAK, I. BOZTOSUN, and H. CIFTCI, "Exact Analytical Solutions to the Kratzer Potential by the Asymptotic Iteration Method," *Int. J. Quantum Chem.*, vol. 107, no. 3, pp. 540–544, 2007.

[21] L. E. Sutton, *Tables of Interatomic Distances and Configu- ration in Molecules and Ions*. London,: Chemical Society of London, 1958.

IOP Conf. Series: Journal of Physics: Conf. Series **820** (2017) 012015 doi:10.1088/1742-6596/820/1/012015

A generalized volumetric dispersion model for a class of two-phase separation/reaction: finite difference solutions

Chairat Siripatana[1,2], Hathaikarn Thongpan[1] and Arwut Promraksa[1]
[1]School of Engineering and Resources, Walailak University, 80161, Nakhon Si Thammarat Thailand
[2]Biomass and Oil-Palm Excellence Center, Walailak University, 80161, Nakhon Si Thammarat Thailand

E-mail: schairat61@gmail.com

Abstract. This article explores a volumetric approach in formulating differential equations for a class of engineering flow problems involving component transfer within or between two phases. In contrast to conventional formulation which is based on linear velocities, this work proposed a slightly different approach based on volumetric flow-rate which is essentially constant in many industrial processes. In effect, many multi-dimensional flow problems found industrially can be simplified into multi-component or multi-phase but one-dimensional flow problems. The formulation is largely generic, covering counter-current, concurrent or batch, fixed and fluidized bed arrangement. It was also intended to use for start-up, shut-down, control and steady state simulation. Since many realistic and industrial operation are dynamic with variable velocity and porosity in relation to position, analytical solutions are rare and limited to only very simple cases. Thus we also provide a numerical solution using Crank-Nicolson finite difference scheme. This solution is inherently stable as tested against a few cases published in the literature. However, it is anticipated that, for unconfined flow or non-constant flow-rate, traditional formulation should be applied.

1. Introduction

Although dispersion models are notorious for complexity it has been proven very useful in describing various dispersive phenomena occurs frequently in chemical and environmental engineering fields [1-11]. However, applying dispersion models in unit operations such as adsorption column, ion-exchange rasin, counter-current extractors, continuous anaerobic digestion such as upflow anaerobic sludge blanket (UASB), industrial flow reactors, or even the spread of contaminants along the water passages can be overkill particularly when the cross-sectional area of the columns/passages or/and column porosity vary with linear position. In most cases researchers choose to simplify the problems by assuming well-mixed (infinitely dispersive) or plug-flow (no axial dispersion) formulation otherwise they have to resort specialists who can deal with the mathematical complexity numerically.

The purpose of this article is to tackle mathematical problems for a class of engineering problems where two- or three-dimensional flow problems can be simplified with little loss of model usefulness by treating them as one-dimensional flow problems. Here a volumetric approach in formulating differential equations will be explored for a class of flow problems involving mass transfer and reactions in chemical and environmental engineering. The resulting model will be applicable for dispersive counter-current, co-current, moving-bed, packed-bed or batch modes where the flow passage is not uniform (so do the fluid/solid velocities) and column porosity varies with position.

2. Model development

IOP Conf. Series: Journal of Physics: Conf. Series **820** (2017) 012015 doi:10.1088/1742-6596/820/1/012015

Although the derivation in this article covers many common cases in industrial unit operations, we will illustrate the analysis for only *counter-current configuration (CC)* as shown in Figure 1. However, the resulting models (referred as volumetric dispersion model, VDM) for other frequent cases will be also summarized.

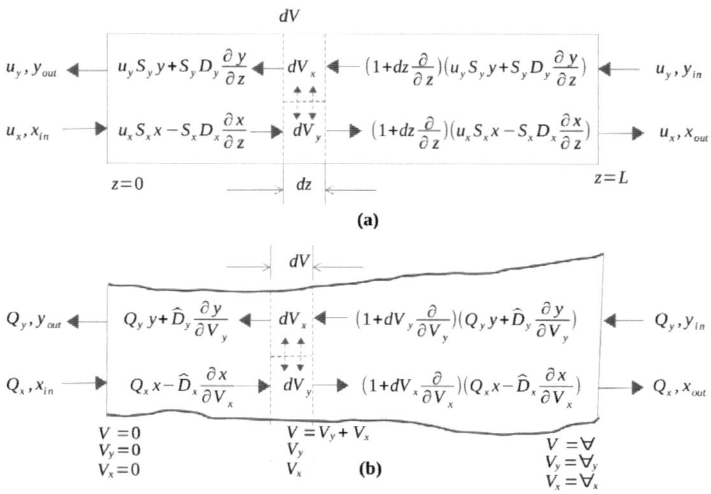

Figure 1. Differential volume analysis a) traditional formulation b) volumetric dispersive formulation

2.1 Traditional Dispersion Formulation (TDF) (sometimes referred "backmixing models")

Conventionally, as elaborated in the monumental work of Mecklenburgh and Hartland [15], dispersive mass transport models are formulated based on linear displacement (z) and velocity as depicted in Figure 1a. Here $x, y, u_x, u_y, S_x, S_y, D_x, D_y$ are solute concentration, the linear velocity, cross-sectional area, and dispersion coefficients, of dispersed (x-phase) and continuous phase (y-phase) at a certain position respectively. x^*, y^* are the concentration which is in equilibrium with the opposite phase at a specific position. k_x, k_y are the mass transfer coefficient based on x- and y-phase respectively. The following mass balances can written for x-phase.

inflow into dV_x - outflow from dV_x + mass transfer into dV_x = accumulation in dV_x

$$u_x S_x x - S_x D_x \frac{\partial x}{\partial z} - (1+dz\frac{\partial}{\partial z})(u_x S_x x - S_x D_x \frac{\partial x}{\partial z}) + k_y dA(y-y^*) = dV_x \frac{\partial x}{\partial t} \quad (1)$$

In general u_x and S_x change with z, thus we have

$$-u_x S_x dz \frac{\partial x}{\partial z} + x\frac{\partial (u_x S_x)}{\partial z} + S_x D_x dz \frac{\partial^2 x}{\partial z^2} + dz \frac{\partial x}{\partial z}\frac{\partial (S_x D_x)}{\partial z} + k_y dA(y-y^*) = dV_x \frac{\partial x}{\partial t} \quad (2)$$

However, if u_x and S_x can be assumed constant, independent of z, we have

$$-u_x S_x dz \frac{\partial x}{\partial z} + u_x S_x dz \frac{\partial}{\partial z}\left(\frac{D_x}{u_x}\frac{\partial x}{\partial z}\right) + k_y dA(y-y^*) = dV_x \frac{\partial x}{\partial t} \quad (3)$$

Equation (2) is more generic than (3) albeit introduces two extra derivative terms while Equation (3) is much simpler but too restrictive for many practical problems where both u_x and S_x vary with z. Certainly we can insist to use Equation (2) to represent non-uniform flow problems but it would be better to start with simpler balance equations that would reduce the complexity in finding solutions considerably. This should be done without significantly compromising model generality.

2.2 Volumetric Dispersion Formulation (VDF)

In searching for simpler approach, we noticed that this class of problems is normally uniform in term of flow rate rather than in term of linear velocities in both phases. Thus we started formulating the balance equation based on constant flow rates instead of velocities. Referring to Figure 1b, we have

For x-phase:

inflow into dV_x - outflow from dV_x + mass transfer into dV_x = accumulation in dV_x

$$Q_x x-\hat{D}_x\frac{\partial x}{\partial V_x}-\left(1+dV_x\frac{\partial}{\partial V_x}\right)\left(Q_x x-\hat{D}_x\frac{\partial x}{\partial V_x}\right)+k_y dA(y-y^*)=dV_x\frac{\partial x}{\partial t} \tag{4}$$

For y-phase:
inflow into dV_y - outflow from dV_y + mass transfer in dV_y = accumulation in dV_y

$$\left(1+dV_y\frac{\partial}{\partial V_y}\right)\left(Q_y y+\hat{D}_y\frac{\partial y}{\partial V_y}\right)-Q_y y-\hat{D}_y\frac{\partial y}{\partial V_y}-k_y dA(y-y^*)=dV_y\frac{\partial y}{\partial t} \tag{5}$$

Here $V_x, V_y, Q_x, Q_y, \hat{D}_x=S_x^2 D_x, \hat{D}_y=S_y^2 D_y$ are accumulative volume, volumetric flow rates and volumetric dispersion coefficients of x- and y-phases respectively. Accumulative volume relative to the left-end is referred as $V=V_x+V_y$, and total volume as $\forall$. Notice that the mass transfer terms can be replaced by different rate relations, equilibrium isotherm or some expression of reaction rates. This is not restrictive although in this article we consider only mass transfer cases exclusively without reaction.

Assuming constant Q_x independent of V_x, we have

$$-Q_x dV_x\frac{\partial x}{\partial V_x}+Q_x dV_x\frac{\partial}{\partial V_x}\left(\frac{\hat{D}_x}{Q_x}\frac{\partial x}{\partial V_x}\right)+k_y dA(y-y^*)=dV_x\frac{\partial x}{\partial t} \tag{6}$$

For y-phase, assuming constant Q_y independent of V_y, we have

$$Q_y dV_y\frac{\partial y}{\partial V_y}+Q_y dV_y\frac{\partial}{\partial V_y}\left(\frac{\hat{D}_y}{Q_y}\frac{\partial y}{\partial V_y}\right)-k_y dA(y-y^*)=dV_y\frac{\partial y}{\partial t} \tag{7}$$

It is more convenient to write Equations (6) and (7) in term of ∂V rather than ∂V_x and ∂V_y. Noting that $dV_x=(1-\epsilon)dV$ and $dV_y=\epsilon dV$, where $\epsilon=dV_y/(dV_x+dV_y)=dV_y/dV$ is column porosity at a specific position. After some algebraic manipulation, we have.

$$-Q_x\frac{\partial x}{\partial V_x}+Q_x\frac{\partial}{\partial V_x}\left(\frac{\hat{D}_x}{Q_x}\frac{\partial x}{\partial V_x}\right)+\frac{k_y a}{1-\epsilon}(y-y^*)=\frac{\partial x}{\partial t} \tag{8}$$

$$Q_y\frac{\partial y}{\partial V_y}+Q_y\frac{\partial}{\partial V_y}\left(\frac{\hat{D}_y}{Q_y}\frac{\partial y}{\partial V_y}\right)-\frac{k_y a}{\epsilon}(y-y^*)=\frac{\partial y}{\partial t} \tag{9}$$

Here $a=dA/dV$ is mass transfer area per unit volume, and $k_y a(y-y^*)=-k_x a(x-x^*)$.

Equation (6-9) have the level of complexity as that of (3) without the being too restrictive as its counterparts. Equation (8) and (9) can be normalized using the following dimensionless groups: $\tau_x=\forall_x/Q_x, \tau_y=\forall_y/Q_y, \theta=t/\tau_y, \kappa=\tau_y/\tau_x, \hat{Pe}_x=\forall_x Q_x/\hat{D}_x, \hat{Pe}_y=\forall_y Q_y/\hat{D}_y, T_y=k_y a\tau_y, Z=V/\forall$. Thus, we obtain

$$\kappa\left(\frac{\partial(1/\hat{Pe}_x)}{\partial Z}-1\right)\frac{\partial x}{\partial Z}+\frac{\kappa}{\hat{Pe}_x}\frac{\partial^2 x}{\partial Z^2}+\frac{T_y}{1-\epsilon}(y-y^*)=\frac{\partial x}{\partial\theta} \tag{10}$$

BCs: At $Z=0$: $x_{in}=x-(\kappa/\hat{Pe}_x)(\partial x/\partial Z)$ and at $Z=1$: $x_{out}=x$ or $\partial x/\partial Z=0$

$$\left(\frac{\partial(1/\hat{Pe}_y)}{\partial Z}+1\right)\frac{\partial y}{\partial Z}+\frac{1}{\hat{Pe}_y}\frac{\partial^2 y}{\partial Z^2}-\frac{T_y}{\epsilon}(y-y^*)=\frac{\partial y}{\partial\theta} \tag{11}$$

BCs: At $Z=0$: $y_{out}=y$, or $\partial y/\partial Z=0$ and at $Z=1$: $y_{in}=y+(1/\hat{Pe}_y)(\partial y/\partial Z)$

Similarly, it can be shown that, if y-phase flows through the column in the direction as portrayed in Figure 1b, the following equations are applied for x-phase for other important cases.

For fixed-bed adsorption and desorption (FB) where solid (x-phase) is not moving $(\kappa=0, 1/\hat{Pe}_x\to\infty)$, equation (10) becomes

$$\frac{T_y}{1-\epsilon}(y-y^*)=\frac{\partial x}{\partial\theta} \tag{12}$$

For moving bed without in- and out-flow in x-phase $(\kappa=0, 1/\hat{Pe}_x\neq 0)$. Equation (10) becomes.

$$\kappa\left(\frac{\partial(1/\hat{Pe}_x)}{\partial Z}\right)\frac{\partial x}{\partial Z}+\frac{\kappa}{\hat{Pe}_x}\frac{\partial^2 x}{\partial Z^2}+\frac{T_y}{1-\epsilon}(y-y^*)=\frac{\partial x}{\partial\theta} \quad \text{where} \quad \kappa\to 0 \tag{13}$$

It seems that, since $\kappa=0$, then Equation (13) would reduce to the same form of (12). However, we observe that $\kappa/\hat{Pe}_x=(\forall_y\hat{D}_x)/(\forall_x^2 Q_y)$ which is non-zero. Thus if we denote $\kappa/\hat{Pe}_x$ as $1/\hat{Pe}'_x$,

Equation (13) becomes

$$\frac{1}{Pe'_x}\frac{\partial^2 x}{\partial Z^2}+\frac{T_y}{1-\epsilon}(y-y^*)=\frac{\partial x}{\partial \theta} \quad (14)$$

BCs: At $Z=0$: $\partial x/\partial Z=0$ and at $Z=1$: $\partial x/\partial Z=0$

And for co-current/batch adsorption and desorption (BCC), Equation (10) becomes.

$$\kappa\left(\frac{\partial(1/\widehat{Pe}_x)}{\partial Z}+1\right)\frac{\partial x}{\partial Z}+\frac{\kappa}{\widehat{Pe}_x}\frac{\partial^2 x}{\partial Z^2}+\frac{T_y}{1-\epsilon}(y-y^*)=\frac{\partial x}{\partial \theta} \quad (15)$$

BCs: At $Z=0$: $x_{out}=x$ or $\partial x/\partial Z=0$ and at $Z=1$: $x_{in}=x+(\kappa/Pe_x)(\partial x/\partial Z)$

It can be shown that in traditional formulation with constant u_x, u_y, S_x and S_y we finally arrive at the same form of differential equations and boundary conditions as in (10-15) simply by substituting Pe_x, Pe_y for $\widehat{Pe}_x, \widehat{Pe}_y$. Table 1 summarized the similarity and differences of these two models.

Table 1. Definition of terms and the differences between TDF and VDF approaches

	Tradtional formulation (TDF)	**Volumetric formulation (VDF)**
Assumptions/ constraints	u_x, u_y are constant, practically only Pe_x and Pe_y can be varied with Z and θ	Q_x, Q_y are constant, practically $\widehat{Pe}_x, \widehat{Pe}_y, S_x, S_y, \epsilon, u_x$ and u_y can be varied with Z and θ
Characteristic position and size	z, L	$V=\int_0^z S\,dz$, $\forall=\int_0^L S\,dz$ $\forall_x=\int_0^L S_x dz$, $\forall_y=\int_0^L S_y dz$
Peclet number of dispersed phase	$Pe_x=\frac{L u_x}{D_x}$	$\widehat{Pe}_x=\frac{\forall_x Q_x}{\widehat{D}_x}=\frac{\forall_x Q_x}{S_x^2 D_x}$
Peclet number of continuous phase	$Pe_y=\frac{L u_y}{D_y}$	$\widehat{Pe}_y=\frac{\forall_y Q_y}{\widehat{D}_y}=\frac{\forall_y Q_y}{S_y^2 D_y}$
Ratio of residence times of the two phases	$\kappa=\frac{\tau_y}{\tau_x}=\frac{u_x}{u_y}$	$\kappa=\frac{\tau_y}{\tau_x}=\frac{Q_x/\forall_x}{Q_y/\forall_y}$
Remarks	This comparison only valid when we assume constant u_x, u_y, S_x, S_y	This comparison only valid when we assume constant Q_x, S_y

3. Solution methods: implicit finite difference

A large class of two-phase separation or reaction can be described by system of partial differential equation of the following form.

$$a_x\frac{\partial x}{\partial Z}+b_x\frac{\partial^2 x}{\partial Z^2}+c_x=\frac{\partial x}{\partial \theta} \quad (16)$$

$$a_y\frac{\partial y}{\partial Z}+b_y\frac{\partial^2 y}{\partial Z^2}+c_y=\frac{\partial y}{\partial \theta} \quad (17)$$

Here, $a_x, b_x, c_x, a_y, b_y, c_y$ are functions of Z and θ. The boundary and initial conditions are to be given for specific cases. For simplicity, clarity and stability, we choose the Crank-Nicolson implicit finite difference method to obtain numerical solutions by replacing the following discrete approximations (Figure 2.) for x derivatives. These are applied for y derivatives in the a similar way.

$$\frac{\partial x}{\partial Z}\approx\frac{1}{2}\left[\frac{x_{i+1}^j-x_{i-1}^j}{2\Delta Z}+\frac{x_{i+1}^{j+1}-x_{i-1}^{j+1}}{2\Delta Z}\right]+O(\Delta x^2) \quad (18)$$

$$\frac{\partial^2 x}{\partial Z^2}\approx\frac{1}{2}\left[\frac{x_{i+1}^j-2x_i^j+x_{i-1}^j}{\Delta Z^2}+\frac{x_{i+1}^{j+1}-2x_i^{j+1}+x_{i-1}^{j+1}}{\Delta Z^2}\right]+O(\Delta x^2) \quad (19)$$

Although in general $a_x, b_x, c_x, a_y, b_y, c_y$ are functions of Z as well as θ, for simplicity we will limit the finite difference solutions only for the cases that they are time independent.

Table 2. Coefficients of Equation (16) and (17) for different flow arrangements (for TDF $\widehat{Pe}_x, \widehat{Pe}_y, \widehat{D}_x$, and $\widehat{D}_y$ must be replaced by Pe_x, Pe_y, D_x and D_y). .

Types	Terms	Values
All flow arrangements	a_y	$\partial(1/\widehat{Pe}_x)/\partial Z+1$
	b_y	$1/\widehat{Pe}_y$
	c_y	$(T_y/\epsilon)(y-y^*)$
	b_x	$1/\widehat{Pe}'_x=\kappa/\widehat{Pe}_x$ Except FB where $b_x=0$
	c_x	$(T_y/(1-\epsilon))(y-y^*)$
Counter-current (CC)	a_x	$\kappa[\partial(1/\widehat{Pe}_x)/\partial Z-1]$
Moving bed (MB)	a_x	$\partial(1/\widehat{Pe}_x)/\partial Z$
Fixed bed (FB)	a_x	0
Batch or co-current (BCC)	a_x	$\partial(1/\widehat{Pe}_x)/\partial Z+1$

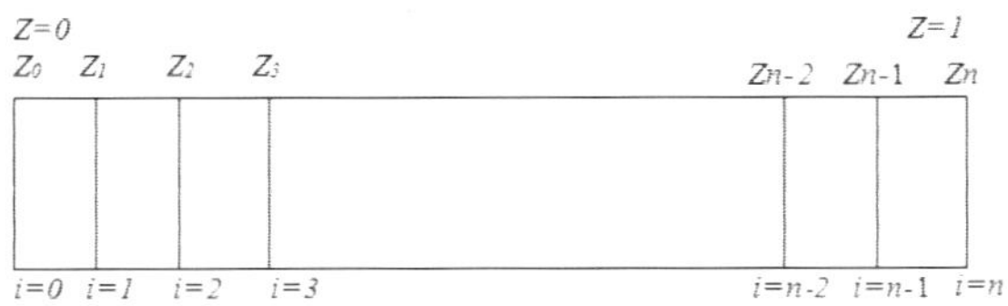

Figure 2. Volumetric channel discretization for the finite difference method.

$$\frac{\partial x}{\partial \theta}=\left[\frac{x_i^{j+1}-x_i^j}{\Delta\theta}\right]+O(\Delta x^2) \tag{20}$$

Here i and j are position and time index respectively. $\Delta Z=1/(n+1)$, And $\Delta\theta$ is chosen to be sufficiently small to ensure stable output. The numerical solution for Equation (16) and (17) can be written in the following matrix forms.

$$\boldsymbol{A}_x\boldsymbol{x}=\boldsymbol{C}_x, \boldsymbol{A}_y\boldsymbol{y}=\boldsymbol{C}_y \tag{21}$$

Thus

$$\boldsymbol{x}=\boldsymbol{A}_x^{-1}\boldsymbol{C}_x \quad \text{and} \quad \boldsymbol{y}=\boldsymbol{A}_y^{-1}\boldsymbol{C}_y \tag{22}$$

where $\boldsymbol{x}=[x_0^{j+1}, x_1^{j+1}, x_2^{j+1}, \ldots, x_{n-1}^{j+1}, x_n^{j+1}, x_{n+1}^{j+1}]^T$ and $\boldsymbol{y}=[y_0^{j+1}, y_1^{j+1}, y_2^{j+1}, \ldots, y_{n-1}^{j+1}, y_n^{j+1}, y_{n+1}^{j+1}]^T$

For $k \in [x, y]$

$$\boldsymbol{A}_k=\begin{bmatrix} F_k & G_k & & & & \\ M_{k,1} & N_{k,1} & O_{k,1} & & & \\ & M_{k,2} & N_{k,2} & O_{k,2} & & \\ & \ddots & \ddots & \ddots & & \\ & & M_{k,n-2} & N_{k,n-2} & O_{k,n-2} & \\ & & & M_{k,n-1} & N_{k,n-1} & O_{k,n-1} \\ & & & & H_k & I_k \end{bmatrix}$$

$\boldsymbol{C}_k=[C_{k,0}, C_{k,1}, C_{k,2}, \cdots, C_{k,n-2}, C_{k,n-1}, C_{k,n}]^T$,

For $0\le i\le n$

$M_{k,i}=\beta_{k,i}-\alpha_{k,i}$, $N_{k,i}=-2\beta_{k,i}-\delta$, $O_{k,i}=\alpha_{k,i}+\beta_{k,i}$

$P_{k,i}=\alpha_{k,i}-\beta_{k,i}$, $Q_{k,i}=2\beta_{k,i}-\delta$, $R=-\alpha_{k,i}-\beta_{k,i}$

$C_{k,i}=-M_{k,i}k_{i-1}^j+P_{k,i}k_i^j-O_{k,i}k_{i+1}^j-\gamma_{k,i}$, $\delta=4(\Delta Z)^2$

$\alpha_{k,i}=\Delta Z\Delta\theta a_{k,i}^j$, $\beta_{k,i}=2\Delta\theta b_{k,i}^j$, $\gamma=4(\Delta Z)^2\Delta\theta c_{k,i}^j$

For y-phase,from BCs we have $Z=0,\rightarrow y_{-1}=y_1$ and $Z=1,\rightarrow y_{n+1}=\rho_{y,n}(y_{in}-y_n)+y_{n-1}$, $\rho_{y,n}=2Pe_{y,n}\Delta Z$. All flow arrangements have same forms of $F_y, G_y, H_y, I_y, C_{y,0}, C_{y,n}$: that is $F_y=N_{y,0}$, $G_y=M_{y,0}+O_{y,0}$ $H_y=M_{y,n}+O_{y,n}$, $I_y=N_{y,n}-\rho_{y,n}O_{y,n}$, $C_{y,n}=(P_{y,n}+R_{y,n})y_{n-1}^j+(Q_{y,n}-\rho_{y,n}R_{y,n})y_n^j+R_{y,n}y_{in}^j-\rho_{y,n}O_{y,n}y_{in}^{j+1}-\gamma_{y,n}$ and $C_{y,0}=Q_{y,0}y_0^j+(P_{y,0}+R_{y,0})y_1^j-\gamma_{y,0}$. Other terms which depend on flow arrangement are summarized in Table 3. The simulation procedure can be summarized schematically as in Figure 3. An illustrative results shown in this article will based on a moving-bed (MB) column depicted in Figure 4 (a section of vertical cone). The results will be compared with the simulation results from the volumetrically equivalent uniform column and that from the inverted column of the same shape.

Table 3. Formulas of terms in the finite difference formulation for different flow types

Type	Terms
CC	BCs: $Z=0,\rightarrow x_{-1}=\rho_{x,0}(x_{\text{in}}-x_0)+x_1,\ \rho_{x,0}=2Pe_{x,0}\Delta Z$ and $Z=1,\rightarrow x_{n+1}=x_{n-1}$ $F_x=N_{x,0}-\rho_{x,0}M_{x,0},\ G_x=O_{x,0}+M_{x,0},\ H_x=M_{x,n}+O_{x,n},\ I_x=N_{x,n}$ $C_{x,0}=\rho_{x,0}P_{x,0}x_{\text{in}}^j-\rho_{x,0}M_{x,0}x_{\text{in}}^{j+1}+(Q_{x,0}-\rho_{x,0}P_{x,0})x_0^j+(R_{x,0}+P_{x,0})x_1^j-\gamma_{x,0}$ $C_{x,n}=(P_{x,n}+R_{x,n})x_{n-1}^j+Q_{x,n}x_n^j-\gamma_{x,n}$
BCC	BCs: $Z=0,\rightarrow x_{-1}=x_1$ and $Z=1,\rightarrow x_{n+1}=\rho_{x,n}(x_{\text{in}}-x_n)+x_{n-1},\ \rho_{x,n}=2Pe_{x,n}\Delta Z.$ $F_y=N_{x,0},\ G_x=M_{x,o}+O_{x,0}\quad H_x=M_{x,n}+O_{x,n},\ I_x=N_{x,n}-\rho_{x,n}O_{x,n}$ $C_{x,0}=Q_{x,0}x_0^j+(P_{x,0}+R_{x,0})x_1^j-\gamma_{x,0},C_{x,n}=(P_{x,n}+R_{x,n})x_{n-1}^j+(Q_{x,n}-\rho_{x,n}R_{x,n})x_n^j+R_{x,n}x_{\text{in}}^j-\rho_{x,n}O_{x,n}x_{\text{in}}^{j+1}-\gamma_{x,n}$
MB/ FB	$x_{-1}=x_1,\ x_{n+1}=x_{n-1}\quad F_x=N_{x,0},G_x=M_{x,0}+O_{x,0},H_x=M_{x,n}+O_{x,n},I=N_{x,n}$ $C_{x,0}=Q_{x,0}x_0^j+(P_{x,0}+R_{x,0})x_1^j-\gamma_{x,0}\quad C_{x,n}=(P_{x,n}+R_{x,n})x_{n-1}^j+Q_{x,n}x_n^j-\gamma_{x,n}$

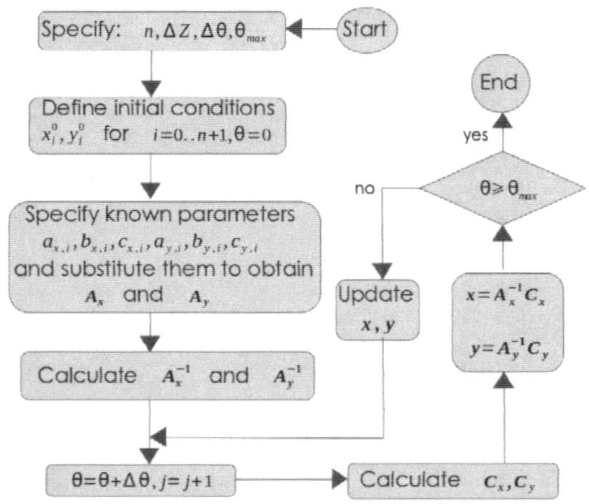

Figure 3. Solution algorithm for simulating concentration profile inside a column based on Equations (10-16)

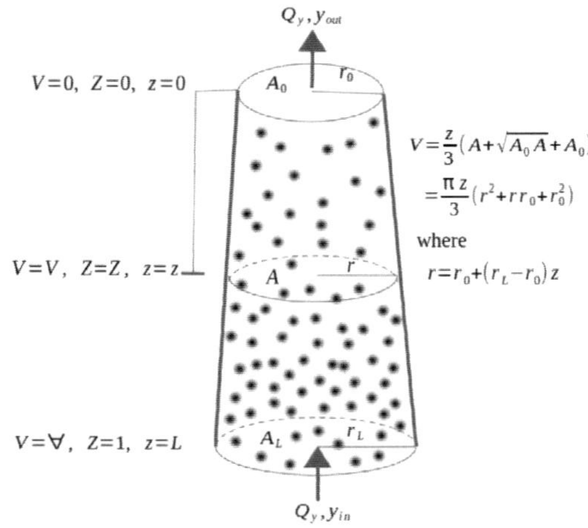

Figure 4. A moving bed with no in-out-flow, having variable (as a function of Z) cross-sections S,ϵ,T_y,Pe_x and Pe_y

4. Verification with the work of Rittirut et al. [13,14]

Rittirut et al [13,14] developed a counter-current diffusion-dispersion model for industrial extraction of fruit juices based on a specific case similar to current volumetric formulation. They verified the numerical solutions of the model by comparing them with the well-established/analytical solutions for steady-state counter-current configuration provided by Mecklenburgh and Hartland [15]. To use their results as a specific case for comparison, we did the following steps.

Firstly, we mapped our models with their models to relate all parameters. Essentially, we found that their model is a special case of the current models which include more popular flow arrangement including counter-current, co-current/batch, fixed and moving bed. One of their inherently-limited assumption, unfortunately, is that the extraction draft (α: the ratio of the flow rate of continuous phase (Q_y) over the flow rate of dispersed phase (Q_x)) determined the porosity (ϵ) such that $\alpha=Q_y/Q_x=V_y/V_x=\forall_y/\forall_x=\epsilon/(1-\epsilon)$. This assumption implies that $\kappa=(\forall_y/Q_y)/(\forall_x/Q_x)=1$, $\tau=\tau_x=\tau_y$ and $\theta=t/\tau=\theta_x=t/\tau_x=\theta_y=t/\tau_y$. Although this assumption may be not serious for their experimental cases, it severely restricted the utility of their formulation for more general counter-current flow configuration. This restriction is largely removed in the current VDF formulation however. That is to say the formulation in this article allows the independence between the hold-up (as indicated by ϵ) in the column and the corresponding flow ratio (as indicated by α).

From their work, the following intermediate differential equations were obtained after direct substituting Q_x, Q_y for S, L (the notations used in Rittirut et al as equivalent to Q_x, Q_y).

$$-\frac{\partial x}{\partial Z}+\frac{\partial}{\partial Z}\left(\frac{1}{P}\frac{\partial x}{\partial Z}\right)-\frac{\forall_x}{Q_x}\frac{k_x dA}{(1-\epsilon)dV}(x-x^*)=-\frac{\partial x}{\partial Z}+\frac{\partial}{\partial Z}\left(\frac{1}{P}\frac{\partial x}{\partial Z}\right)-\frac{\tau k_x a}{(1-\epsilon)}(x-x^*)=\frac{\partial x}{\partial \theta_x}=\frac{\partial x}{\partial \theta} \quad (23)$$

Similarly for continuous phase,

$$\frac{\partial y}{\partial Z}+\frac{\partial}{\partial Z}\left(\frac{Q_x}{Q_y R}\frac{\partial y}{\partial Z}\right)+\frac{\forall_y}{Q_y}\frac{k_x dA}{\epsilon dV}(x-x^*)=\frac{\partial y}{\partial Z}+\frac{\partial}{\partial Z}\left(\frac{1}{\alpha R}\frac{\partial x}{\partial Z}\right)+\frac{\tau k_x a}{\epsilon}(x-x^*)=\frac{\partial y}{\partial \theta_y}=\frac{\partial y}{\partial \theta} \quad (24)$$

Matching Equation (23) and (24) to (10) and (11), assuming equilibrium constant (m) as unity, and noting that $k_y a(y-y^*)=-k_x a(x-x^*)$, we have

$P=\widehat{Pe}_x, \alpha R=\widehat{Pe}_y, \kappa=1, \alpha=Q_y/Q_x=V_y/V_x=\forall_y/\forall_x=\epsilon/(1-\epsilon)$, $T_S=T_y/(1-\epsilon)$, and $T_L=T_y/\epsilon$.

Table 4. Parameters' values used for stability test and verification as compared to those of Riitirut et al [13,14] and Mecklenburgh and Hartland [15].

Parameters	**Values**									
$\alpha=Q_y/Q_x=\forall_y/\forall_x$	0.5	0.5	0.5	0.5	1.0	1.5	1.5	1.5	1.5	2.0
$P=\widehat{Pe}_x$	2.0	5.0	20	100	10	20	20	20	100	100
$R=\widehat{Pe}_y/\alpha$	0.1	5.0	20	100	20	0.01	1.0	20	100	100
$T_S=T_y/(1-\epsilon)$	0.01	1.0	1.0	1.0	10	1.0	1.0	1.0	1.0	20

For counter-current extraction, the numerical solutions for this work give identical solutions as their works for all cases given in their articles. The steady-state solutions in their results (and thus our results) agreed well (within 1% relative error) with the well-established solution given by Mecklenburgh and Hartland [15]. Since the formulation of Riitirut et al [13,14] is simply a special case of the current formulation, the current VDM is verified only partially. However, to verify the simulation results for unsteady state cases thoroughly, we need much more elaborative effort since only few solutions for (only) plug-flow cases are available. This is not treated in this article.

5. Illustrative simulation: Batch and Moving-bed (MB) adsorption

In order to illustrate the applicability of VDM, we will consider a special case of a MB/FB adsorption as depicted in Figure 4. It should be noted that a FB adsorber is a special case of MB which has $\kappa=0$ and $\widehat{Pe}'_x\rightarrow\infty$ or the bed is completely at rest. This means $a_x=b_x=0$. This flow (MB) configuration covers many unit operations including: fluidized-bed adsorbers, fluidized-dryers, fluidized-bed reactors, ion-exchange column, packed-bed reactor, up-flow anaerobic sludge blanket (UASB) reactors etc.

5.1 Batch direct yellow ARLE dye adsorption

Suppose we want to adsorb ARLE dye onto coconut palm shell-based activated carbon (CPS-AC) in a moving-bed absorber depicted in Figure 4. The illustration presented here is based on the equilibrium sorption isotherm of CPS-AC versus the ARLE dye given by Módenes et al. [16]. In their papers, the best isotherm representation was the famous Langmuir equation of the following form.

$$q_e=\frac{Q_0 b C_e}{1+bC_e}=\frac{101.56\times 0.044\, C_e}{1+0.044\, C_e} \quad (25)$$

where C_e (mg dm^{-3}) the concentration of ARLE dye at equilibrium, q_e (mg g^{-1}) the amount of dye adsorbed onto CPS-AC at equilibrium, $Q_0=101.56$ (mg g^{-1}) the maximum capacity of the absorbent and $b=0.044$ (dm^3 mg^{-1}) a Langmuir parameter. Given the density of water-saturated CPS-AC as approximately equal to that of water (1,000 g dm^{-3}), equation (25) can be translated as follows. That is $y^*=C_e$, and $x^*=1000q_e$. If we define $x(t=0)=0$ and x_m as the maximum absorbent capacity in mg dm^{-3}. Thus we have.

$$x^*=1000q_e=\frac{1000Q_0 b y^*}{1+by^*}=\frac{x_m b y^*}{1+by^*}=\frac{4468y^*}{1+0.044y^*} \quad \text{or} \quad y^*=\frac{x^*}{4468-0.044x^*}=\frac{x^*}{b(x_m-x^*)} \tag{26}$$

Then we can obtain the numerical solution from the following PDEs and their respective boundary conditions.

$$\left(\frac{\partial(1/\widehat{Pe}_y)}{\partial Z}+1\right)\frac{\partial y}{\partial Z}+\frac{1}{\widehat{Pe}_y}\frac{\partial^2 y}{\partial Z^2}+\frac{k_x a\tau_y}{\epsilon}\left(x-\frac{4468y}{1+0.044y}\right)=\frac{\partial y}{\partial\theta} \tag{27}$$

BCs: At $Z=0$: $x_{in}=x-(\kappa/\widehat{Pe}_x)(\partial x/\partial Z)$ and at $Z=1$: $x_{out}=x$ or $\partial x/\partial Z=0$

$$\frac{1}{\widehat{Pe}'_x}\frac{\partial^2 x}{\partial Z^2}-\frac{k_x a\tau_y}{1-\epsilon}\left(x-\frac{4468y}{1+0.044y}\right)=\frac{\partial x}{\partial\theta} \tag{28}$$

BCs: At $Z=0$: $x_{-1}=x_1$ or $\partial x/\partial Z=0$ and at $Z=1$: $x_{n+1}=x_{n-1}$ or $\partial x/\partial Z=0$

Módenes et al [16] also studied the kinetics of ARLE dye batch adsorption onto CPS-AC (0.177 mm particle size). An experiment was carried out for 96 h at pH 2.0, temperature 30 °C, stirring speed 120 rpm, the initial dye concentration 70 mg dm^{-3} and used 0.3 g of CPS-AC suspending in 0.125 dm^3 dye solutions. We analysed their data by assuming well-mixed in both solid and liquid phases $(1/\widehat{Pe}'_x=1/\widehat{Pe}_y=0)$ of which equation (27) and (28) reduced to the following most simplified form.

$$\frac{dx}{dt}=k_y a\left(\frac{\epsilon}{1-\epsilon}\right)(y-y^*)=k_y a\left(\frac{\epsilon}{1-\epsilon}\right)\left(y-\frac{x}{4468-0.044x}\right)=-k_x a\left(\frac{\epsilon}{1-\epsilon}\right)\left(x-\frac{4468y}{1+0.044y}\right) \tag{29}$$

$$\frac{dy}{dt}=k_y a(y-y^*)=-k_y a\left(y-\frac{x}{4468-0.044x}\right)=k_x a\left(x-\frac{4468y}{1+0.044y}\right) \tag{30}$$

Initial conditions: $x(0)=0$ an $y(0)=70$ (mg dm^{-3})

In deriving equations (29) and (30) a few assumptions were made including constant overall mass transfer coefficeint $(K_x=k_x a)$, Langmuir equilibrium isotherm and constant solid-liquid ratio (or external porosity, ϵ). This two first-order ODEs were solved numerically using standard methods provided in Scipy: a scientific module for Python programming language. Our task here is to estimate the magnitudes of $k_x a$ and ϵ by matching some experimental data from the article to the simulation produced from Equation (29) and (30). As shown in Figure 4, although the model did not fit the data well in the initial period (t <10 h) due to surface adsorption, K_x was found to be 5.x10^{-5} h^{-1} with $\epsilon=0.9945$ approximately. Figure 5a demonstrates the good fit between experimental data provided by Módenes et al. [16] and the best estimated profile simulated by Equation (29) and (30). Figure 5b is the sorption isotherm taken from the same source to be used in our demonstration.

5.2 Direct yellow ARLE dye adsorption in moving-bed and fixed-bed adsorbers

Now let us proceed to MB illustrative simulation which based on the column geometry as depicted in Figure 4 given r_0, r_L and L as 0.05, 0.1 and 1 m respectively. This provides the total volume of 0.018326 m^3 (or 18.326 dm^3) which equivalent to the uniform column having r of 0.076376 m. In the subsequent simulations K_x of 5x10^{-5} h^{-1} was used exclusively. Firstly we will show how variable $\epsilon(t)$ affects concentration profile in the absorber (uniform cross-sectional area, r = 0.076376 m) (Figure 6) and secondly we illustrate how VDM is used for non-uniform column (Figure 7).

It is illustrated that both external porosity and variable column cross-section have strong effects on concentration profiles in both solid and liquid phases. In the examples given, however, we assume that the solid phase is essential fixed $\widehat{Pe}'_x\to\infty$ and the liquid phase disperses such that $\widehat{Pe}_y=5$. Low porosity (thus high density of absorbent) at the liquid inlet tends to make the concentration profile in solid phase flat while variable cross-sectional area causes the profile skew depending on the direction of increasing (or decreasing) cross-sectional area. Similar trends occur for liquid phase but adsorption time only weakly affects the liquid profile in the simulation ranges.

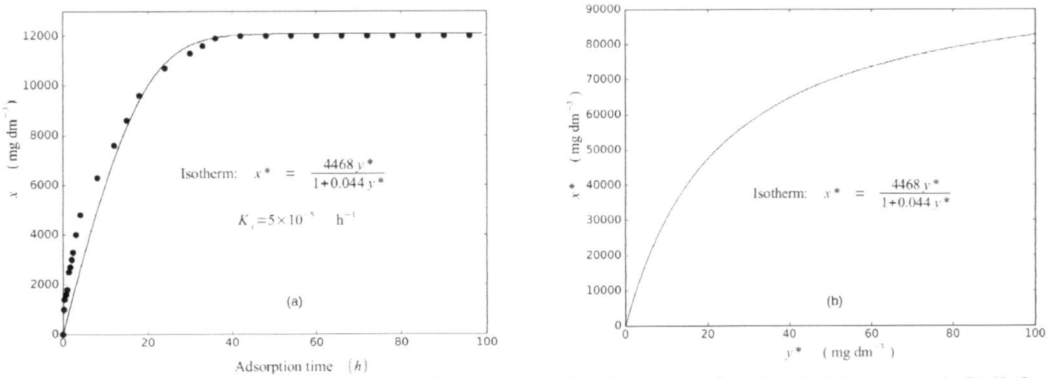

Figure 5. (a) Fitting Equation (29) and (30) to a batch adsorption data in Módenes et al. [16] from which K_x was estimated as $5x10^{-5}$ h^{-1} (b) Its corresponding sorption isotherm.

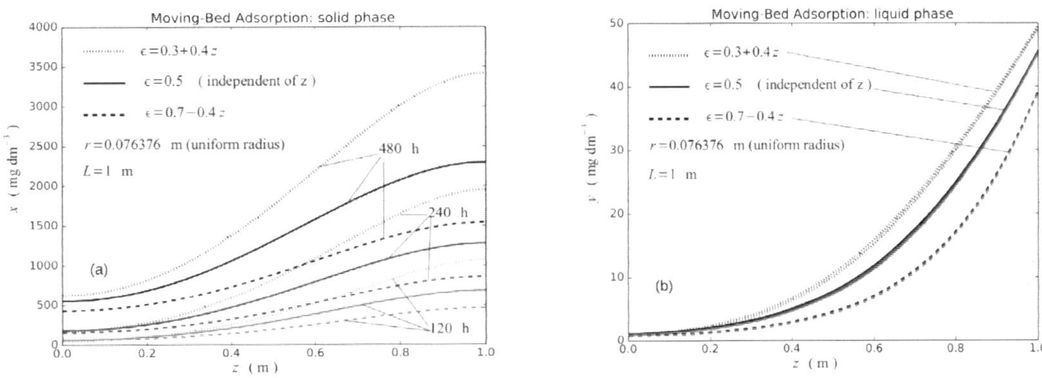

Figure 6. Illustrated numerical simulation of a moving-bed adsorption having variable external porosity (ϵ) (a) concentration profile in solid phase (b) concentration profile in liquid phase. $(\widehat{Pe}'_x=200, \widehat{Pe}_y=5)$

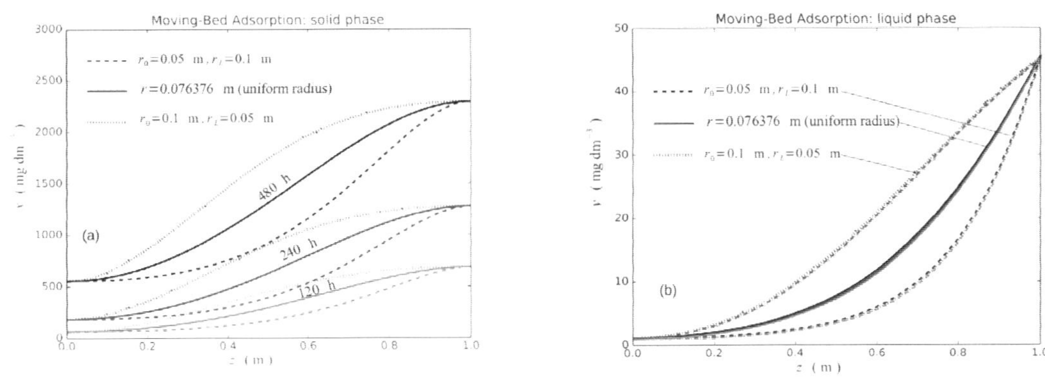

Figure 7. Illustrated numerical simulation of a moving-bed adsorption having variable cross-sectional area(a) concentration profile in solid phase (b) concentration profile in liquid phase. $(\widehat{Pe}'_x=200, \widehat{Pe}_y=5)$

In the specific cases illustrated here, it seems that, although both variable porosity and cross-section affected strongly the concentration profiles in dispersed phase, the adsorption effectiveness was essentially the same as judged from the concentration profiles in liquid phase. This can be explained that CPS-AC had very high adsorption capacity such that, in the range of our simulated experiments, its unused capacity did not decrease low enough to have a significant effect on the profile of the liquid phase.

Although it is not shown here, we also conducted a similar simulation experiments based on batch adsoption data provided by Yasemin and Zeki [17] and Kujundziski et al [18] and had similar results albeit there were some instability for the simulation cases based on the data from [18] due to very high non-linearity of the corresponding adsorption isotherm.

6. Conclusions

Volumetric dispersion models developed in this work can simplify a class of problems appeared in many continuous and batch mass-transfer operations where the cross-sectional area and porosity are varied with position (thus so too the velocities of dispersed and continuous phases) as long as the flow-rates of both dispersed and continuous phases are essentially constant. This work also suggests that many problems in chemical and environmental engineering can be simplified and extended by using volumetric approaches in model formulation to achieve better models in terms of both simplicity and generality. However, it is anticipated that for unconfined flow where flow rate is not constant TDF should be applied because the constant flow-rate assumption is invalid for VDF.

7. Acknowledgment

The authors would like to thanks Walailak University and Ministry of Science and Technology of Thailand grant for financial support.

8. References

[1] Mohanty S 2000 *Rev. Chem. Eng.* **16** 3 199-248
[2] Pirdashti M, and Kompany R. 2009 *Iranian J. Chem. Eng.* **6** 3 46-56
[3] Rados R, Al-Dahhan M H, and Dudikovic M P 2003 *Catalysis Today.* **79-80** 211-218
[4] Mahdi I H 2009 *Al-Khwarizmi Eng. J.* **5** 4 1-12
[5] Ganaie I A , Arora S, and Kukreja V K 2013 *Int. J. of Differential Equations.* 2013:ID 875298
[6] Choi J-W, Min J, Lee W-H, and Lee S-B 2000 *Biotechnol. and Bioproc. Eng.* **5** 65-70
[7] Zeng Y, Mu S J, Lou S J, Tartakovsky B, Guiot S R and Wu P 2005 *Biochem. Eng. J.* **25** 113-123
[8] Iliuta I, Bildea S C, Iliuta, M C and Larachi F 2002 *Brazilian J. Chem. Eng.* **19** 1 1-19
[9] Pérez J, Aldana G, Useche M, Rincón N, Bracho N, and Mesa J 2010 *Rev. Téc. Ing. Univ. Zulia.* **33** 3 213-222
[10] Nigam K D P, Iliuta I, and Larachi F 2001 *Chem. Eng. and Processing* **41** 365-371
[11] Ren T-T, Mu Y, Yu H-Q, Harada H, and Li Y-Y 2008 *Chem. Eng. J.* **142** 182-189
[12] Korsak L 2008 Anaerobic Treatment of Wastewater in a UASB Reactor. Master Thesis, Department of Chemical Engineering and Technology, Royal Institute of Technogy, Stockholm, Sweden.
[13] Rittirut W, Thongurai C and Siripatana C 2010 *Int. J. of Chem. Reactor Eng.* **8** A112. 1-35.
[14] Rittirut W, Thongurai C and Siripatana C. 2010 *Int. J. of Chem. Reactor Eng.* **8** A113. 1-30.
[15] Mecklenburgh J C and Hartland S 1975 The Theory of Backmixing: The Design of Continuous Flow Chemical Plant with Backming, John Wiley & Son, London.
[16] Módenes A N, Scheufele F B, Espinoza-Quiñones F R, Carraro de Souza P S, Cripa C R B, dos Santos J, Steffen V, and Kroumov A D 2015 *Int. J. Bioautomation*, **19** 2 187-206
[17] Yasemin B, and Zeki T 2007 *J. Environ. Sc.* **19** 160-166
[18] Kujundziski A P, Markovska L and Meshko V 2015 *Zaštita Materijala*, **56** 2 179-190

IOP Conf. Series: Journal of Physics: Conf. Series **820** (2017) 012016 doi:10.1088/1742-6596/820/1/012016

Large-Scale Simulation of Multi-Asset Ising Financial Markets

Tetsuya Takaishi[1]

[1]Hiroshima University of Economics, Hiroshima 731-0192, Japan

E-mail: tt-taka@hue.ac.jp

Abstract. We perform a large-scale simulation of an Ising-based financial market model that includes 300 asset time series. The financial system simulated by the model shows a fat-tailed return distribution and volatility clustering and exhibits unstable periods indicated by the volatility index measured as the average of absolute-returns. Moreover, we determine that the cumulative risk fraction, which measures the system risk, changes at high volatility periods. We also calculate the inverse participation ratio (IPR) and its higher-power version, IPR6, from the absolute-return cross-correlation matrix. Finally, we show that the IPR and IPR6 also change at high volatility periods.

1. Introduction

Statistical properties of asset prices have been intensively studied, and some pronounced properties such as the fat-tailed return distribution and volatility clustering have been investigated. These properties are now classified as "stylized facts" , e.g., see [1]. Possible dynamics for asset returns have been given by Clark [2]. He suggested that return dynamics follow a Gaussian random process with time-varying volatility, also known as the mixture-of-distribution hypothesis (MDH). Let r_t be the return at time t; the return is described by $r_t = \sigma_t \epsilon_t$, where σ_t^2 is the time-varying volatility and ϵ_t is a standard normal value $\sim N(0,1)$. Under the MDH, the volatility varies according to the rate of information arrival to the market. Since it is difficult to estimate the rate of information arrival in real financial markets, volume was used as a proxy [2].

Determining whether the return dynamics follow a Gaussian random process with time-varying volatility can be verified by examining the returns standardized by σ_t. The standardized returns are given by $\bar{r}_t = r_t/\sigma_t$. If the return is described by $r_t = \sigma_t \epsilon_t$, then the standardized returns will be ϵ_t, and the normality for $\bar{r}_t$ should be observed, e.g., the variance is equal to one and the kurtosis is equal to three. A drawback of this verification is that the volatility cannot be directly obtained in real financial markets. Recent availability of high-frequency intraday returns enables us to construct realized volatility (RV) [3, 4, 5], which converges to the integrated volatility at the infinite sampling frequency. Empirical studies using the RV claim that the return dynamics are approximately consistent with a Gaussian random process with time-varying volatility [4, 5, 6, 7, 8, 9].

The RV has different properties from other volatility measures such as absolute volatility, which is often used by econophysists. The clustering and memory effects of the RV have been investigated, and the RV has also been compared with the absolute volatility [10]. The

IOP Conf. Series: Journal of Physics: Conf. Series **820** (2017) 012016 doi:10.1088/1742-6596/820/1/012016

distribution of the RV has also been studied, and it has been argued that the RV distribution is well described by an inverse gamma distribution [11].

To better understand the price dynamics observed in real financial markets, Bornholdt proposed a simple and minimalistic Ising-based spin model that includes only two interactions that conflict with each other [12]. One of the interactions corresponds to the majority effect that agents imitate their neighbors; the other is the effect that agents tend to join minority groups. The model with these conflicting interactions shows non-equilibrium dynamics in return time series and successfully exhibits major stylized facts such as fat-tailed return distributions and volatility clustering [12, 13, 14, 15]. The return dynamics with time-varying volatility can be verified by calculating R_t/σ_t; moreover, it has been shown that R_t/σ_t recovers the same standard normality as real financial markets [16, 17].

The original model by Bornholdt can only simulate the dynamics of one asset price. Real financial markets include many stocks correlated with each other. Measuring correlations between asset returns is especially important to the study of financial market stability, and quit a few empirical studies have been conducted that unveil properties of correlations among asset returns in real financial markets, e.g., [18, 19, 20, 21, 22, 23, 24].

To simulate multiple stock time series, the Bornholdt model was extended in [25, 26, 27], and it is shown that the extended model exhibits major stylized facts similar to the original Bornholdt model. In this study, to further investigate the properties of the extended model, we construct a large-scale simulation including 300 assets and study dynamical properties of correlations and the instability of the system simulated by the extended model.

2. Extended Ising-based Financial Model

Originally, the Ising-based model was introduced by Bornholdt [12] and then extended to simulate multiple time series in [25, 26, 27]. We consider a financial market where N stocks are traded and assume that spin agents are located on sites of an $L \times L$ square lattice as in [12]. The number of total sites on a lattice is $P = L \times L$. Each site of the lattice has a spin agent s_i that takes the value $+1$ or -1, where i stands for the i-th agent; $s_i = +1$ (-1) denotes that the agent is assigned the "Buy" ("Sell") state.

The agents flip their spin states probabilistically according to a local field. The local field of the i-th spin $h_i^{(k)}(t)$ at time t for the k-th stock is defined by

$$h_i^{(k)}(t) = \sum_{\langle i,j \rangle} J s_j^{(k)}(t) - \alpha^{(k)} s_i^{(k)}(t)|M^{(k)}(t)| + \sum_{j=1}^{N} \gamma_{jk} M^{(j)}(t), \tag{1}$$

where $\langle i,j \rangle$ stands for a summation over the nearest neighbor pairs and J is the nearest neighbor coupling. In this study, we set $J = 1$, and $M^{(k)}(t)$ is the magnetization defined by $M^{(k)}(t) = \frac{1}{P}\sum_{l=1}^{P} s_l^{(k)}(t)$.

The first term on the right-hand side of (1) with $J > 0$ introduces the ferromagnetic effect similar to the original Ising model that tends to align nearest neighbor spins with the same sign. The second term introduces the effect that promotes a spin-flip, which corresponds to the minority effect. In this study, we assume that all $\alpha^{(k)}$ have the same value; that is, $\alpha^{(k)} \equiv \alpha$. The third term that is not present in the Bornholdt model describes the interaction with other stocks and introduces the effect of imitating the states of other stocks. We assume that the magnitude of the interaction is given by the interaction parameters that form a matrix γ_{lm} that has zero diagonal elements, i.e., $\gamma_{ll} = 0$. We update the spin states according to the following probability p:

$$s_i^{(k)}(t+1) \quad = +1 \qquad p = 1/(1 + \exp(-2\beta h_i^{(k)}(t))), \tag{2}$$

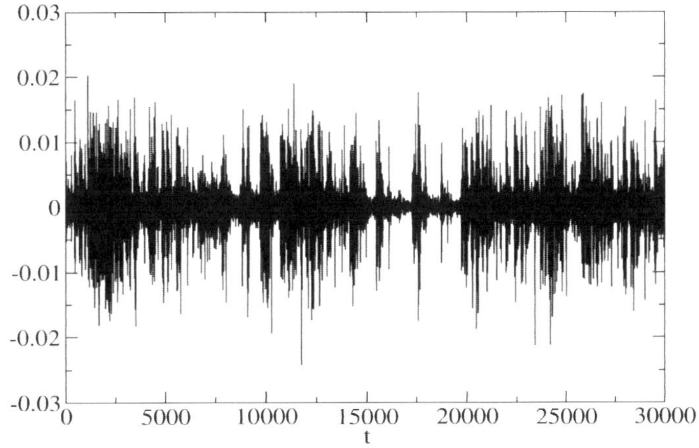

Figure 1. Representative return time series $R_k(t)$ simulated from the extended model, where t is incremented in units of one update.

$$s_i^{(k)}(t+1) \quad = -1 \qquad 1-p,$$

where β is a parameter that corresponds to the inverse temperature in the original Ising model.

3. Simulation

In this study, we consider a financial system that trades 300 stocks. Each stock is traded on a 100×100 square lattice with a periodic boundary condition. The simulation parameters are set to $(\beta, \alpha) = (2.3, 60)$. Since the situation where all elements γ are non-zero is unrealistic, we assume that ten percent of the off-diagonal elements in γ are non-zero and set to non-zero values drawn from Gaussian random numbers with average 0.05 and variance 0.01. The remaining elements are set to zero.

Spins are updated according to (2) in random order. We begin the simulation on a lattice with ordered spins and then discard the first 5×10^3 sweeps as thermalization. Then, we collect data from 3×10^4 updates for analysis.

Following [13], the return of the k-th stock is defined by the difference of the magnetization,

$$R_k(t) = (M^{(k)}(t+1) - M^{(k)}(t))/2, \tag{3}$$

where t is incremented in units of one update. In Fig. 1, we show a representative return time series that exhibits volatility clustering, which is one of the stylized facts often seen in real financial markets [28, 29]. Figure 2 shows the autocorrelation functions of return, absolute-return, and squared-return. Notice that the autocorrelation of return quickly disappears with time. On the other hand, the autocorrelations of absolute-return and squared-return remain finite at large time and slowly converge to zero within the noise level. The integrated autocorrelation times are estimated to be 223 ± 112 for the absolute-return and 93 ± 33 for the squared-return. These properties of autocorrelation are also observed in real financial markets [1].

Figure 3 shows the return distribution constructed from the normalized returns. Each return time series is normalized according to $R_k(t) = (R_k(t) - A_k)/\sigma_k$, where A_k is the average value of $R_k(t)$ and σ_k is the standard deviation of $R_k(t)$. Notice that the return distribution is fat-tailed. In real financial markets, this fat-tailed nature of return distributions has been investigated in literature such as in [29, 30, 31].

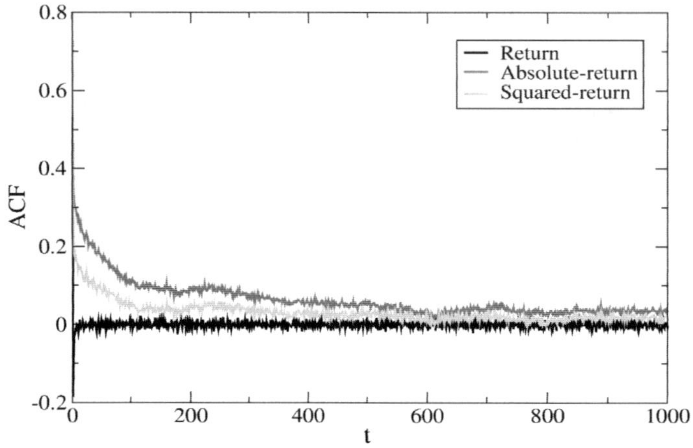

Figure 2. Autocorrelation functions of return, absolute-return, and squared-return as a function of t, where t is incremented in units of one update.

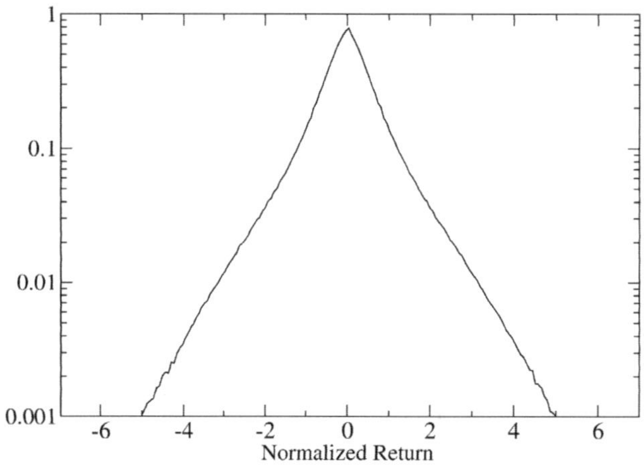

Figure 3. Normalized return distribution.

To dynamically quantify the volatility level of the financial system simulated according to the model, we define a volatility index by the average of $|R_k(t)|$: $I(t) = \frac{1}{N}\sum_{k=1}^{N}|R_k(t)|$. We determined that the volatility index $I(t)$ shown in Fig. 4 exhibits high and low periods. We expect the system to experience unstable periods when the volatility index is high.

4. Cross-correlation Matrix

To investigate the instability of the system, we calculate the cross-correlation matrix. Let $R_k(t)$ be a return for stock k ($k = 1, ..., N$) at time t ($t = 1, ..., T$) defined by (3), where $N = 300$ and

IOP Conf. Series: Journal of Physics: Conf. Series **820** (2017) 012016 doi:10.1088/1742-6596/820/1/012016

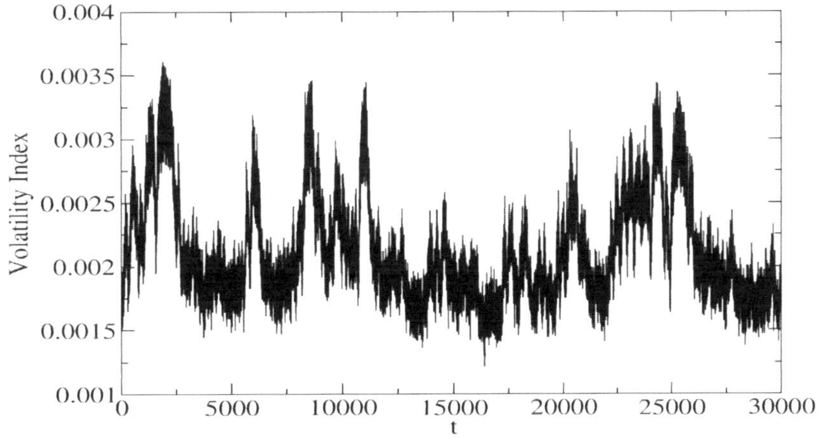

Figure 4. Volatility index $I(t)$.

$T = 30000$. We also define the normalized return $m_k(t)$ by

$$m_k(t) = \frac{R_k(t) - \langle R_k \rangle}{\sigma_k}, \tag{4}$$

where $\langle ... \rangle$ indicates the time series average and σ_k is the standard deviation of $R_k(t)$. Using the normalized return $m_k(t)$, an equal-time cross-correlation matrix $C(t)$ is defined by

$$C_{kj}(t) = \frac{1}{M} \sum_{i=0}^{M-1} m_k(t-i) m_j(t-i), \tag{5}$$

where $M = 400$, i.e., the average in (5) is taken over a period of 400. In this study, we also calculate the equal-time cross-correlation matrix from the absolute-returns.

5. Cumulative Risk Fraction

In order to investigate the systemic risk in the financial system, Billio *et al.* [32] suggested using principal component analysis (PCA) and introduced the cumulative risk fraction (CRF) as a risk measure. The CRF has also been studied in [33, 34, 35, 36]. Here, we use the CRF to study the dynamical evolution of instability of the system.

The CRF is calculated as follows. First, we compute the eigenvalues of the cross-correlation matrix, denoted by $\lambda_1, \lambda_2, \ldots, \lambda_N$, such that $\lambda_1 > \lambda_2 > \ldots > \lambda_N$. Then, the CRF is defined by [32]:

$$CRF_m = \frac{\omega_m}{\Omega}, \tag{6}$$

where Ω is the total variance of the system given by $\Omega = \sum_{j=1}^{N} \lambda_j$ and ω_m is the risk associated with the first m principal components, given by $\omega_m = \sum_{j=1}^{m} \lambda_j$. Here, note that for cross-correlation matrices, $\Omega = 1$. The CRF quantifies the portion of the system variance explained by the first m principal components over the total variance [33]. In periods of financial crisis, many stocks are highly correlated with each other and their prices easily move together. As a result, the volatility of stocks also increases, and the CRF is expected to increase in such financial crisis periods.

Figure 5(a) shows CRF_1 to CRF_5 calculated from the return cross-correlation matrix. Compared to the volatility index $I(t)$ in Fig. 4, notice that the CRF increases in high volatility

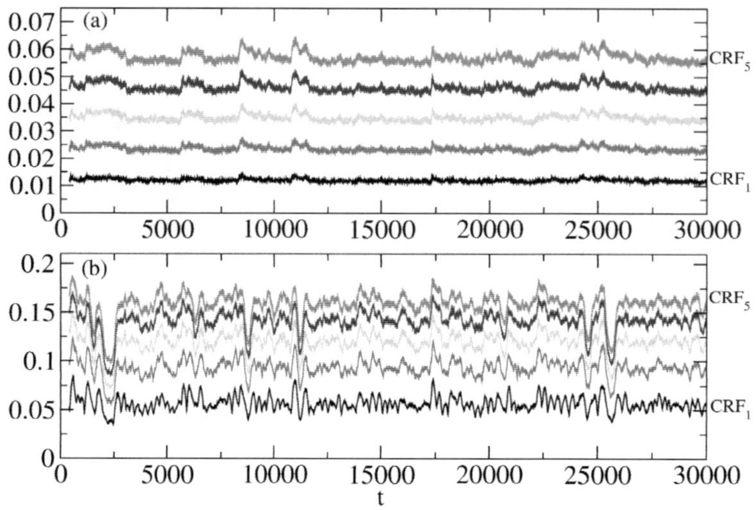

Figure 5. Cumulative risk fraction from the return (a) and absolute-return (b).

periods. Figure 5(b) is the same as Fig. 5(a) but for the absolute-return cross-correlation matrix. Contrary to the CRF from the return cross-correlation matrix, the CRF from the absolute-return cross-correlation matrix decreases in high volatility periods.

6. Random Matrix Theory

To further investigate the dynamical instability of the system, we utilize results from random matrix theory (RMT). Let $y_i(t)$ be an independent, identically distributed random variable with $i = 1, ..., N$ at time $t = 1, ..., T$. Then, we define the normalized variable

$$w_i(t) = \frac{y_i(t) - \langle y_i \rangle}{\sigma_{y_i}}, \tag{7}$$

where σ_{y_i} is the standard deviation of y_i. The equal time cross-correlation between variables $y_i(t)$ is given by $W_{ij} = \langle w_i w_j \rangle$. The matrix W is called the Wishart matrix. For $N \to \infty$ and $T \to \infty$ with $Q = T/N > 1$, the eigenvalue distribution of the matrix W is known in RMT [37, 38] as

$$\rho(\lambda) = \frac{Q}{2\pi} \frac{\sqrt{(\lambda_+ - \lambda)(\lambda - \lambda_-)}}{\lambda}, \quad \lambda_\pm = 1 + \frac{1}{Q} \pm 2\sqrt{\frac{1}{Q}}. \tag{8}$$

For $T = 400$ and $N = 300$, we obtain $\lambda_+ = 3.482$ and $\lambda_- = 0.01795$.

The inverse partition ratio (IPR) that characterizes the eigenvectors is defined by $IPR(l) = \sum_{j=1,}^{N} (v_l^j)^4$, where v_l^j is the j-th component of the eigenvector for the l-th eigenvalue. RMT predicts that the eigenvector components are de-localized and distributed as a Gaussian distribution $\sim \sqrt{\frac{N}{2\pi}} \exp\left(-\frac{N}{2}(v_l^j)^2\right)$. In such a case, the expectation of the IPR is given by $3/N$. On the other hand, when the eigenvector components are localized, for example, if only one component has a non-zero value, the expectation of the IPR is 1.

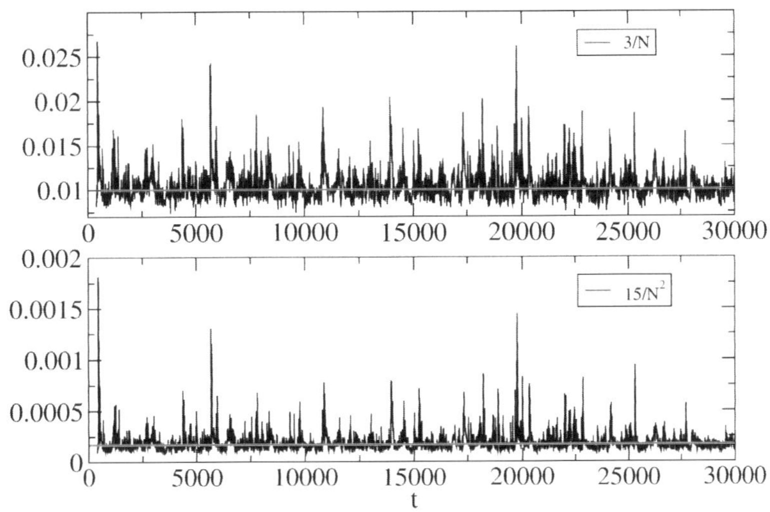

Figure 6. $IPR(1)$ and $IPR6(1)$ from return eigenvectors.

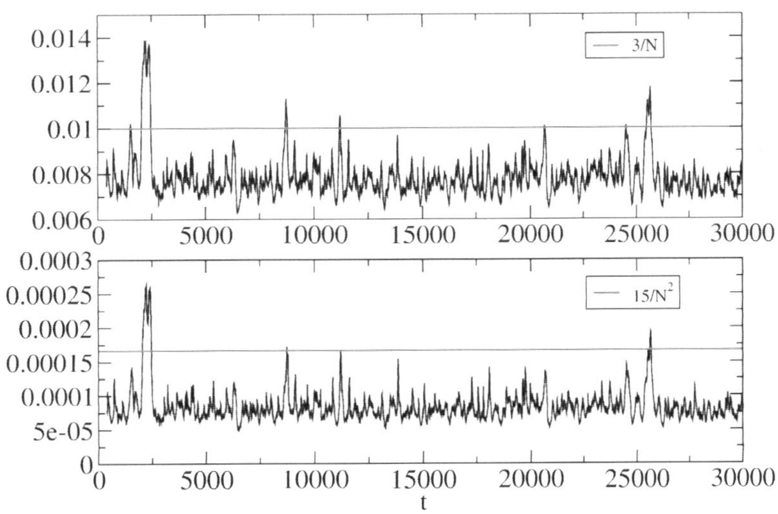

Figure 7. $IPR(1)$ and $IPR6(1)$ from absolute-return eigenvectors.

The IPR can be extended to its higher-power version, IPR6. In this study, we also focus on IPR6 defined by $IPR6(l) = \sum_{j=1,}^{N} (v_l^j)^6$. The expectation of IPR6 in RMT is given by $15/N^2$.

Figure 6 shows the time evolution of $IPR(1)$ and $IPR6(1)$ for the return cross-correlation matrix. $IPR(1)$ and $IPR6(1)$ vary around the values expected from RMT, i.e., $3/N$ and $15/N^2$ respectively, and no clear correspondence between the volatility index and the IPRs is observed.

On the other hand, $IPR(1)$ and $IPR6(1)$ from the absolute-return cross-correlation matrix in Fig. 7 exhibit different behaviors from those observed in the return cross-correlation matrix. Namely, the IPRs vary below the expected values from RMT when the volatility index is low,

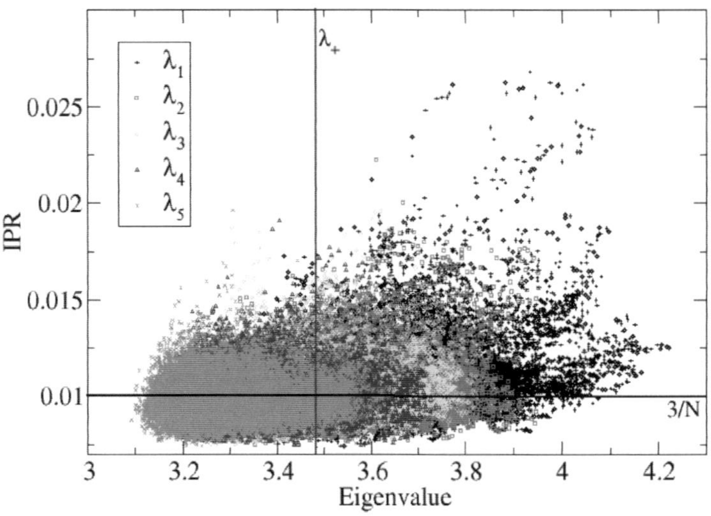

Figure 8. $IPR(l)$ versus return eigenvalues λ_l, $l = 1, \ldots, 5$.

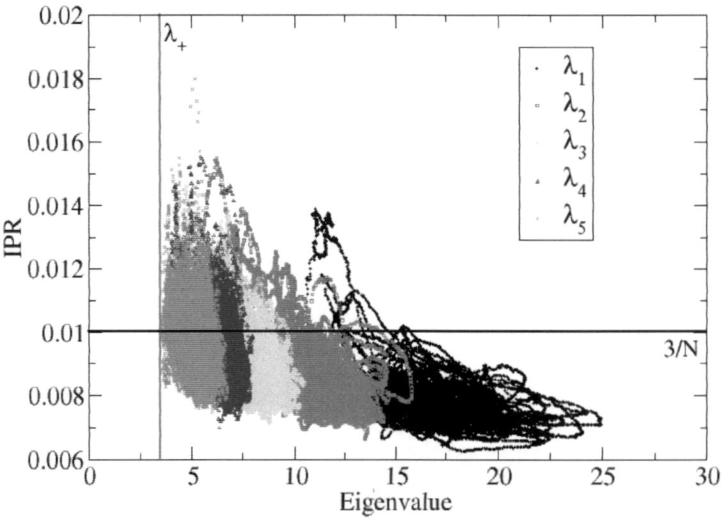

Figure 9. $IPR(l)$ versus absolute-return eigenvalues λ_l, $l = 1, \ldots, 5$.

and the IPRs increase when the volatility index is high.

Figure 8 shows the scatter plot of $IPR(l)$ versus eigenvalue λ_l for $l = 1, \ldots, 5$ for the return cross-correlation matrix. It seems that the return cross-correlation matrix is already close to the random matrix in this model, and thus, it might be understood that the IPRs do not clearly show a correspondence with the volatility index.

Figure 9 shows the scatter plot of $IPR(l)$ versus eigenvalue λ_l for $l = 1, \ldots, 5$ for the absolute-return cross-correlation matrix. For the largest eigenvalue λ_1, $IPR(1)$ deviates below $3/N$, which is the value expected from RMT. Moreover, the value of λ_1 is far beyond λ_+ from RMT. Furthermore, $IPR(l)$ gradually approaches the expectation from RMT for the smaller eigenvalues $\lambda_l, l = 2, \ldots, 5$.

7. Conclusion

We performed a large-scale simulation of an artificial financial market on a 100×100 lattice with 300 stocks using the extended Ising-based model. The financial system simulated by the model exhibited a fat-tailed return distribution and volatility clustering similar to real financial markets. To estimate the volatility level or stability of the financial system, the volatility index was defined by the average of the absolute-return. This measure showed that the financial system in the simulation underwent several unstable periods. We also calculated the CRF introduced as a risk measure [32]. We showed that the CRF from the return cross-correlation matrix increases at high volatility periods. On the other hand, the CRF from the absolute-return cross-correlation matrix decreases at high volatility periods. It is advantageous to measure the magnitude of changes in the CRF to identify unstable periods rather than to measure the actual changes [34].

RMT was used to estimate the stability of the financial system. We calculated the IPR and its higher-power version, IPR6, and compared them with the results from RMT. We showed that the IPR and IPR6 from the return cross-correlation matrix were not sensitive to the instability of the financial system. On the other hand, the IPR and IPR6 from the absolute-return cross-correlation matrix increased at high volatility periods.

While the extended Ising-based model exhibits major stylized facts such as a fat-tailed return distribution and volatility clustering, there exist some inconsistencies with real financial markets. For instance, in the real financial market, the IPR from the return cross-correlation matrix changes considerably at unstable periods. In addition, the IPR shows a decrease rather than an increase when the market is unstable [36]. Therefore, the model introduced in this study remains a prototype and further studies are needed to build a satisfactory model for real financial markets.

Acknowledgement

Numerical calculations in this work were carried out at the Yukawa Institute Computer Facility and at the facilities of the Institute of Statistical Mathematics.

[1] Cont R 2001 *Quantitative Finance* **1** 223
[2] Clark P K 1973 *Econometrica* **41** 135
[3] Andersen T G and Bollerslev T 1998 *International Economic Review* **39** 885
[4] Andersen T G, Bollerslev T, Diebold F X and Labys P 2001 *J. Am. Statist. Assoc.* **96** 42
[5] Andersen T G, Bollerslev T, Diebold F X and Ebens H 2001 *Journal of Financial Economics* **61** 43
[6] Andersen T G, Bollerslev T, Diebold F X and Labys P 2000 *Multinational Finance Journal* **4** 159
[7] Takaishi T, Chen T T and Zheng Z 2012 *Prog. Theor. Phys. Supplement* **194** 43
[8] Takaishi T 2012 *Procedia - Social and Behavioral Sciences* **65** 968
[9] Takaishi T and Watanabe T 2016 *J. Phys. Conf. Ser.* **710** 012010
[10] Zheng Z, Qiao Z, Takaishi T, Stanley H E and Li B 2014 *PLoS ONE* **9(7)** e102940 doi:10.1371/journal.pone.0102940
[11] Takaishi T 2010 *Evolutionary and Institutional Economics Review* **7** 89

[12] Bornholdt S 2001 *Int. J. Mod. Phys.* C **12** 667
[13] Kaizoji T, Bornholdt S and Fujiwara Y 2002 *Physica* A **316** 441
[14] Takaishi T 2005 *Int. J. Mod. Phys.* C **16** 1311
[15] Krause S M and Bornholdt S 2013 *Physica* A **392** 4049
[16] Takaishi T 2013 *J. Phys.: Conf. Ser.* **454** 012041
[17] Takaishi T 2014 JPS Conf. Proc., **1** (2014) 019007. *JPS Conf. Proc.* **1** 019007
[18] Plerou V, Gopikrishnan P, Rosenow B, Amaral L A N and Stanley H E 1999 *Phys. Rev. Lett.* **83** 1471
[19] Laloux L, Cizeau P, Bouchaud J and Potters M 1999 *Phys. Rev. Lett.* **83** 1467
[20] Plerou V, Gopikrishnan P, Rosenow B, Amaral L A N and Stanley H E 2000 *Physica* A **287** 374
[21] Plerou V, Gopikrishnan P, Rosenow B, Amaral L A N, Guhr T and Stanley H E 2002 *Phys. Rev.* E **65** 066126
[22] Utsugi A, Ino K and Oshikawa M 2004 *Phys. Rev.* E **70** 026110
[23] Kim D and Jeong H 2005 *Phys. Rev.* E **72** 046133
[24] Wang D, Podobnik B, Horvatić D and Stanley H E 2011 *Phys. Rev.* E **83** 046121
[25] Takaishi T 2015 *J. Phys. Conf. Ser.* **574** 012149
[26] Takaishi T 2015 *IEEE international conference on cyber technology in automation, control, and intelligent systems (CYBER)* 104 doi:10.1109/CYBER.2015.7287918
[27] Takaishi T 2016 *Evolutionary and Institutional Economics Review* **13** 455
[28] Mantegna R N and Stanley H E 1997 *Physica* A **239** 255
[29] Gopikrishnan P, Plerou V, Amaral L A N, Meyer M and Stanley H E 1999 *Phys. Rev.* E **60(5)** (1999) 5305.
[30] Lux T 1996 *Appl. Financ. Econ.* **6(6)** 463
[31] Gopikrishnan P, Meyer M, Plerou V, Amaral L A N and Stanley H E 1998 *Eur. Phys. J.* B **3** 139
[32] Billio M, Getmansky M, Lo A W and Pelizzon L 2012 *J. Financ Econ.* **104** 535
[33] Kritzman M, Li Y Z, Page S and Rigobon R 2011 *J. Portf. Manag.* **37** 112
[34] Zheng Z, Podobnik B, Feng L and Li B 2012 *Sci. Rep.* **2** 888
[35] Ren F and Zhou W Z 2014 *PLoS ONE* **9(5)** e97711
[36] Takaishi T 2016 *J. Phys. Conf. Ser.* **738** 012077
[37] Edelman A 1998 *SIAMJ. Matrix Anal. Appl.* **9** 543
[38] Sengupta A M and Mitra P P 1999 *Phys. Rev.* E **60** 3389

IOP Conf. Series: Journal of Physics: Conf. Series **820** (2017) 012017 doi:10.1088/1742-6596/820/1/012017

Using $Ag/Ag_2O/SnO_2$ Nanocomposites to Remove Malachite Green by a Photocatalytic Process

A. Taufik[1,2], V. Paramarta[1,2], S.P. Prakoso[2], R. Saleh[1,2,a]
[1]Departemen Fisika, Fakultas MIPA-Universitas Indonesia, 16424 Depok, Indonesia

[2]Integrated Laboratory of Energy and Environment, Fakultas MIPA-Universitas Indonesia, 16424 Depok, Indonesia

[a]Corresponding's author: rosari.saleh@gmail.com, rosari.saleh@ui.ac.id

Abstract. Silver/silver oxide/tin oxide nanocomposites of various weight ratios were synthesized using a microwave-assisted method. The $Ag/Ag_2O:SnO_2$ nanoparticle weight ratios used were 25:75, 50:50, and 75:25. All samples were characterized using X-ray diffraction, UV-Vis spectroscopy, Differential Thermal Analysis (DTA) and Thermogravimetric Analysis (TGA). The $Ag/Ag_2O/SnO_2$ nanocomposites contained cubic structures provided by the Ag and Ag_2O and tetragonal structures provided by the SnO_2. The silver resulted in surface plasmon resonance (SPR) at a wavelength of about 435 nm. The silver oxide material was transformed into pure Ag at a temperature of about 370 °C. The photocatalytic activity was tested on the degradation of malachite green (MG) from an aqueous solution. The results showed that $Ag/Ag_2O/SnO_2$ at a ratio of 50:50 exhibited the best photocatalytic performance for degrading MG under visible-light irradiation. The degradation of MG using $Ag/Ag_2O/SnO_2$ nanocomposites followed pseudo first-order kinetic reactions, and electron holes were found to be the main species acting on the degradation process.

1. Introduction

Semiconductor photocatalytics is a potential application for solving energy and pollution problems [1]. One of these candidate semiconductors is SnO_2 nanoparticles, which are n-type semiconductors having a wide band gap (3.6 eV), good photosensitivity and transparency, lower cost and no toxicity [1-4]. However, in practical applications, SnO_2 nanoparticles still have several limitations, including a high recombination rate of electrons and holes and the ability to be activated only under UV light irradiation.

To enhance the photocatalytic efficiency of SnO_2 nanoparticles, previous studies have applied several techniques, including doping with transition metal, coupling with other semiconductors, deposition of noble metals, and others [1, 5, 6]. Recently, silver and its oxide have been considered as promising materials to support photocatalytics. Ag_2O is a visible-light driven photocatalyst with a band gap of about (1.3–15 eV) [7, 8]. In addition, silver nanoparticles, Ag, can use the surface plasmon resonance (SPR) effect to enhance optical absorption and improve photocatalytic activity [9]. When silver nanoparticles interact with a semiconductor, it creates a metal-semiconductor Mott-Schottky (M-S) heterojunction, which promotes charge transfer from the SnO_2 to the Ag; this, in turn, leads to the concentration of electrons into the Ag in built-in electric fields [7]. In addition, the coexistence of SnO_2, Ag_2O, and Ag maintains the photocatalytic stability of the Ag_2O.

As one of organic pollutant, Malachite green has been widely used in various industries such as a dye in silk; as a food colouring additive; and as a dye in wool, leather, cotton, jute, and paper [10]. Therefore, in this research investigate the formation of $Ag/Ag_2O/SnO_2$ nanocomposites as visible-light photocatalysts for removing Malachite Green (MG), a model organic pollutant. The research also proposed and investigated a photocatalytic mechanism.

2. Experimental

2.1. Materials
Tin Chloride ($SnCl_2$), Silver Nitrate ($AgNO_3$), Sodium Hydroxide (NaOH), and Sodium Dodecyl Sulphate (SDS) were purchased from Merk and used without further purification.

2.2. Synthesizing SnO_2 nanoparticles
SnO_2 nanoparticles were synthesized using the sol-gel method, in which $SnCl_2$ was dissolved in 40 mL of deionized (DI) water and 15 mL of ethanol by magnetic stirring until a homogenous solution was obtained. Then, an NaOH solution was added to the above solution slowly and stirred for 30 min. Next, the mixture solution was heated at 180 °C for 3 h. After heat treatment, the mixture solution was cooled to room temperature naturally. The precipitates were collected by centrifugation, washed with deionized water and ethanol several times each, and dried in a vacuum drying oven at 80 °C for 8 h. Finally, the powder was calcinated in a furnace at 700 °C for 3 h to obtain the SnO_2 nanoparticles.

2.3. Synthesizing $Ag/Ag_2O/SnO_2$
$AgNO_3$ was dissolved in DI water to which SnO_2 nanoparticles were then added. An amount of SDS solution was added to the above mixture and stirred for 5 min. Then, 10 mM of NaOH was added to the solution and stirred for 10 min. The solution was irradiated by 800 W of microwaves for 2 min and then stirred for 30 min. The precipitated samples were collected using centrifugation and washed in ethanol and DI water several times. The final product was dried at 120 °C for 5 h. The molar ratios of $Ag/Ag_2O:SnO_2$ were 25:75, 50:50, and 75:25.

2.4. Photocatalytic measurement
A photocatalytic experiment was used to test the degradation of MG, a model organic pollutant. Typically, 20 mg/L of MG was poured into a 100-mL glass beaker. Then, 0.03 g of $Ag/Ag_2O/SnO_2$ nanocomposite was added to the solution and kept in the dark for 30 min to reach adsorption/desorption equilibrium. When this was achieved, the solution was irradiated for 2 h using a 40-W Xe lamp. The MG solution was collected every 15 min and analyzed using UV-Vis spectroscopy to investigate the MG degradation process. The decolorization and percent degradation of MG were calculated using the following equations.

$$Decolorization: \frac{C_t}{C_0} \tag{1}$$

$$\%\,Degradation: \frac{C_0 - C_t}{C_0} x100\% \tag{2}$$

where C_0 is the initial dye concentration and $C_t$is the dye concentration after time t.
To investigate the role of the photocatalytic mechanism, various scavengers were added, including Sodium Sulfate, Di-Ammonium Oxalate, and Tert-Butyl alcohol as electron, hole, and OH radicals, respectively.

3. Result and Discussion
Figure 1 shows the XRD spectra of $Ag/Ag_2O/SnO_2$ with various weight percentages. For comparison, the figure also plots the XRD spectra of pure SnO_2. The $Ag/Ag_2O/SnO_2$ XRD results included several peaks corresponding to the combined cubic structures of the Ag and Ag_2O and the tetragonal structures of the SnO_2. As Figure 1 shows, increased SnO_2 loading resulted in greater peak intensity of the

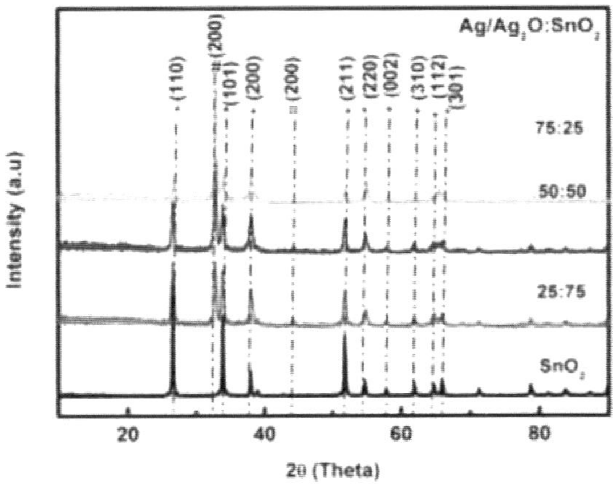

(a)

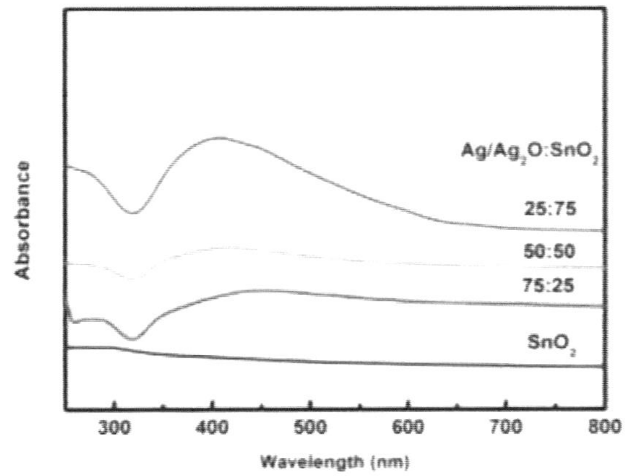

(b)

Figure 1. (a) XRD spectra of $Ag/Ag_2O/SnO_2$ with different weight ratio (black (SnO_2), red (25:75), blue (50:50), and green (75:25)). (b) UV-vis absorbance spectra of $Ag/Ag_2O/SnO_2$. Blue (SnO_2), red (75:25), green (50:50), dark cyan (25:75).

Table 1. Lattice parameters and grain size of $Ag/Ag_2O/TiO_2$ with different weight ratio.

Sample	Lattice Parameter				Grain size (nm)		
	Ag	Ag_2O	SnO_2		Ag	Ag_2O	SnO_2
	a=b=c	a=b=c	a=b	c			
SnO_2	-	-	4.738	3.187	-	-	44
$Ag/Ag_2O:SnO_2$ (25:75)	4.090	4.731	4.736	3.189	20	19	44
$Ag/Ag_2O:SnO_2$ (50:50)	4.088	4.727	4.734	3.186	22	17	44
$Ag/Ag_2O:SnO_2$ (75:25)	4.083	4.712	4.710	3.182	24	14	44

tetragonal structures of the SnO_2, which indicated increased SnO_2 quantities in the samples. The grain size and lattice parameters of the samples were investigated using Debye-Scherrer equations and the Rietveld refinement method. Table 2 shows the grain-size and lattice-parameter results, which indicated that the lattice parameters of the $Ag/Ag_2O/SnO_2$ did not change significantly. The results also indicated that formation of the $Ag/Ag_2O/SnO_2$ nanocomposites did not change the corresponding Ag, Ag_2O, and SnO_2 crystal structures.

To analyze the surface plasmon resonance (SPR) spectra of the silver nanoparticles, the optical properties of the samples were investigated. Figure 1(b) shows the absorbance spectra of SnO_2 and $Ag/Ag_2O:SnO_2$ with various weight ratios. The results showed that the SnO_2 nanoparticles did not exhibit the typical absorbance of the SPR spectra. However, after incorporating Ag/Ag_2O, the typical SPR spectra of Ag nanoparticles appeared at the 435 nm wavelength [9], and higher SnO_2 loads resulted in higher SPR peaks. This suggested that SnO_2 could the Ag_2O materials into Ag nanoparticles.

Figure 2 shows the differential thermal analysis (DTA) and thermogravimetric analysis (TGA) measurement of the $Ag/Ag_2O/SnO_2$ with a weight ratio of $Ag/Ag_2O:SnO_2$ of 75:25. The figure shows that the TGA measurement indicated a loss of weight of about 0.3% from 100–200 °C due to the removal of adsorbed water from the samples. An interesting feature of the DTA/TGA measurements of the $Ag/Ag_2O/SnO_2$ nanocomposites was the exothermic process from 336–379 °C, which was due to the transformation of Ag_2O to Ag nanoparticles and the release of oxygen into the atmosphere [11].

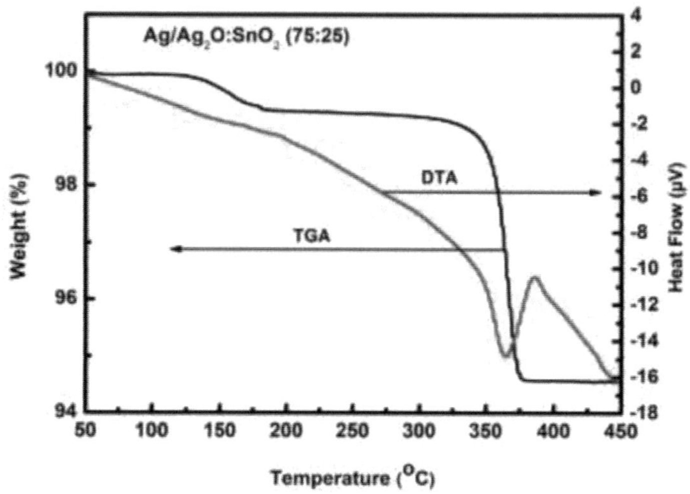

Figure 2. DTA-TGA measurement of $Ag/Ag_2O/SnO_2$ (75:25) nanocomposites.

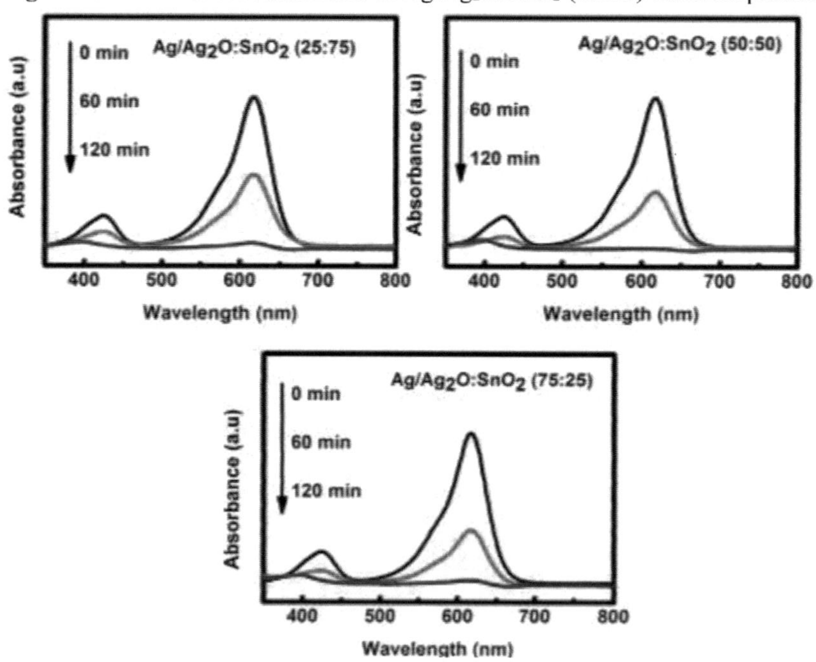

Figure 3. Malachite green absorbance spectra after and before photocatalytic reactions.

Figure 3 shows the MG absorbance spectra before and after the photocatalytic performance of $Ag/Ag_2O/SnO_2$ nanocomposites with various weight percentages. The results indicated that the absorbance spectra of the MG showed a maximum intensity at 600 nm. After the photocatalytic reaction, the absorbance spectra of the MG gradually decreased. This indicated

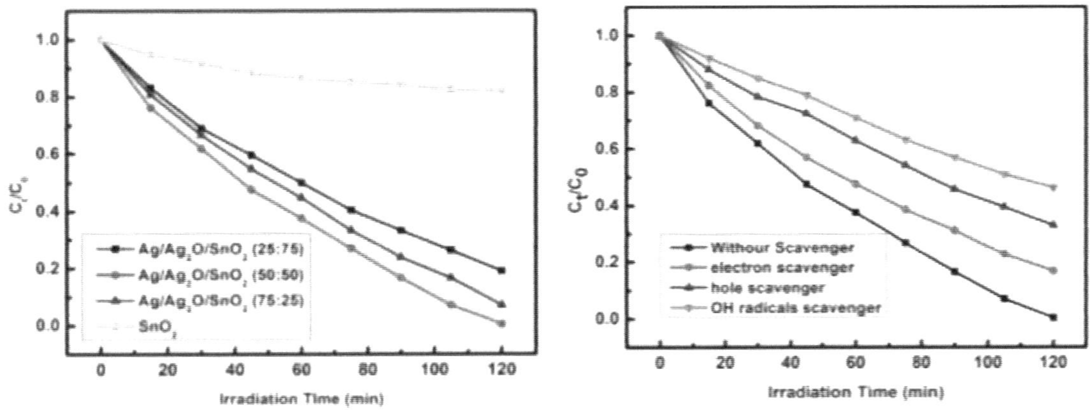

(a) (b)

Figure 4. (a) decolorization process of MG. (b) The effect of addition scavenger on photocatalytic performance of $Ag/Ag_2O/SnO_2$ nanocomposites.

Table 2. % degradation and rate constant of malachite green degradation using $Ag/Ag_2O/SnO_2$ photocatalyst.

Sample	% Degradation	Rate constant (min^{-1})
SnO_2	18	0.00157
$Ag/Ag_2O:SnO_2$ (25:75)	80	0.01327
$Ag/Ag_2O:SnO_2$ (50:50)	99	0.03521
$Ag/Ag_2O:SnO_2$ (75:25)	92	0.01984

that the concentration of MG was reduced due to the photocatalytic experiments using $Ag/Ag_2O/SnO_2$ nanocomposites.

Figure 4(a) shows the decolorization process by which the $Ag/Ag_2O/SnO_2$ nanocomposites removed the MG from the aqueous solution. The results showed that as the weight ratio of the $Ag/Ag_2O:SnO_2$ increased from 25:75 to 50:50, the decolorization of the MG increased. However, as the weight ratio increased to 75:25, the decolorization decreased, indicating that the 50:50 weight ratio exhibited the optimum conditions for removing MG from the aqueous solution using photocatalytic activity under visible-light irradiation. Table 2 shows the percentages of degradation of the MG.

The kinetic reactions of the photocatalytic process of $Ag/Ag_2O/SnO_2$ nanocomposites on the degradation of MG were calculated using pseudo first-order kinetic reactions.

$$K_{app}t = -\ln\frac{C_t}{C_0} \quad (3)$$

where K_{app} is the rate constant of the photocatalytic reactions. Table 2 shows the apparent rate constants of all samples. The results showed that the apparent rate constant of the $Ag/Ag_2O/SnO_2$ nanocomposite increased from 0.01327 min^{-1} at a weight ratio of 25:75 to 0.03521 min^{-1} at a weight ratio of 50:50, and then decreased to 0.01984 min^{-1} at a weight ratio of 75:25.

Figure 4(b) shows the influence of various radical scavengers on the photocatalytic activity of the $Ag/Ag_2O/SnO_2$ nanocomposite. The results showed that after adding any radical

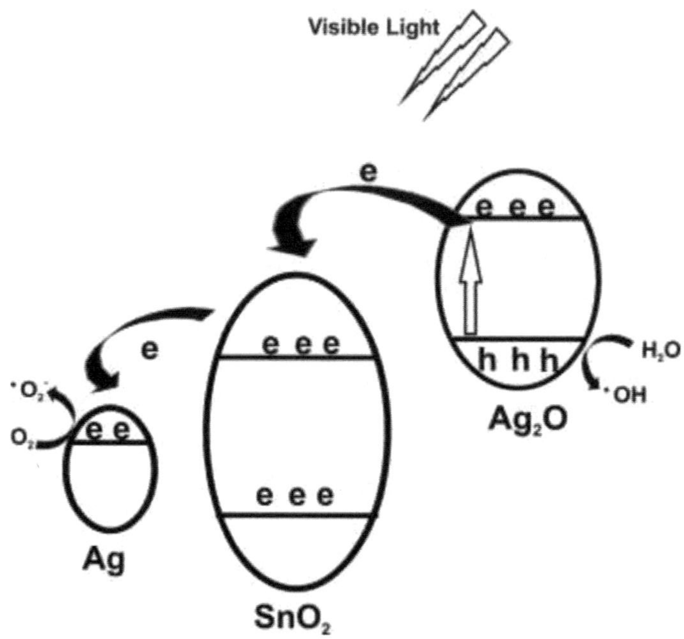

Figure 5. Photocatalytic mechanism.

scavenger, the photocatalytic activity of the $Ag/Ag_2O/SnO_2$ nanocomposite decreased due to the elimination of active species from the photocatalytic process. As the figure shows, adding Di-Ammonium Oxalate produced the lowest photocatalytic activity, indicating that it was the most active free radical species, and meaning that electron holes were the species having the most effect on the degradation process.

The proposed photocatalytic mechanism can be described as follows (see Figure 5): under visible-light irradiation, only Ag_2O can create electron hole pairs, due to its suitable band-gap energy [7, 12]. The electrons are transferred into the conduction band, leaving holes in the valence band. The electrons in the conduction band tend to recombine with the electron holes; however, the addition of another trapping state, including SnO_2 and Ag, can reduce the recombination of electrons and holes because the electrons from the conduction band of the Ag_2O may be transferred into the conduction band of the SnO_2. Furthermore, added Ag nanoparticles, which have lower conduction-band positions, can accept electrons from the SnO_2. Thus, the electron-transfer process leaves electron holes to react with water molecules, producing OH radicals and attacking the organic dye.

4. Conclusions

$Ag/Ag_2O/SnO_2$ with various weight ratios was synthesized using a microwave-assisted method. The existence of Ag in the samples was proved by XRD and UV-Vis absorbance spectroscopy. The thermal behavior of the samples revealed that all Ag_2O nanoparticles were transformed into Ag material. The samples with the $Ag/Ag_2O:SnO_2$ weight ratio of 50:50 exhibited the highest photocatalytic performance of the $Ag/Ag_2O/SnO_2$. The higher photocatalytic performance of $Ag/Ag_2O/SnO_2$ nanocomposites compare to SnO_2 nanoparticles is due to the inhibition of recombination of electron and holes. For further study $Ag/Ag_2O/SnO_2$ nanocomposites will be combined with graphene for enhance the photocatalytic performance.

Reference

[1] X. Wang, H. Fan, P. Ren 2013 *Colloids Surf., A,* **419** 140–146.
[2] R. Yang, Y.G. Gu, Y.Q. Li, J. Zheng, X.J. Li 2010 *Acta Mater.* **58** 866–874.
[3] W.C. Zhou, R.B. Liu, Q. Wan, Q.L. Zhang, A.L. Pan, L. Guo, B.S. Zou 2009 *J. Phys. Chem. C* **113** 1719–1726.
[4] H. Yuan, J.Q. Xu 2010 *Int. J. Eng. Appl. Chem.* **1** 241–246.
[5] Z. Lin, N. Li, Z. Chen, P. Fu 2017 *Sensors and Actuator B* **239** 501–510.
[6] S.G. Ghugal, S.S. Umare, R. Sasikala 2015 *Appl. Catal., A,* **496** 25–31.
[7] H. Yang, J. Tian, T. Li, H. Cui 2016 *Catal. Commun.* **87** 82–85.
[8] X. Wang, S. Li, H. Yu, J. Yu, S. Liu 2011 *Chem. Eur. J.* **17** 7777–7780.
[9] R.A. Rather, S. Singh, B. Pal 2017 *Sol. Energy Mater. Sol. Cells* **160** 463–469.
[10] E.S. Baeissa 2016 *J. Alloys Compd.* **672** 564–570.
[11] A. Sakulkalavek, R. Sakdanuphab 2016 *Mater. Sci. Semicond. Process.* **56** 313–323.
[12] Q. Luo, L. Wang, D. Wang, R. Yin, X. Li, J. An, X. Yang 2015 *Journal of Environmental Chemical Engineering* **3** 622–629.

IOP Conf. Series: Journal of Physics: Conf. Series **820** (2017) 012018 doi:10.1088/1742-6596/820/1/012018

The Synthesis and Characterization of $Ag/Ag_2O/TiO_2$/NGP Composites as Adsorbents and Photocatalysts for Wastewater Removal

A. Taufik[1,2], **S.P. Prakoso**[2], **R. Saleh**[1,2,a]
[1]Departemen Fisika, Fakultas MIPA-Universitas Indonesia, 16424 Depok, Indonesia

[2]Integrated Laboratory of Energy and Environment, Fakultas MIPA-Universitas Indonesia, 16424 Depok, Indonesia

[a]Corresponding's author: rosari.saleh@gmail.com, rosari.saleh@ui.ac.id

Abstract. An $Ag/Ag_2O/TiO_2$/NGP with different nanographene platelets (NGP) weight percents was synthesized using the hydrothermal method. The NGP weight percent was 5, 10, and 15 weight percent, respectively. The structural properties of the samples were characterized by X-ray diffraction. The optical properties of the samples were investigated by UV-Vis spectroscopy and their thermal properties were investigated by Thermo Gravimetric Analysis and Differential Thermal Analysis. The existence of a cubic structure from Ag and Ag_2O, an anatase from TiO_2, and a graphitic like structure from NGP indicated that the samples were successfully formed. The maximum light absorption at 435 nm indicated the existence of surface plasmon resonance (SPR) from silver nanoparticles. The phase transformation of Ag_2O to Ag occurred at 435 °C. Adsorption and the photocatalytic performance were investigated in the degradation of methylene blue from the aqueous solution. The adsorption ability of $Ag/Ag_2O/TiO_2$/NGP composites increased by increasing the NGP weight percent. The adsorption process of MB followed pseudo-second-order kinetic reactions. The photocatalytic activity of $Ag/Ag_2O/TiO_2$/NGP showed that $Ag/Ag_2O/TiO_2$/NGP with a 10 weight percent NGP exhibited the highest photocatalytic performance, with a removal ability of approximately 96 percent.

1. Introduction

Recently, organic dyes have been widely used in many fields, such as the textile, rubber, paper, and printing industries [1]. However, the wide range of applications of organic dyes caused environmental problems due to their hazardous properties. A portion of used organic dyes is discharged into the natural environment via industrial wastewaters, posing severe threats to aquatic organisms and human health due to the high toxicity and low biodegradability of synthetic organic dyes [2-4]. Therefore, it is important to treat organic dyes in industrial waste before they are discharged into the environment. Adsorption has been widely used in wastewater treatment due to its simple design, low-cost, and easy operation [5].

Many adsorbent materials, such as zeolite, clay, activated carbon, and graphene, have been used to remove organic dye from the environment. [6]. Graphene has been considered to be a promising material as an adsorbent due to its high specific surface area, high adsorption capacities, rapid adsorption rates, and tunable functionality of surface properties along with the intriguing mechanical, thermal, and electrical properties of the constitutive materials [6]. However, adsorption techniques merely transfer dye pollutants from one phase to another, and the pollutants are not degraded well or removed [7]. Therefore, it is very important to find an effective method to enhance the removal of organic dyes.

Photocatalytic technology is a desirable method for energy generation and wastewater treatment, because it is an inexpensive and convenient, and can completely decompose dye molecules into inorganic small molecules [8]. Numerous studies have addressed the photocatalytic performance of an oxide semiconductor [8]. TiO_2 has been considered as a photocatalyst over past two decades due to its high photocatalytic performance, high chemical and thermal stability and it was inexpensive [9]. However, TiO_2 still has several drawbacks, such as a high-rate recombination electron and holes, and it could be activated only under UV light irradiation due to its wide bandgap (3.2 eV) [9,10]. Several studies reported that many methods could improve the photocatalytic performance of a catalyst; methods included the deposition of noble metal, such as silver, coupled with a narrow bandgap semiconductor, such as Ag_2O [10,11].

Based on the aforementioned explanation, this research evaluates the comparison between adsorption and the photocatalytic performance of a combination of TiO_2, graphene, Ag, and Ag_2O in $Ag/Ag_2O/TiO_2$/nanographene platelets (NGP) composites. The structural, thermal, and optical properties of the samples were investigated to confirm the existence of $Ag/Ag_2O/TiO_2$/NGP. This study will investigate the adsorption ability of the samples, the kinetic study and photocatalytic mechanism.

2. Experimental

2.1. Synthesis $Ag/Ag_2O/TiO_2$/NGP

All reagents used were analytical grade and had no further purifications. Ag/Ag_2O/TiO2/NGP with different NGP weight percent were synthesized using microwave assisted method. Typically, $AgNO_3$ (silver nitrate), TiO_2, and nanographene were dispersed in 40 mL DI water using magnetic stirrer. Then, 80 mL SDS (sodium dodecyl sulfate) was added into the above solution. After that, 10 mM of NaOH were added into the solution and stirred for 10 min. After the homogeneous solution was achieved, the solution was irradiated by a microwave wave operated at 800 W for 2 min. The solution was then centrifuged and washed using ethanol and DI water to remove the unnecessary substances. The precipitated was dried under air condition at 120 °C for 5 hours. The NGP amounts were 5, 10, and 15 weight percent, respectively, on the $Ag/Ag_2O/TiO_2$ nanocomposites.

2.2. Characterization

The structural properties of the samples were characterized by an X-ray diffraction (XRD) Rigaku Miniflex 600 operated at 30 kV and 15 mA and monitored in the range $2\theta = 10° - 90°$. The optical properties of the samples were investigated by a UV-Vis spectrometer Hitachi UH5300 from 200 nm to 800 nm. The thermal properties of the samples were investigated using DTA/TGA (Rigaku Thermo Plus EVO2 TG-8121), with alumina as a reference and operated at room temperature until 500 °C.

2.3. Adsorption and Photocatalytic measurement

Adsorption and the photocatalytic performance of $Ag/Ag_2O/TiO_2$/NGP were tested on the degradation of methylene blue (MB) from the aqueous solution. Typically, 20 mg/L of MB were poured into a 100 mL beaker glass. After that, 0.03 g of $Ag/Ag_2O/TiO_2$/NGP were added to the solution and kept in the dark for 4 hours. At the specified time interval, the methylene blue solution was collected and monitored by UV-Vis spectroscopy to investigate the degradation of methylene blue. Its degradation was calculated using the following equation:

$$Degradation: \frac{C_t}{C_0}, \quad \text{(eq.1)}$$

IOP Conf. Series: Journal of Physics: Conf. Series **820** (2017) 012018 doi:10.1088/1742-6596/820/1/012018

where C_0 was the initial dye concentraton and C_t was the dye concentration after time t. The absorption capacity of $Ag/Ag_2O/TiO_2$/NGP composites was calculated using:

$$q_e = \frac{(C_o - C_e)V}{m}, \qquad \text{(eq.2)}$$

where q_e represents the equilibrium adsorption capacity of the Fe_3O_4 nanocomposite (mg/g); C_0 and C_e are the initial and equilibrium concentrations of methylene blue (mg/L) in aqueous solutions, respectively; V is the volume of the aqueous MB (mL); and m is the amount of $Ag/Ag_2O/TiO_2$/NGP used (g). After 4-hour adsorption process, the solution was irradiated using a 40 W Xe lamp for 2 hours, and every 15 minutes, the MB solution was taken and monitored by UV-vis spectroscopy to investigate the degradation process.

3. Result and Discussion

Figure 1 shows the XRD spectra of $Ag/Ag_2O/TiO_2/NGP$ with different NGP weight percents. The comparison the XRD spectra of $Ag/Ag_2O/TiO_2$ without the addition of NGP, was also plotted in the figure. The XRD result from $Ag/Ag_2O/TiO_2/NGP$ revealed several diffraction patterns that corresponded to the existence of a cubic structure from Ag_2O and Ag nanoparticles, an anatase structure from TiO_2 and a graphiticlike structure from NGP materials. The result showed that the diffraction pattern at $2\theta = 26°$ increases with an increase in the NGP weight percent. It indicated the higher NGP loading on the samples. The lattice parameter and grain size of the samples were

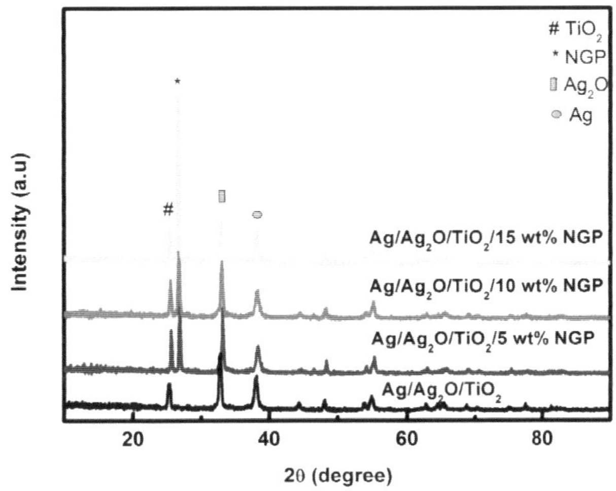

Figure 1. XRD spectra of $Ag/Ag_2O/TiO_2/NGP$ with different NGP weight percent. Black (0 wt%), blue (5 wt%), red (10 wt%), green (15 wt%).

Table 1. Lattice parameter and grain size of $Ag/Ag_2O/TiO_2/NGP$ with different NGP weight percent.

	Lattice Parameter					Grain size (nm)			
	Ag	Ag_2O	TiO_2		NGP				
Sample	a=b=c	a=b=c	a=b	c	a=b=c	Ag	Ag_2O	TiO_2	NGP
$Ag/Ag_2O/TiO_2$	4.094	4.736	3.791	9.529	-	20	19	31	-
$Ag/Ag_2O/TiO_2$/5 wt% NGP	4.072	4.683	3.747	9.277	3.600	13	36	37	32
$Ag/Ag_2O/TiO_2$/10 wt% NGP	4.070	4.696	3.749	9.469	3.491	13	27	32	33
$Ag/Ag_2O/TiO_2$/15 wt% NGP	4.068	4.706	3.762	9.455	3.627	16	23	31	33

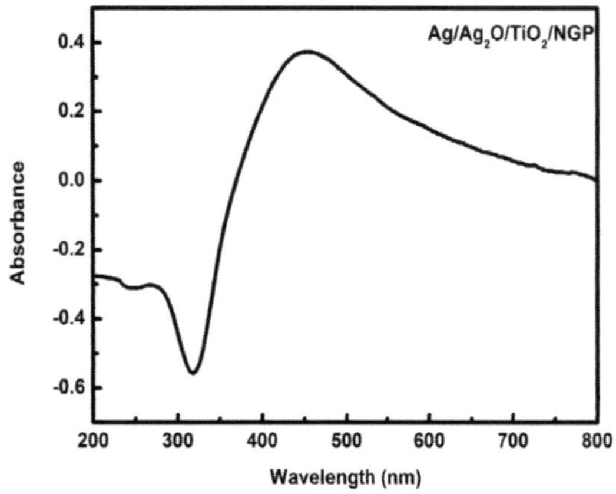

(a)

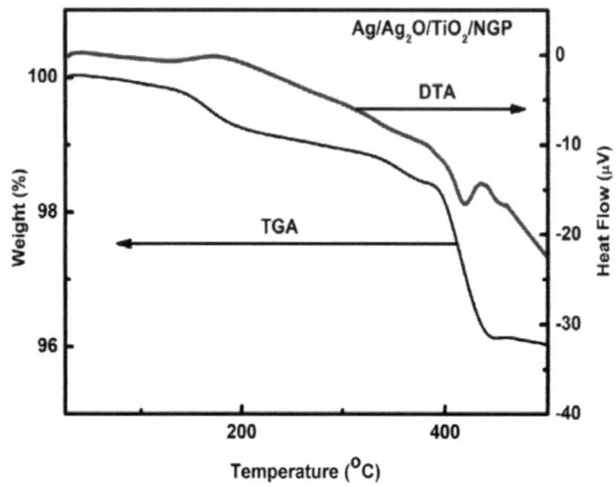

(b)

Figure 2. (a) UV-Vis absorbance and (b) DTA/TGA measurement of $Ag/Ag_2O/TiO_2$/NGP composites.

calculated using the Retveld Refinement method and Debye-Scherer Equation, respectively. The lattice parameter and grain size of the corresponding materials in $Ag/Ag_2O/TiO_2$/NGP composites are tabulated in Table 1. The table shows that the corresponding lattice parameter from Ag, Ag_2O, TiO_2, and NGP materials does not change significantly. It indicates that the all prepared samples have the same structural properties. It could be observed from the XRD result that the incorporation of NGP could enhance the Ag loading on the samples. The figure shows a diffraction peak at $2\theta = 38°$ that corresponds to the cubic structure from a silver increase with increased NGP loading.

To understand the behavior of silver nanoparticles, the optical properties of $Ag/Ag_2O/TiO_2$/NGP composites were measured (Figure 2(a)). The figures show that the prepared sample has broad light absorption, with the maximum absorbance intensity of 435 nm. Rather et al. reported that the absorbance behavior corresponded to the existence of Surface Plasmon Resonance (SPR) from silver nanoparticles [12].

The thermal properties of the samples were characterized by the DTA/TGA measurement, and the results are shown in Figure 2(b). The DTA/TGA measurement of $Ag/Ag_2O/TiO_2$ nanocomposites revealed several weight losses, such as at 100 °C to 200 °C, that corresponded to the removal of the absorbed water from the samples. The significant weight lost from $Ag/Ag_2O/TiO_2$ occurs at 400 °C to 437 °C due to the phase transformation from Ag_2O into Ag nanoparticles [13]. The DTA measurement of $Ag/Ag_2O/TiO_2$ showed that the transformation of Ag_2O into Ag nanoparticles was an exothermic

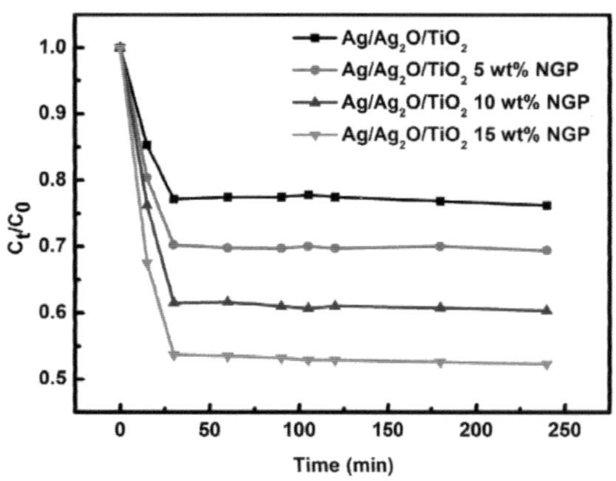

(a)

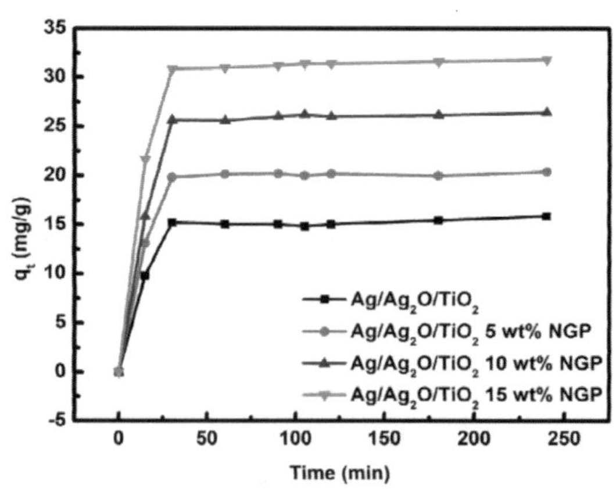

(b)

Figure 3. (a) Decolorization of NGP using $Ag/Ag_2O/TiO_2$/NGP composites through adsorption process. (b)Adsorption capacity of $Ag/Ag_2O/TiO_2$/NGP composites.

process.

Figure 3(a) shows the decolorization of MB after an absorption process that uses $Ag/Ag_2O/TiO_2/NGP$ composites with different NGP weight percents. The result shows that the absorption ability of $Ag/Ag_2O/TiO_2$ nanocomposites increases with the incorporation of NGP materials. Higher NGP loading also resulted in a higher absorption ability. Samples with higher NGP loading exhibit the highest absorbance ability. Figure 3(b) shows the adsorption capacity of all prepared samples. As can be seen from the figure that the adsorption capacity of $Ag/Ag_2O/TiO_2/NGP$ increase with increasing NGP weight percent. It indicates that NGP materials are very important in the absorption ability of the samples.

The kinetic model of the adsorption process was investigated using pseudo-first-order and second-order kinetic reactions [14]: The derivation of pseudo first-order-kinetic reactions is given by the following equations:

$$q_t = q_e(1 - \exp(-K_1 t)) \quad \text{(eq.3)}$$
$$q_t = q_e - q_e \exp(-K_1 t)) \quad \text{(eq.4)}$$
$$\ln q_t = \ln q_e - \ln(q_e \exp(-K_1 t)) \quad \text{(eq.5)}$$
$$\ln q_t = \ln q_e - (\ln q_e + (-K_1 t)) \quad \text{(eq.6)}$$
$$\ln(q_e - q_t) = \ln q_e - K_1 t \quad \text{(eq.7)}$$

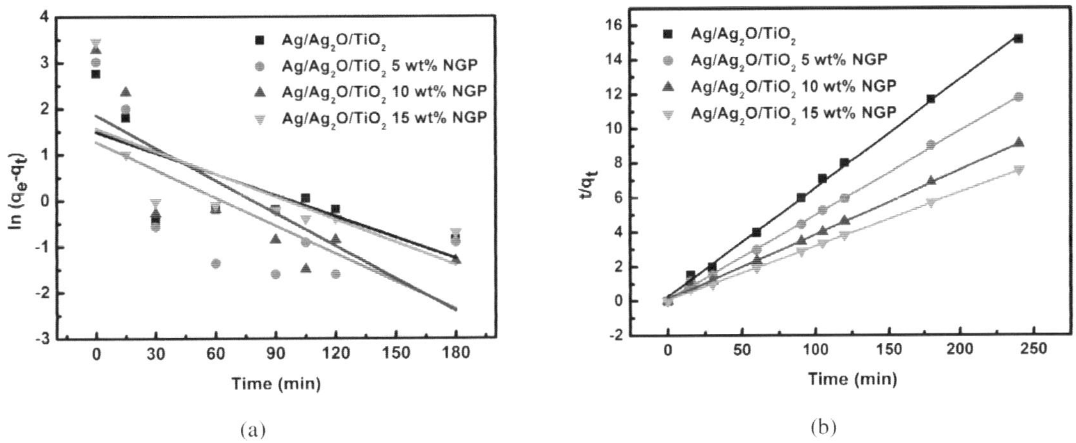

Figure 4. (a) Pseudo first and (b) second order kinetic reactions model.

Table 2. Fitting result from pseudo first and second order kinetic reactions.

Sample	qe (mg/g) exp	The pseudo-First-order rate qe (mg/g) cal	K_1	R^2	The pseudo-second-order rate qe (mg/g) cal	K_2	R^2
$Ag/Ag_2O/TiO_2$	15.85	1.91	0.00660	0.47	15.93	0.013406	0.99
$Ag/Ag_2O/TiO_2$/5 wt% NGP	20.38	1.74	0.00877	0.40	20.54	0.018022	0.99
$Ag/Ag_2O/TiO_2$/10 wt% NGP	26.41	2.24	0.01027	0.60	26.78	0.010835	0.99
$Ag/Ag_2O/TiO_2$/15 wt% NGP	31.78	2.44	0.01076	0.68	32.11	0.011616	0.99

while the pseudo second order kinetic reactions were derived by:

$$q_t = \frac{q_e^2 K_2 t}{1+q_e K_2 t} \quad \text{(eq.9)}$$

$$\frac{1}{q_t} = \frac{1+q_e K_2 t}{q_e^2 K_2 t} \quad \text{(eq.10)}$$

$$\frac{1}{q_t} = \frac{1}{q_e^2 K_2 t} + \frac{q_e K_2 t}{q_e^2 K_2 t} \; multiple\ by\ t \quad \text{(eq.11)}$$

$$\frac{t}{q_t} = \frac{1}{q_e^2 K_2} + \frac{t}{q_e} \quad \text{(eq.12)}$$

q_e and q_t are the adsorption capacities at equilibrium and time t, respectively. K_1 is the pseudo-first-order kinetic constant (min^{-1}), and K_2 is the pseudo-second-order constant (g/(mg min)). The pseudo-first-order and pseudo-second-order models are plotted in Figures 4(a) and (b), respectively. The fitting parameters are given in Table 2. It can be seen from the tables that the fitting experimental data is more fitted with pseudo-second-order kinetic reactions than pseudo-first-order kinetic reactions.

The photocatalytic activity of all samples is shown in Figure 5. The figures show that the incorporation of NGP could enhance the photocatalytic activity. The degradation percentage of all samples is tabulated in Table 3. The degradation efficiency increase from 76% to 95% with increasing NGP from 0, 5, until 10 wt% NGP. The enhancement of photocatalytic performance $Ag/Ag_2O/TiO_2$/NGP compare tp $Ag/Ag_2O/TiO_2$ is due to the existence of NGP could prevent recombination electron and holes and therefore enhance the photocatalytic efficiency. However, the increasing NGP loading until 15 wt% the photocatalytic activity of $Ag/Ag_2O/TiO_2$/NGP decrease. It is probably because the higher NGP loading could have prevented the light from penetrating the samples.

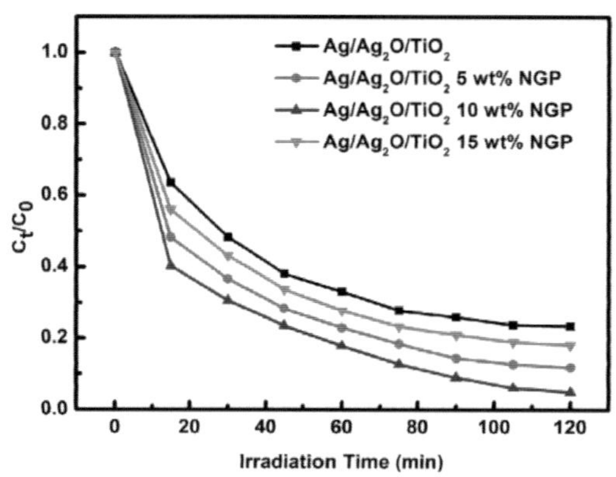

Figure 5. Photocatalytic activity of $Ag/Ag_2O/TiO_2$/NGP with different NGP weight percent.

Table 3. Percentage degradation of photocatalytic activity of $Ag/Ag_2O/TiO_2$/NGP with different NGP weight percent.

Sample	Photocatalytic degradation (%)
$Ag/Ag_2O/TiO_2$	76
$Ag/Ag_2O/TiO_2$/5 wt% NGP	88
$Ag/Ag_2O/TiO_2$/10 wt% NGP	95
$Ag/Ag_2O/TiO_2$/15 wt% NGP	82

4. Conclusion

$Ag/Ag_2O/TiO_2/NGP$ composites with different NGP weight percents were successfully synthesized using the hydrothermal method. The existence of silver nanoparticles induced the surface plasmon resonance to improve the charge separation process in photocatalytic activity. The incorporation of NGP could enhance the adsorption ability and photocatalytic performance for remove organic dye from the aqueous solution. The higher NGP loading from 5, 10, and 15 wt% in $Ag/Ag_2O/TiO_2/NGP$ composites resulted in higher adsorption performance and the photocatalytic performance of $Ag/Ag_2O/TiO_2/NGP$ composites with 10 wt% NGP resulted in higher photocatalytic performance. The adsorption performance of $Ag/Ag_2O/TiO_2/NGP$ composites followed pseudo-second-order kinetic reaction.

Reference

[1] F. Zhang, X. Chen, F. Wu, Y. Ji 2016 *Colloids Surf.*, ***509*** 474–483.
[2] Y. Qi, M. Yang, W. Xu, S. He, Y. Men 2017 *J. Colloid Interface Sci.* **486** 84–96.
[3] L. Liu, Z.Y. Gao, X.P. Su, X. Chen, L. Jiang, J.M. Yao 2015 *ACS Sustain. Chem. Eng.* **3** 432–442.
[4] G. Crini 2006 *Bioresour. Technol.* **97** 1061–1085.
[5] S. Yang, Y. Wu, Y. Wu, L. Zhu 2015 *J. Taiwan Inst. Chem. E* **50** 205–214.
[6] H. Kim, S.O. Kang, S. Park, H.S. Park 2015 *Ind. Eng. Chem. Res.* **21** 1191–1196.
[7] P. Nuengmatcha, S. Chanthai, R. Mahachai, W.C. Oh 2016 *Journal of Environmental Chemical Engineering* **4** 2170–2177.
[8] G. Hai, J. Huang, L. Cao, Y. Jie, J. Li, X. Wang, G. Zhang 2017 *J. Alloys Compd.* **690** 239–248.
[9] L. Liu, C. Luo, J. Xiong, Z. Yang, Y. Zhang, Y. Cai, H. Gu 2017 *J. Alloys Compd.* **690** 771–776.
[10] N. Wei, H. Cui, Q. Song, L. Zhang, X. Song, K Wang, Y. Zhang, J. Li, J. Wen, J. Tian 2016 Appl. Catal., B **198** 83–90.
[11] Y.C Yao, X.R. Dai, X.Y. Hu, S.Z. Huang, Z. Jin 2016 *Appl. Surf. Sci.* **387** 469–476.
[12] R.A. Rather, S. Singh, B. Pal 2017 *Sol. Energy Mater. Sol. Cells* **160** 463–469.
[13] A. Sakulkalavek, R. Sakdanuphab 2016 *Mater. Sci. Semicond. Process.* **56** 313–323.
[14] A.N.M. Salem, M.A. Ahmed, M.F.E. Shahat 2016 *J. Mol. Liq.* **219** 780–788.

IOP Conf. Series: Journal of Physics: Conf. Series **820** (2017) 012019 doi:10.1088/1742-6596/820/1/012019

A Computational Review of The Solar Wind in Oblate Spheroidal Coordinates Perspective

F W Wibowo[1], Wihayati[2]
[1]Department of Informatics Engineering, STMIK AMIKOM Yogyakarta, Jl. Ring Road Utara, Condong Catur, Depok, Sleman, Yogyakarta 55283, Indonesia
[2]Astrophysics and Mathematics Group, Sekip Utara Bulaksumur, Yogyakarta 55281, Indonesia

E-mail: ferry.w@amikom.ac.id[1], wihayati@gmail.com

Abstract. The solar as theoretical is known not perfectly round. Many types of research have been trying to approach the states and properties of the solar. Complexities of these issues are very interesting to study and observe. In this paper presented the review of a solar wind that blows from its surface based on Parker's model of solar wind that approached using continuity and momentum equations in the fluid. But in this case, the solar wind also approached in oblate spheroidal coordinates. This paper didn't describe the effect of the magnetic field and only calculated in the radial direction. The solar wind equation in oblate spheroidal coordinates has been observed in around the pole and equator. These approached equations have been solved computationally using parameters of G, M_*, a and c_s.

1. Introduction

It has been shown theoretically that the shape of the sun isn't perfectly round, but suffered flattening caused by the rotation. The symptom that attracted the most attention in the review of a stellar is the wind that blows from the surface of the stellar. This is due to several reasons, namely wind interactions with the environment to produce shock, which creates an impression amazing [1]. The presence of a stellar wind is a major supplier of mechanical energy source in an interstellar medium (ISM). The stellar wind is an important medium for the repatriation of stellar nucleosynthesis results into the ISM. A nucleosynthesis is a process of making a novel atomic from nucleons.

The stellar wind is a case of reversal of the process of the stellar accretion. A both of them involve identical equations i.e. the continuity equation, momentum equation, M_* is the value of the missing mass, and u is the velocity flow. When it shows acceleration, the forces involved in generating additional variables in the momentum equation. In practice, the stellar wind flows as time-dependent hydrodynamics problems [2]. Symptoms of the stellar wind aren't detected directly. Most of the stellar are bigger, hotter, denser, spins faster, and has a magnetic field that is stronger, older or younger than the solar [3]. The best study of the stellar wind is the solar wind.

At the time of the gravitational field of the solar isn't able to hold the shocks of the hot gas, the corona radiates out continuously. The corona with a high temperature, all the elements are ionized and since the chances of recombination are low, the gas that comes out of the solar intangible ion up to a distance of the astronomical unit. The steady flow of ionized gas particles emitted from the solar is highly conductive plasma giving a rise to a magnetic field or also known as the solar wind [4]. The solar wind consists of a hot gas of charged particles emitted from the sun flowing radially outward through the solar system and into interstellar space. It takes several days for its journey to the earth [5]. The photoionization has an important role in the charge state composition of the solar wind [6]. The differences in ion temperature properties are affected by their charge and mass [7].

Aims of this paper are approaching the solar wind in oblate spheroidal coordinates computationally. In this paper also defines the mechanism of the solar wind model and its behavior in the oblate spheroidal. This paper concludes the effect of the magnetic field in solving of the solar wind within the oblate spheroidal and the solar wind calculated is only in the radial direction. This paper also defines fluid basic equation namely continuity and momentum equations in the oblate spheroidal coordinate system which is implemented in the fluid equation to obtain solar wind equation in the oblate spheroidal coordinate.

2. Related works

Contents of the universe are dominated by gas matter. Each gram unit of it consists of particles ordered 10^{24} such as atoms, ions, protons, electrons, etc. The state of the gas system is very dynamic, each constituent particle moves freely assumed not to collide and no interaction of force on each particle. It is seen that the universe is a complex system and not easy to describe the patterns of events on it. In many cases, the complexity of the system in the universe is able to be approached with another concept [8]. In this paper to approach this phenomenon is using a fluid concept [9].

The depiction of the dynamics of complex systems isn't easy as a look and complicated. Some cases due to the complex system could be approached a matter using matter approaching as a fluid. In the implementation of the fluid study involves many macroscopic physics concepts. The model of dynamic fluid analysis is able to perform a hassle of microphysics contained into a single parameter i.e. a state equation. The state equation obtained could be resolved easily with including to dynamic fluid equations.

2.1. Fluids

The term of fluid relates to gasses and liquids. Gasses and liquids have more relation than with solids. Since both of them have molecules that are free to move around and not isolated in a place as in a solid. The hotter the fluid, the faster its molecules move on and the more area the fluid will flow. Fluid constituent atomic particles differ from atoms structuring the solid matter. Inter fluid constituent particle drifted to one another while the solid constituent particles bound to the crystal lattice. Practically the fluid is always composed of microscopic particles, however, the equations of the hydrodynamics have imposed the fluid as a continuous medium with macroscopic properties such as well define and having pressure (p), density (ρ), and velocity (u) given at each point. A medium is described as a fluid at any point defined physical quantities such as density $\rho(r, t)$, velocity $u(r, t)$, where r is the vector position of the time t. The magnitude of the fluid dynamic is a well defined on the range of such areas.

2.2. Dynamic Fluid

There is a special perspective when such a system enacted as a fluid, namely number of macroscopic representing averages defined locally in such element fluid within region range. A fluid dynamic equation is a form of a statement of conservation law depicting fluid element motion based on Newton's law of motion i.e. conservation equation of mass, momentum, and energy. Presentation of the quantities of fluid that can change over time is able to be viewed in two ways namely Lagrangian and Euler. According to Euler, a partial differential $d\rho/dt$ is implemented to measure changing of ρ at t on position r. Meanwhile, according to Lagrangian, the density changing value measured belongs to element fluid move together with fluid. From this case, the changing value is known as Lagrangian differential and symbolized as a total differential. The basic equations to depict stellar wind are Euler equation and momentum equation.

2.2.1. A mass conservation. In the fluid, consider that there is a volume *V* limited by surface S so the fluid mass included in the volume given by equation 1.

$$m = \int_V \rho dV \tag{1}$$

The mass is in volume *V* could be changed only through fluid flow penetrating surface S, thus the mass conservation written as equation 2.

$$\frac{d}{dt}\int_V \rho dV = -\int_S \rho u \, dS \tag{2}$$

where *dS* is an area element on the surface *S*. That's way the volume is fixed, so the left side could be derived and with implementing divergent theorem on the right side it would be obtained equation 3.

$$\int_V \left(\frac{d\rho}{dt} + div(\rho u)\right) dV = 0 \tag{3}$$

Volume V exists in any value but not zero, so the integrant could be neglected in this case it would form as equation 4.

$$\frac{d\rho}{dt} + div(\rho u) = 0 \tag{4}$$

This case could be simplified as equation 5.

$$\frac{d\rho}{dt} + \frac{d}{dx_j}(\rho u_j) = 0 \tag{5}$$

2.2.2. A momentum conservation. The momentum equation is obtained with the same way using an average calculation of total momentum changing in volume *V* as given in equation 6.

$$\frac{d}{dt}\int_V \rho u \, dV \tag{6}$$

In this case, the momentum flux penetrating the surface *S* couldn't be neglected with calculating of force at each unit of the volume known as force flux f_i at the fluid and surface strain, in this case given as a strain tensor T_{ij}. Thus obtained momentum equation as given in equation 7.

$$\frac{d}{dt}(\rho u_i) + \frac{d}{dx}(\rho u_i u_j) = f_i + \frac{d}{dx_j}[T_{ij}] \tag{7}$$

The supporting forces that are often considered in this case. One of it is a gravitation force as shown in equation 8.

$$f_i = -\rho \frac{d\Phi}{dx_i} \tag{8}$$

The gravitation potential equation due to the density presented by Poisson equation (see equation 9)

$$\nabla^2 \Phi = 4\pi G\rho \tag{9}$$

where G is a gravitation constant. The other force is the magnetic force that is shown in equation 10.

$$f_i = \left(j \wedge B\right)_i \tag{10}$$

where j is a current and B is a magnetic field while for electrical force written as equation 11.

$$f_i = \rho_Q E_i \tag{11}$$

where ρ_Q is an electrical current density and E_i is a magnetic field.

A strain tensor defined as infinitesimal surface vector elements dS at fluid. A vector unit is an extent of surface element and vector direction as a normalization on the surface element. So the surface force F_i is given according to equation 12.

$$F_i = T_{ij} dS_j \tag{12}$$

where dS and F are vectors so T_{ij} is a second ordered tensor. In this case, the strain tensor donated by pressure p as shown in equation 13.

$$T_{ij} = -p\delta_{ij} \tag{13}$$

where δ_{ij} is a Kronecker delta. The magnetic force as strain tensors written in equation 14.

$$m_{ij} = B_i B_j - \frac{1}{2}\delta_{ij} B_k B_k \tag{14}$$

Although it seems not to calculate the viscosity effect, but in this case, the strain donated by the viscosity effect is still showing from du_i/dx_j. The strain tensor is the average including information in the element fluid relative flow. Physically, this shows a microscopic motion especially the temperature in the gas particles. These particles will take a momentum as long as the average free path. By implementing equation 4 and to replace a partial differential $d\rho/dt$ obtained a momentum equation or a fluid motion equation as shown in equation 15.

$$\frac{du_i}{dt} + u_j \frac{du_i}{dx_j} = -\frac{1}{\rho}\frac{d\rho}{dx_i} - \frac{d\Phi}{dx_i} + \frac{dm_{ij}}{dx_j} \tag{15}$$

3. Tensor Approaching

Coordinates and differential operators against time in the fluid equation are contravariant vectors, whereas differentials in Cartesian coordinates replaced with covariant differentials using metric tensor g_{ij} with a line element dS that is given in equation 16 [10].

$$ds^2 = g_{ij} dx^i dx^j \tag{16}$$

The continuity equation could be written in the tensor notation as written in equation 17.

$$\frac{d\rho}{dt} + \left(\rho \upsilon^i\right)_{;i} = 0 \tag{17}$$

with coordinate axis x^i, namely a sum of all space components with i=1,2,3 and the contravariant velocity vector υ^i is a space coordinate differential against time $\upsilon^i = dx^i/dt$. The symbol of semicolon shows a covariant differential so the contravariant vector random T^i is able to enact the equation 18.

$$T^i_{;j} = \frac{dT^i}{dx^j} + \Gamma^i_{kj} T^k \tag{18}$$

with Γ^i_{kj} is the second form of Christoffel symbol and for a scalar number ϕ, its covariant differential could be written as equation 19.

$$\psi_{;j} = \frac{d\psi}{dx^j} \tag{19}$$

So the momentum equation will be as equation 20.

$$\frac{dv^i}{dt} + \upsilon^j \upsilon^i_{;j} = -g^{ij}\left(\frac{1}{\rho} p_{;j} + \psi_{;j}\right) \tag{20}$$

where p is a pressure and ψ is a gravitation potential.

4. Fluid equation in oblate spheroidal coordinates

Oblate spheroidal coordinates are coordinate systems of a confocal ellipsoid or known as planetary ellipsoid because the earth shape is close to this. The confocal ellipsoid coordinate system is often used to solve this problem in mechanics, electrodynamic potential, and hydrodynamic. General coordinates (ξ, υ, ϕ) are chosen in oblate spheroidal coordinates and the Cartesian coordinate transformation to oblate spheroidal coordinate is shown in equation 21.

$$\begin{aligned} x &= a\cosh\xi \sin\upsilon \cos\phi, \\ y &= a\cosh\xi \sin\upsilon \sin\phi, \\ z &= a\sinh\xi \cos\upsilon \end{aligned} \tag{21}$$

with $0<\xi<\infty$, $0<\upsilon<\pi$, $x=r\cos\phi$, $y=r\sin\phi$, $0<\phi<2\pi$. The coordinate axis z as a rotation axis and curves focus on the plane spot x=0, the centre is zero (0), and radius a for ξ is a fixed value. When the value of ξ is changed, it will be obtained a confocal oblate spheroidal. The scale factor is valued $h_\xi = h_\upsilon = a(\sinh^2\xi + \cos^2\upsilon)^{1/2}$, $h_\phi = a\cosh\xi \sin\upsilon$. Oblate spheroidal coordinates are shown in Figure 1.

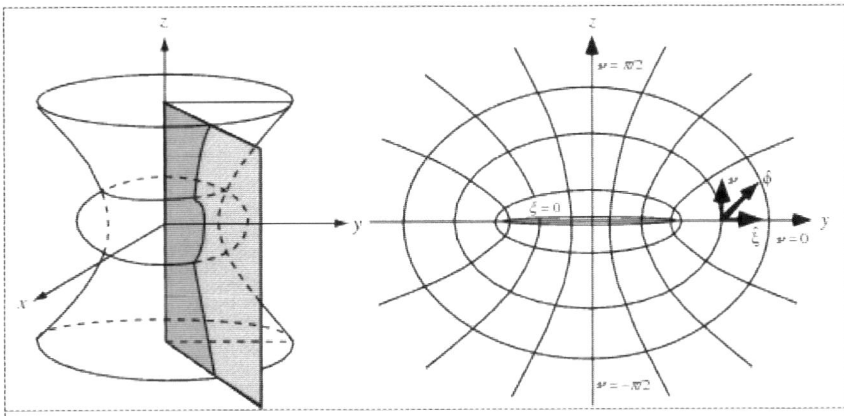

Figure 1. Oblate spheroidal coordinates.

So the metric space for oblate spheroidal coordinates (ξ, υ, ϕ) is shown in equation 22.

$$ds^2 = a^2\left(\sinh^2\xi+\cos^2\upsilon\right)\left(d\xi^2+d\upsilon^2\right)+a^2\cosh^2\xi\sin^2\upsilon d\phi^2 \tag{22}$$

A diagonal metric of g_{ij} for i = j = k and there is no re-added index could be calculated using Christoffel Γ^i_{kj} using equation 23.

$$\Gamma^i_{kj}=0, \Gamma^i_{jj}=-\frac{1}{2g_{ii}}\frac{dg_{jj}}{dx^i}, \Gamma^i_{ji}=\Gamma^i_{ij}=-\frac{1}{2g_{ii}}\frac{dg_{ii}}{dx^j}, \Gamma^i_{ii}=-\frac{1}{2g_{ii}}\frac{dg_{ii}}{dx^i} \tag{23}$$

It could be obtained the Christoffel symbol existed as below,

$$\Gamma^1_{11}=\frac{\sinh\xi\cosh\xi}{\left(\sinh^2\xi+\cos^2\upsilon\right)} \qquad \Gamma^2_{12}=\Gamma^2_{21}=\frac{\sinh\xi\cosh\upsilon}{\left(\sinh^2\xi+\cos^2\upsilon\right)}$$

$$\Gamma^1_{12}=\Gamma^1_{21}=-\frac{1}{2}\frac{\sin 2\upsilon}{\left(\sinh^2\xi+\cos^2\upsilon\right)} \qquad \Gamma^2_{22}=-\frac{1}{2}\frac{\sin 2\upsilon}{\left(\sinh^2\xi+\cos^2\upsilon\right)}$$

$$\Gamma^1_{22}=-\frac{\sinh\xi\cosh\xi}{\left(\sinh^2\xi+\cos^2\upsilon\right)} \qquad \Gamma^2_{33}=-\frac{1}{2}\frac{\cosh^2\xi\sin 2\upsilon}{\left(\sinh^2\xi+\cos^2\upsilon\right)}$$

$$\Gamma^1_{33}=-\frac{\sinh\xi\cosh\xi\sin^2\upsilon}{\left(\sinh^2\xi+\cos^2\upsilon\right)} \qquad \Gamma^3_{13}=\Gamma^3_{31}=\tanh\xi$$

$$\Gamma^2_{11}=\frac{1}{2}\frac{\sin 2\upsilon}{\left(\sinh^2\xi+\cos^2\upsilon\right)} \qquad \Gamma^3_{23}=\Gamma^3_{32}=\cot\upsilon$$

From the continuity equation in the oblate spheroidal coordinate system could be written as equation 24.

$$\frac{d\rho}{dt}+\frac{d\rho u^{\xi}}{d\xi}+\frac{d\rho u^{\upsilon}}{d\upsilon}+\frac{d\rho u^{\phi}}{d\phi}+\rho u^{\xi}\left(\frac{2\sinh\xi\cosh\xi}{\left(\sinh^{2}\xi+\cos^{2}\upsilon\right)}+\tanh\xi\right)+\rho u^{\upsilon}\left(\cot\upsilon-\frac{\sin 2\upsilon}{\left(\sinh^{2}\xi+\cos^{2}\upsilon\right)}\right)=0 \quad (24)$$

and from the other fluid equation resulted in momentum equation as shown in equation 25, 26, and 27.

$$\begin{aligned}&\frac{du^{\xi}}{dt}+u^{\xi}\left(\frac{du^{\xi}}{d\xi}+\frac{\sinh\xi\cosh\xi}{\left(\sinh^{2}\xi+\cos^{2}\upsilon\right)}u^{\xi}-\frac{1}{2}\frac{\sin 2\upsilon}{\left(\sinh^{2}\xi+\cos^{2}\upsilon\right)}u^{\upsilon}\right)+\\&u^{\upsilon}\left(\frac{du^{\upsilon}}{d\upsilon}-\frac{1}{2}\frac{\sin 2\upsilon}{\left(\sinh^{2}\xi+\cos^{2}\upsilon\right)}u^{\xi}-\frac{\sinh\xi\cosh\xi}{\left(\sinh^{2}\xi+\cos^{2}\upsilon\right)}u^{\upsilon}\right)+\\&u^{\phi}\left(\frac{du^{\upsilon}}{d\phi}-\frac{\sinh\xi\cosh\xi\sin^{2}\upsilon}{\left(\sinh^{2}\xi+\cos^{2}\upsilon\right)}u^{\phi}\right)=-\frac{1}{a^{2}\left(\sinh^{2}\xi+\cos^{2}\upsilon\right)}\left(\frac{1}{\rho}\frac{dp}{d\xi}+\frac{d\psi}{d\xi}\right)\end{aligned} \quad (25)$$

$$\begin{aligned}&\frac{du^{\upsilon}}{dt}+u^{\xi}\left(\frac{du^{\upsilon}}{d\xi}+\frac{1}{2}\frac{\sinh 2\upsilon}{\left(\sinh^{2}\xi+\cos^{2}\upsilon\right)}u^{\xi}+\frac{\sinh\xi\cosh\xi}{\left(\sinh^{2}\xi+\cos^{2}\upsilon\right)}u^{\upsilon}\right)+\\&u^{\upsilon}\left(\frac{du^{\upsilon}}{d\upsilon}-\frac{1}{2}\frac{\sin 2\upsilon}{\left(\sinh^{2}\xi+\cos^{2}\upsilon\right)}u^{\xi}-\frac{\sinh\xi\cosh\xi}{\left(\sinh^{2}\xi+\cos^{2}\upsilon\right)}u^{\upsilon}\right)+\\&u^{\phi}\left(\frac{du^{\upsilon}}{d\phi}-\frac{1}{2}\frac{\cosh^{2}\xi\sin 2\upsilon}{\left(\sinh^{2}\xi+\cos^{2}\upsilon\right)}u^{\phi}\right)=-\frac{1}{a^{2}\left(\sinh^{2}\xi+\cos^{2}\upsilon\right)}\left(\frac{1}{\rho}\frac{dp}{d\upsilon}+\frac{d\psi}{d\upsilon}\right)\end{aligned} \quad (26)$$

$$\begin{aligned}&\frac{du^{\phi}}{dt}+u^{\xi}\left(\frac{du^{\phi}}{d\xi}+\tanh\xi u^{\phi}\right)+u^{\upsilon}\left(\frac{du^{\phi}}{d\upsilon}+\cot\upsilon u^{\phi}\right)+\\&u^{\phi}\left(\frac{du^{\phi}}{d\phi}+\tanh\xi u^{\xi}+\cot\upsilon u^{\upsilon}\right)=-\frac{1}{a^{2}\left(\cosh^{2}\xi+\sin^{2}\upsilon\right)}\left(\frac{1}{\rho}\frac{dp}{d\phi}+\frac{d\psi}{d\phi}\right)\end{aligned} \quad (27)$$

Assumed that the wind flow directs radially (ξ), $u^{\upsilon}= u^{\phi} = 0$, so every rate $d/d\upsilon$ and $d/d\phi$ are neglected or equal to zero (0). So it will obtain continuity and momentum equations more simple as shown in equation 28 and 29 respectively.

$$\frac{d\rho}{dt}+\frac{d\rho u^{\xi}}{d\xi}+\rho u^{\xi}\left(\frac{2\sinh\xi\cosh\xi}{\left(\sinh^{2}\xi+\cos^{2}\upsilon\right)}+\tanh\xi\right)=0 \quad (28)$$

$$\frac{du^{\xi}}{dt}+u^{\xi}\left(\frac{du^{\xi}}{d\xi}+\frac{\sinh\xi\cosh\xi}{\left(\sinh^{2}\xi+\cos^{2}\upsilon\right)}u^{\xi}\right)=-\frac{1}{a^{2}\left(\sinh^{2}\xi+\cos^{2}\upsilon\right)}\left(\frac{1}{\rho}\frac{dp}{d\xi}+\frac{d\psi}{d\xi}\right) \quad (29)$$

5. Solar wind in oblate spheroidal coordinates

In around the equator, the colatitude degree is about 90° and written as $\upsilon \approx \pi/2$ so the value of $\cos\upsilon$ is zero (0) and $\sin\upsilon$ is one (1). In this case, this approaching is used to implement in around the equator and pole.

5.1. Solar wind in around the equator
Based on the momentum equation if enacted in around the equator with $\upsilon \cong \pi/2$ so it will be obtained as equation 30.

$$\left(\frac{1}{\rho}\frac{dp}{d\upsilon}+\frac{d\psi}{d\upsilon}\right)=0 \tag{30}$$

In this case, the pressure (p) and potential (ψ) aren't dependent on υ. The equation for a stellar radius presented in the oblate spheroidal coordinate system as shown in equation 31.

$$\begin{aligned} r &= a^2\cosh^2\xi\sin^2\upsilon\cos^2\varphi + a^2\cosh^2\xi\sin^2\upsilon\sin^2\varphi + a^2\sinh^2\xi\cos^2\upsilon \\ &= a^2\cosh^2\xi\sin^2\upsilon + a^2\sinh^2\xi\cos^2\upsilon \end{aligned} \tag{31}$$

So the stellar radius from the stellar centre until around the equator, $r_e = a\cosh\xi$, could be found ψ as equation 32.

$$\psi = -\frac{GM_*}{a\cosh\xi} \tag{32}$$

The momentum equation thus substituted $c_s^2 = \frac{p}{\rho}$ and $\frac{d\psi}{d\xi} = \frac{GM_*\sinh\xi}{a\cosh^2\xi}$ become equation 33.

$$u^{\xi}\frac{du^{\xi}}{d\xi} + \left(u^{\xi}\right)^2\frac{\cosh\xi}{\sinh\xi} = -\frac{1}{a^2\sinh^2\xi}\left(\frac{c_s^2}{\rho}\frac{dp}{d\xi} + \frac{GM_*\sinh\xi}{a^2\cosh^2\xi}\right) \tag{33}$$

The equation 33 is a momentum equation in the first order partial differential equation with independent variable ρ and u^{ξ} so that the continuity equation could be shown in equation 34.

$$\frac{d\rho}{d\xi} = -\rho\left(\frac{1}{u^{\xi}}\frac{du^{\xi}}{d\xi} + \frac{2\cosh\xi}{\sinh\xi} + \tanh\xi\right) \tag{34}$$

The equation 34 then substituted into the equation 33 obtained the momentum equation in the differential equation with independent variable u^{ξ} as shown in equation 35.

$$\left(u^{\xi} - \frac{c_s^2}{a^2\sinh^2\xi}\frac{1}{u^{\xi}}\right)\frac{du^{\xi}}{d\xi} + \left(u^{\xi}\right)^2\frac{\cosh\xi}{\sinh\xi} = -\frac{c_s^2}{a^2\sinh^2\xi}\left(\frac{2\cosh\xi}{\sinh\xi} + \tanh\xi\right) - \frac{GM_*\sinh\xi}{a^3\sinh^2\xi\cosh^2\xi} \tag{35}$$

The stellar wind equation in around the equator is shown in Figure 2 and 3.

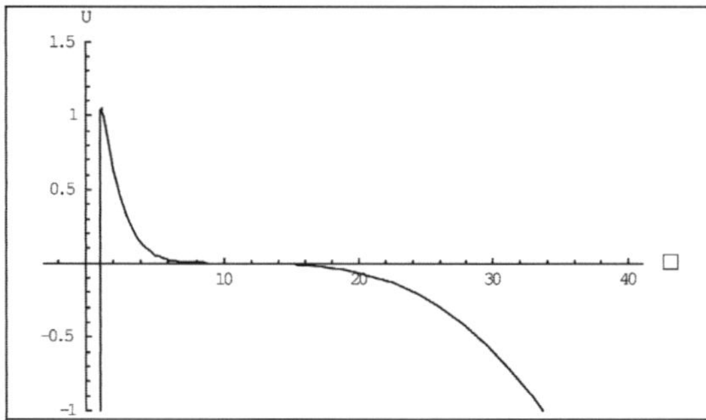

Figure 2. The stellar wind in around the equator with G=2, M_*=1, a=1, and c_s=1.

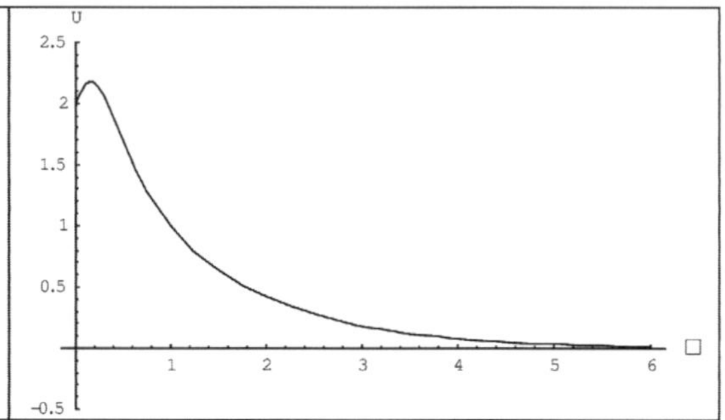

Figure 3. The stellar wind in around the equator with G=2, M_*=1, a=2, and c_s =1

5.2. Solar wind in around the pole

In around the pole, the colatitude degree is assumed 0 and written as $\upsilon \cong 0$ so the value of cos υ is one (1) and sin υ is zero (0). Because the wind velocity is dependent only on the direction ξ so the continuity equation is written as shown in equation 36.

$$\frac{d\rho}{dt}+\frac{d\rho u^{\xi}}{d\xi}+\rho u^{\xi}\left(3\tanh\xi\right)=0 \qquad (36)$$

Thus the momentum equation is shown in equation 37.

$$\frac{du^{\xi}}{dt}+u^{\xi}\frac{du^{\xi}}{d\xi}+\left(u^{\xi}\right)^{2}\tanh\xi=-\frac{1}{a^{2}\cosh^{2}\xi}\left(\frac{1}{\rho}\frac{dp}{d\xi}+\frac{d\psi}{d\xi}\right) \qquad (37)$$

Approaching for state around a pole is obtained a stellar radius r_p=a sinh ξ so assumed the stellar potential as shown in equation 38.

$$\psi=-\frac{GM_{*}}{a\sinh\xi} \qquad (38)$$

So $\frac{d\psi}{d\xi}=\frac{GM_{*}\cosh\xi}{a\sinh^{2}\xi}$ and the steady wind flow is $\frac{d}{dt}=0$. The continuity is shown in equation 39.

$$\frac{d\rho}{d\xi}=-\rho\left(\frac{1}{u^{\xi}}\frac{du^{\xi}}{d\xi}+3\tanh\xi\right) \qquad (39)$$

and the momentum equation is shown in equation 40.

$$u^{\xi}\frac{du^{\xi}}{d\xi}+\left(u^{\xi}\right)^{2}\tanh\xi=-\frac{1}{a^{2}\cosh^{2}\xi}\left(\frac{c_{s}^{2}}{\rho}\frac{d\rho}{d\xi}+\frac{GM_{*}\cosh\xi}{a\sinh^{2}\xi}\right) \qquad (40)$$

with substituting the equation 39 into the equation 40, it will result as shown in equation 41.

$$\left(u^{\xi}-\frac{c_s^2}{a^2\cosh^2\xi}\frac{1}{u^{\xi}}\right)\frac{du^{\xi}}{d\xi}+\left(u^{\xi}\right)^2\tanh\xi=-\frac{3c_s^2}{a^2}\frac{\sinh\xi}{\cosh^3\xi}-\frac{GM_*\sinh\xi}{a\cosh\xi\sinh^2\xi} \quad (41)$$

The stellar wind equation in around the pole is shown in Figure 4 and 5.

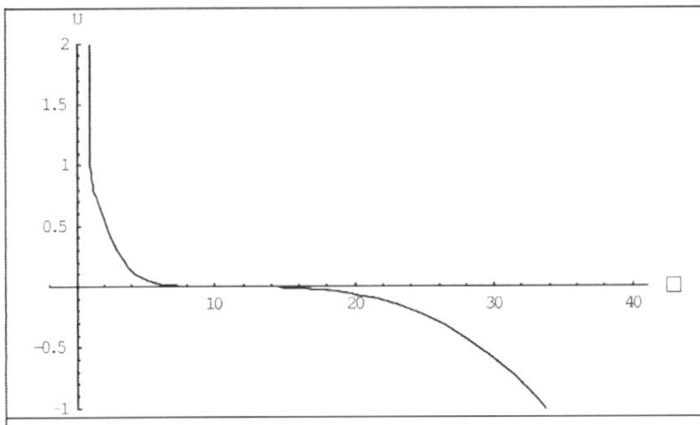

Figure 4. The stellar wind in around pole with G=2, M_*=1, a=1, and c_s =1.

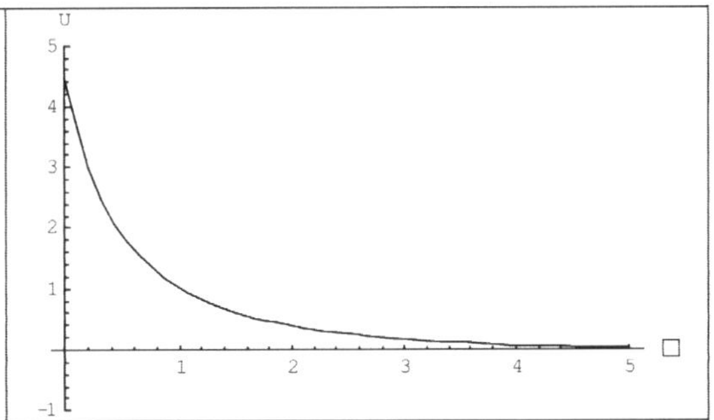

Figure 5. The stellar wind in around pole with G=2, M_*=1, a=2, and c_s =1

6. Conclusion

The computational approach to solving differential equations in this paper has been done. The stellar wind equation in oblate spheroidal coordinates has been evaluated in around the pole and the equator. Both of them are same type namely one ordered differential equation as a velocity function at the direction ξ to the distance $\xi(U^{\xi}(\xi))$. These equations have been solved by using a fixed value for G, M_*, a and c_s. Future work of our paper needs more completion research on other colatitude angle and it will also underlie the study of solar wind dynamics process more complex within a magnetic field, viscosity, etc. So it will have an enhanced interpretation.

7. References

[1] Parker E N 2014 Reminiscing my sixty year pursuit of the physics of the sun and the galaxy *Research in Astronomy and Astrophysics* **14** 1 pp1-14

[2] Feng X, Zhang M and Zhou Y 2014 A new three-dimensional solar wind model in spherical coordinates with a six-component grid *The Astrophysical Journal Supplement Series* **214** 6 (12pp)

[3] Holst B V D, Manchester IV W B, Frazin R A, Văsquez A M, Tóth G, and Gombosi T I 2010 A data-driven, two temperature solar wind model with alfvén waves *The Astrophysical Journal* **725** pp1373-1383

[4] Borovsky J E 2016 Relativity and the solar wind: the Maxwell-equation origins of the solar-wind motional electric field Journal of Electromagnetic Analysis and Applications 8 pp133-151

[5] Chiao M 2008 Solar wind: Thar she blows *Nature Physics* **4** 347

[6] Landi E and Lepri S T 2015 Photoionization in the solar wind *The Astrophysical Journal Letters* **812** L28 6pp

[7] Tracy P J, Kasper J C, Zurbuchen T H, Raines J M, Shearer P, and Gilbert J 2015 Thermalization of heavy ions in the solar wind *The Astrophysical Journal* **812** 170 9pp

[8] Parker E N 2001 Solar activity and classical physics *Chin. J. Astron. Astrophys.* **1** 2 pp99-124

[9] Echim M M, Lemaire J, and Lie-Svendsen Ø 2011 A Review on solar wind modeling: kinetic and fluid aspects *Surv Geophys* **32** pp1-70

[10] Wihayati, Wibowo F W 2017 Metric Tensor Approach in Solar Wind for Rapidly Rotating Stellar *J. Phys.: Conf. Ser.*

Dye adsorption into transition metal-doped zinc oxide nanoparticles supported on natural zeolites to solve wastewater issue

Margaretha Indra Pratiwi [1,2], Nur Afifah[1,2] and Rosari Saleh[1,2,a]
[1]Departemen Fisika, Fakultas MIPA-Universitas Indonesia, 16424 Depok, Indonesia
[2]Integrated Laboratory of Energy and Environment, Fakultas MIPA-Universitas Indonesia, 16424 Depok, Indonesia
[a]Corresponding's author: rosari.saleh@gmail.com, rosari.saleh@ui.ac.id

Abstract. In this work, Fe-doped zinc oxide/natural zeolite (Fe:ZnO/NZ) nanocomposites were prepared using the co-precipitation method with various NZ amounts. The nanocomposites were characterized by X-ray diffraction (XRD), Brunauer-Emmett-Teller (BET) surface area analysis and thermogravimetric analysis (TGA). The nanocomposites were used to remove methylene blue (MB) dye from an aqueous solution. The effect of various NZ amounts and initial MB concentration were tested. The Pseudo-first-order and pseudo-second-order kinetic models were used to explain the adsorption mechanism. The Langmuir and the Freundlich isotherm models were used to fit the adsorption isotherms of the nanocomposites.

1. Introduction

Population growth worldwide has led to an increasing demand for clean water as well as increasing production of wastewater. Wastewater may contain a significant amount of pollutants and microbes that are hazardous for the environment and human health [1-3]. Industrial wastewater is one of the main contributors to water pollution. In particular, dyes used in the textile, paper, and cosmetics industries are a common organic pollutant in wastewater, and they are toxic to aquatic life [1,4,5]. Dye-containing wastewater should be treated before being discharged into the environment. Adsorbtion is considered an effective treatment for wastewater because of its low cost and simplicity [4].

Natural zeolites (NZ) are good adsorbents and are widely used for environmental remediation owing to their low cost, abundant availability in many parts of the world, good mechanical and thermal properties, presence of individual micropores, and nontoxicity [3,6]. Unfortunately, they show limited capacity for dye adsorption [7]. NZ has been used as a supporter in pure and transition-metal-doped semiconductor nanocatalysts to remove dyes from water [8-11].

Zinc oxide (ZnO) nanoparticles are known for their adsorption capability toward cationic and anionic dyes [8]. In our previous work, we had studied the photocatalytic activity of transition metal iron-doped ZnO (Fe:ZnO) [9] and Fe:ZnO/NZ 10 wt% with various atomic percentage (at%) of metal iron (Fe^{3+}) in degrading the cationic dye methylene blue (MB) [12]. In this work, we are going to study the adsorption ability of Fe:ZnO/NZ with constant at% of Fe^{3+} and various amounts of NZ toward MB, in hope that degradation of the dye will increase with increasing amount of NZ.

2. Experimental Details

2.1. Materials
ZnSO4.7H2O, FeSO4.7H2O, and NaOH were purchased from Merck. NZ were purchased from Sukabumi, Indonesia. All materials were of analytical grade and were used without purification. Distilled water was used to prepare the solutions in the experiment.

2.2. Fe:ZnO/Natural Zeolite Nanocomposite Preparation
The Fe:ZnO/natural zeolite nanocomposites were synthesized by the co-precipitation method. The required molar ratio of $ZnSO_4.7H_2O$ and $FeSO_4.7H_2O$ was mixed in distilled water to form a solution. NZ was then added to the solution with a constant amount (10 wt%, 20 wt%, 30 wt%, and 40 wt%). The mixed solutions were placed in an ultrasonic cleaner operating at 57 kHz for 2 h. After sonication, the mixed solutions were magnetically stirred for 1 h while NaOH solution was added until a pH of 13 was reached. The solutions were then allowed to stand at room temperature for 18 h. Then, the solutions were centrifuged and washed several times with distilled water and ethanol to remove impurities. The final products were dried in a vacuum oven for 4 h at 100 °C.

2.3. Characterization
The structural properties of the samples were characterized using X-ray diffraction (XRD; Rigaku Miniflex 600) operated at 30 kV and 15 mA and monitored in the range 2θ = 10°–90°. The specific surface area of the samples was investigated by Brunauer-Emmett-Teller (BET; Nova Quantachrome 2000) surface area analysis. The samples' thermal stability was investigated using differential thermal analysis/thermogravimetric analysis (DTA/TGA; Rigaku Thermo Plus EVO2 TG-8121) with alumina as a reference and operated from room temperature to 500 °C.

2.4. Adsorption Test
The organic pollutant model used was MB. The degradation of 100 mL aqueous solution of MB was conducted using Fe:ZnO/NZ nanocomposites under room temperature for 4 h in the dark. Around 10 mL of the solution was removed at given time intervals to observe the degradation of MB using a UV-Vis Hitachi UH5200 spectrophotometer. The degradation percentage and adsorption capacity of the samples were calculated using the following equations:

$$E(\%) = \frac{(C_0 - C_t)}{C_0}\ x\ 100 \quad (1)$$

$$q_e = \frac{(C_0 - C_e)V}{m} \quad (2)$$

where C_0 denotes the initial concentration of MB and C_t is the concentration of MB at time t. q_e represents the equilibrium adsorption capacity of the Fe:ZnO/NZ (mg/g); C_0 and C_e are the initial and equilibrium concentrations of methylene blue (mg/L) in aqueous solutions, respectively; V is the volume of the aqueous MB (mL); and m is the amount of Fe:ZnO/NZ (g).

3. Results and Discussion
XRD measurement is a useful method for studying the structural properties of the samples. Figure 1 shows the XRD patterns of the samples with various NZ amounts. As seen in the figure, the samples show the hexagonal wurtzite structure of ZnO [13]. Increasing the NZ content in the samples does not increase the NZ intensity significantly at around 2θ=20°. No undesirable peaks are observed as a secondary phase. This indicates that the nanocomposites were successfully formed. The specific surface area of the samples was determined using BET surface area analysis. The results are shown in Figure 2. They indicate that with increasing NZ content, the specific surface area of the samples also increases.

TGA analysis was used to check the samples' thermal stability and weight loss with temperature. Fig. 3 shows TGA curves of the Fe:ZnO nanoparticle and Fe:ZnO/NZ nanocomposite, respectively. The weight loss of Fe:ZnO after increasing the temperature to 500 °C was ~3.5%, whereas that of Fe:ZnO/40 wt%NZ was ~7.3%. The weight loss of Fe:ZnO was due to boundwater elimination [14,15]. The weight

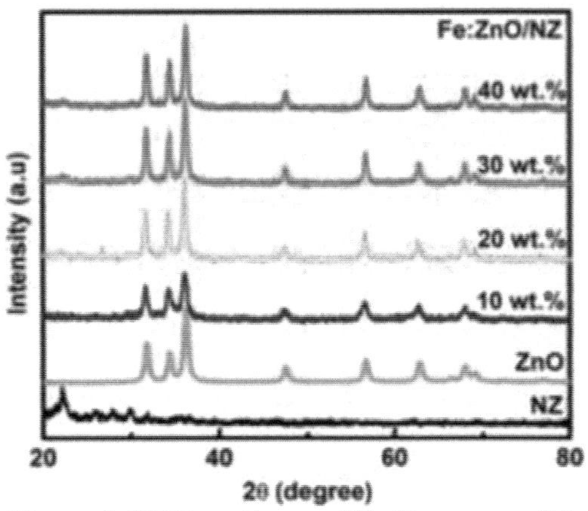

Figure 1. XRD patterns of ZnO nanoparticle and Fe:ZnO/NZ nanocomposites with various amount of NZ.

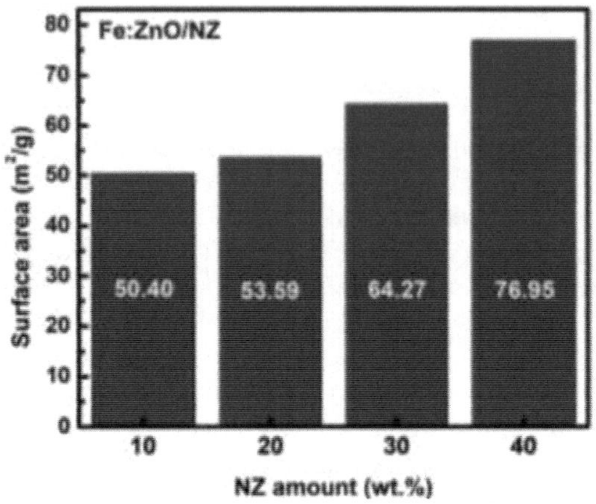

Figure 2. BET Surface area of the samples.

loss of Fe:ZnO/40 wt%NZ was due to the loss of the adsorbed water and iron dopant and water loss associated with silanol nests [15,16].

Figure 4 shows the degradation of MB using the Fe:ZnO catalyst with various NZ amounts (wt%) and initial dye concentration of 20 mg/L. The adsorption of MB in the catalysts reaches equilibrium at 120 min. The degradation of MB also increases as the incorporation of NZ increases. This is attributable to the increasing specific surface area of the samples, indicating that the higher the NZ loading in the samples, the larger is the number of sorption sites on the adsorbent's surface [17]. NZ also has anionic layers, and these interact well with the positively charged MB [1,18].

To understand the adsorption behavior of the samples, two kinetic adsorption models were used. The first is a pseudo-first-order kinetic model and the second, a pseudo-second-order one. The equations for these models are respectively given as follows [19]:

$$\log(q_e - q_t) = \log q_e - \frac{k_1}{2.303} t \quad (3)$$

$$\frac{t}{q_t} = \frac{1}{k_2 q_e^2} + \frac{t}{q_e} \quad (4)$$

where q_e and q_t are the sorption capacity at equilibrium and at time t (mg/g), respectively; k_1 and k_2 are the rate constant of the pseudo-first-order kinetic model (1/min) and pseudo-second-order kinetic model (g/(mg min)), respectively; and t is time (min). The fitted data of the pseudo-first-order and pseudo-second-order kinetic reactions are shown in Figure 5 and 6, respectively. The corresponding kinetic parameters from the pseudo-first-order and pseudo-second-order kinetic models are listed in Table 1. As seen from this table, the equilibrium adsorption capacity of the pseudo-second-order-kinetic model was similar to that obtained experimentally, and the R^2 value of the pseudo-second-order kinetic model was higher than that of the pseudo-first-order one. These results indicate that the pseudo-second-order kinetic model had a better fit than the pseudo-first-order one. This means that the adsorption of MB using the Fe:ZnO/NZ followed the pseudo-second-order kinetic model.These results indicate that the adsorption mechanism depends on the amount of solute adsorbed on the surface of the samples and the amount adsorbed at equilibrium [5].

Figure 7 shows the effect of the initial dye concentration in the adsorption process. The figure shows that dye degradation decreases as the initial dye concentration increases. Figure 8 shows that with increasing initial dye concentration, the adsorption capacity of the catalyst improves. This may be due to increasing driving force of the concentration gradient as the initial dye concentration increases [20]. Both Figure 7 and 8 show that the adsorption reached equilibrium at 120 min.

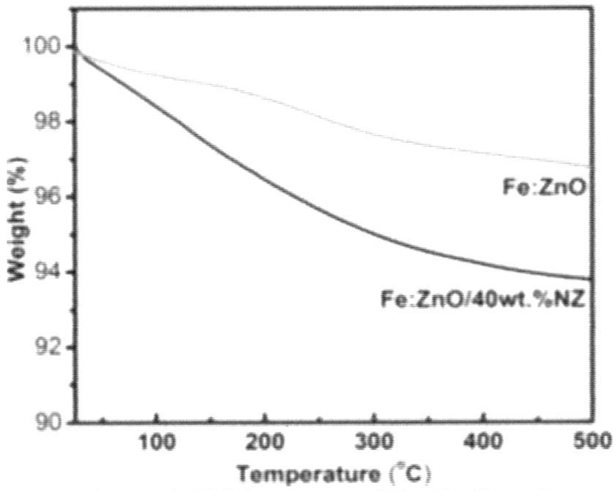

Figure 3. TGA curves of Fe:ZnO and Fe:ZnO/NZ.

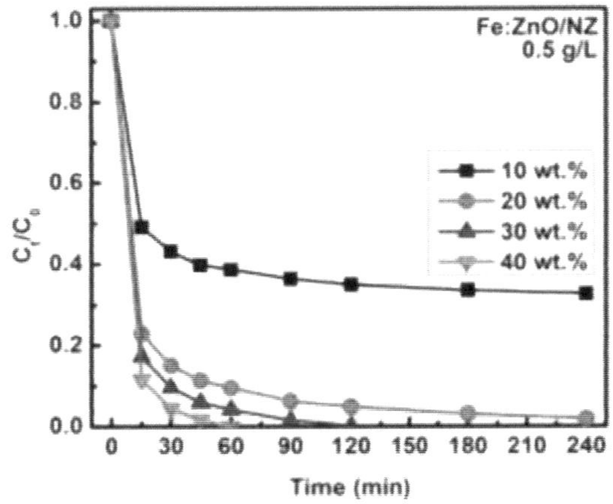

Figure 4. The effect of NZ amount in the MB adsorption process.

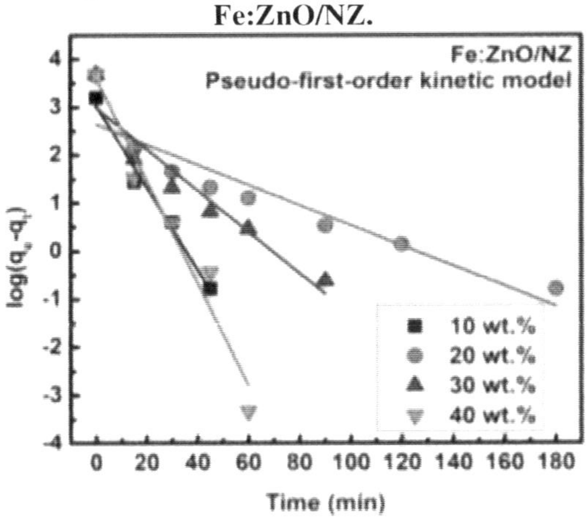

Figure 5. MB adsorption in pseudo-first-order kinetic model.

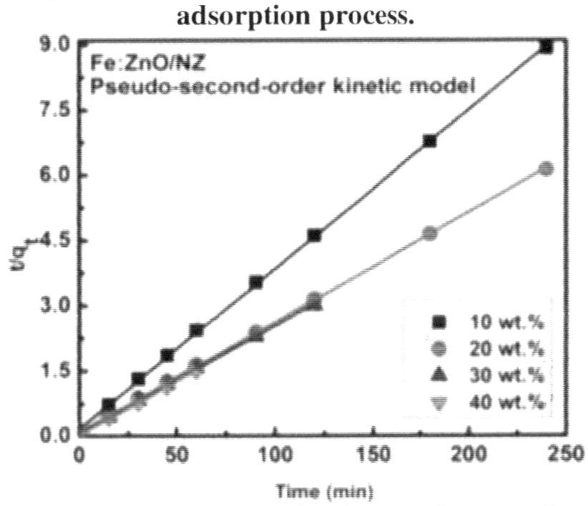

Figure 6. MB adsorption in pseudo-second-order kinetic model.

Table.1 Fitting result from pseudo first and second order kinetic reactions.

		The pseudo-First-order rate			The pseudo-second-order rate		
Sample	**qe (mg/g) exp**	**qe (mg/g) cal**	**K_1**	**R^2**	**qe (mg/g) cal**	**K_2**	**R^2**
10 wt%	24.54	20.76	0.1964	0.97	27.29	0.00748	0.99
20 wt%	39.23	14.00	0.04841	0.85	39.66	0.00611	0.99
30 wt%	39.92	19.91	0.09931	0.89	40.40	0.01009	0.99
40 wt%	40	36.82	0.24548	0.94	40.30	0.02338	0.99

To understand the adsorption mechanism, the experimental data was fitted using the Langmuir and the Freundlich isotherm models [4], as shown in Figure 9 and 10, respectively. The Langmuir model represents chemisorption on a set of distinct localized adsorption sites [4], whereas the Freundlich model describes the relation between the solute concentration on the adsorbent surface and the solute concentration in the liquid. The Langmuir and Freundlich equations are as follows [21]:

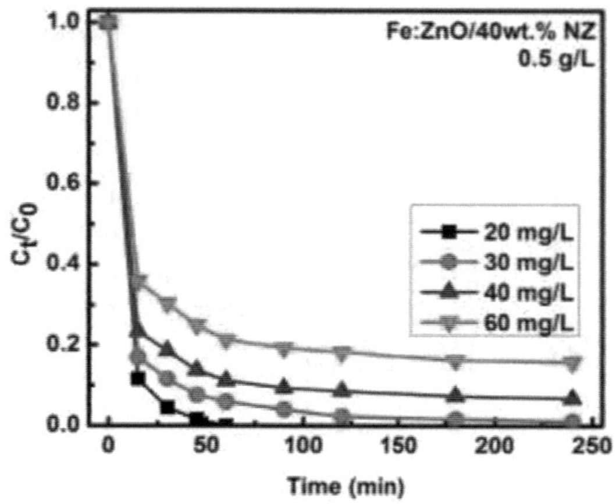

Figure 7. The effect of initial dye concentration in the MB adsorption process.

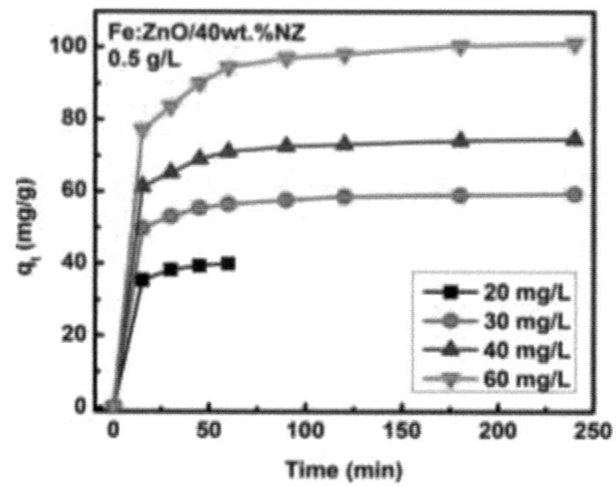

Figure 8. Amount of MB adsorbed on Fe:ZnO/NZ nanocomposites with different initial dye. concentration.

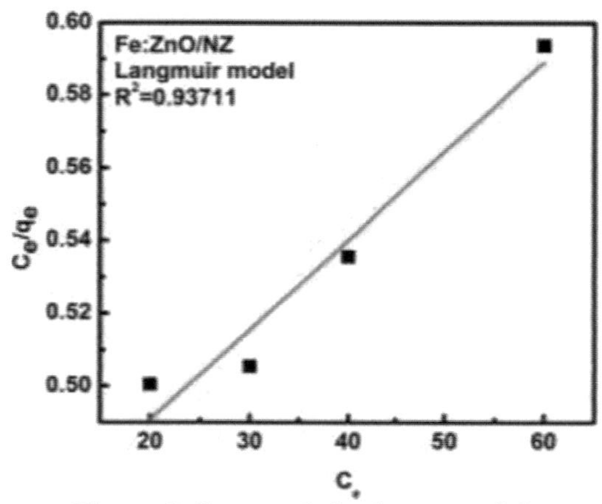

Figure 9. Langmuir isotherm model.

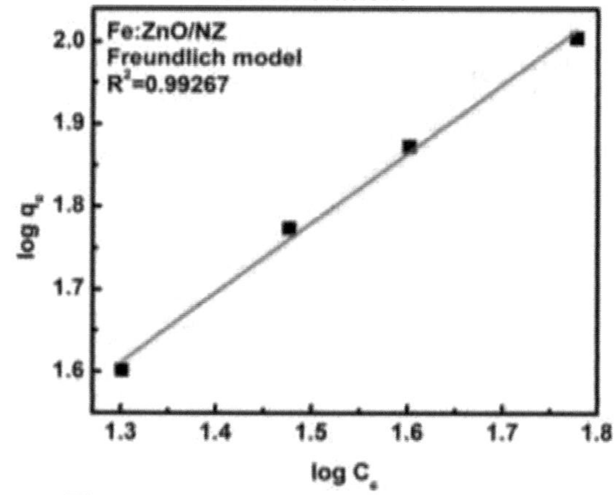

Figure 10. Freundlich isotherm model.

Table 2. Adsorption isotherm parameter.

Sample	Langmuir adsorption-isotherm model			Freundlich adsorption-isotherm model		
	q_m (mg/g)	K_L	R^2	n	K_f	R^2
Fe:ZnO:NZ	406.5	0.006	0.94	1.19	1.68	0.99

$$\frac{C_e}{q_e} = \frac{1}{q_m K_L} + \frac{C_e}{q_m} \tag{5}$$

$$log q_e = log K_f + \frac{1}{n} log C_e \tag{6}$$

The isotherm parameters obtained from the Langmuir and Freundlich adsorption isotherm models are listed in Table 2. As seen from this table, the maximum adsorption capacity of Fe:ZnO/NZ is 32 mg/g, and R^2 of the Freundlich isotherm model is higher than that of the Langmuir isotherm model. These results indicate that the equilibrium data would be better described by the Freundlich model.

4. Conclusion

Fe:ZnO/NZ nanocomposites were successfully synthesized by the co-precipitation method with no undesirable peaks are observed as a secondary phase. With increasing amount of NZ, the specific surface area of the samples also increased. TGA measurement showed that Fe:ZnO/NZ lost more weight (~3.8%) than Fe:ZnO. MB adsorption into the nanocomposites increased as the amount of NZ increased. The kinetic data of the samples fitted the pseudo-second-order kinetic model where the equilibrium adsorption capacity was similar to that obtained experimentally, and the R^2 value of the pseudo-second-order kinetic model was higher than that of the pseudo-first-order one. The adsorption equilibrium data fitted the Freundlich isotherm model better than Langmuir isotherm model because the R^2 value of the Freundlich isotherm model is higher than that of the Langmuir isotherm model. In the future, we intend to couple other materials with Fe:ZnO to degrade organic dyes in aqueous solution.

References

[1] Hor K Y, Chee J M C, Chong M N, Jin B, Saint C P, Poh P E and Aryal R 2016 *J Clean Prod* **118** 197-209.
[2] Chong M N, Jin B, and Saint C P 2011 *Chem Eng J* **171** 16-23.
[3] Bahrami M and Nezamzadeh-Ejhieh A 2015 *Mater Sci Semicond Process* **30** 275-284.
[4] Wang S B and Ariyanto E 2007 *J Colloid Interf Sci* **314** 25-31.
[5] Aysan H, Edebali S, Ozdemir C, Karakaya M C and Karakaya N 2016 *Microporous Mesoporous Mater* **235** 78-86.
[6] Misaelides P 2011 *Microporous Mesoporous Mater* **144** 15-18.
[7] Liu S G, Ding Y Q, Li P F, Diao K S, Tan X C, Lei F H, Zhan Y H, Li Q M, Huang B and Huang Z Y 2014 *Chem Eng J* **248** 135-144.
[8] Zhang F, Chen X, Wu F H and Ji Y F 2016 *Colloids Surf, A* **509** 474-483.
[9] Afifah N, Adriani S, Djaja N F and Saleh R 2015 *Adv Mater Res* **1123** 295-302.
[10] Guesh K, Mayoralc Á, Márquez-Álvarezd C, Chebudea Y and Díaz I 2016 *Microporous Mesoporous Mater* **225** 88-97.
[11] Azmi R, Djaja N F and Saleh R 2014 *Adv Mater Res* **911** 120-125.
[12] Saleh R and Djaja N F 2014 *Superlattices Microstruct* **74** 217-233.
[13] Wang J W, Wan J Q and Chen K Z 2010 *Mater Lett* **64** 2373-2375.
[14] Gui Y, Wang H, He C, Qiu L and Cao X 2009 *Langmuir* **25** 4678.
[15] Ates A 2014 *Powder Technol* **264** 86-95.
[16] Ba-Abbad M M, Kadhum A A H, Mohamad A B, Takriff M S and Sopian K 2013 *Chemosphere* **91** 1604-1611.
[17] Ghaedi M, Hassanzadeh A and Kokhdan S N 2011 *J Chem Eng Data* **56** 2511-2520.
[18] Gürsesa A, Doğarc Ç, Yalçına M, Açıkyıldıza M, Bayraka R and Karacab S 2006 *J Hazard Mater* **131** 217-228.
[19] Panic V V and Velickovic S J 2014 *Sep Purif Technol* **122** 384-394.
[20] Chiou MS and Li HY 2002 *J Hazard Mater* **B93** 233-248.
[21] Sohrabnezhad Sh and Pourahmad A 2010 *Desalination* **256** 84-89.

Multifunctional Photocatalytic Degradation of Methylene Blue Using $LaMnO_3/Fe_3O_4$ Nanocomposite on Different Types of Graphene

Yulia Dwi Susanti[1,2], Nur Afifah[1,2] and Rosari Saleh[1,2,a]
[1]Departemen Fisika, Fakultas MIPA, Universitas Indonesia, Kampus UI Depok 16424, Indonesia
[2]Integrated Laboratory of Energy and Environment, Fakultas MIPA, Universitas Indonesia, kampus UI Depok, 16424, Indonesia
[a]Corresponding's author: rosari.saleh@gmail.com, rosari.saleh@ui.ac.id

Abstract. This study added different types of graphene (GN-1 and GN-2) in to $LaMnO_3/Fe_3O_4$ nanocomposite to degrade MB as an organic pollutant. $LaMnO_3/Fe_3O_4$/GN-1 and $LaMnO_3/Fe_3O_4$/GN-2 composites were synthesized using the co-precipitation method in a constant 5 wt% of GN. The as-prepared samples were characterized using XRD and TGA. The $LaMnO_3/Fe_3O_4$/GN-1 and $LaMnO_3/Fe_3O_4$/GN-2 composites showed the orthorhombic structure of $LaMnO_3$ nanoparticles and the cubic spinel of Fe_3O_4 nanoparticles, as well as of GN-1 and GN-2. The graphite structure of NGP in $LaMnO_3/Fe_3O_4$/GN-1 composites was confirmed. However, the peak of graphene in $LaMnO_3/Fe_3O_4$/GN-2 could not be identified. The photocatalytic efficiency of the $LaMnO_3/Fe_3O_4$/GN-2 was higher than that of $LaMnO_3/Fe_3O_4$/GN-1. The enhancement of the photocatalytic light activity can be attributed to the high separation efficiency of electron-hole recombination and to the large surface contact between $LaMnO_3/Fe_3O_4$ and graphene, which can improve the transfer efficiency of the photocatalytic process. In addition, the effects of catalyst dosage, rate constant, and scavengers were investigated and discussed.

1. Introduction

Organic dyes are produced from textile and other industrial processes that are known to be one of the major of pollutants in wastewater. Removal of organic dyes from wastewater has become priority for researchers recent decades. There are many effective removal methods in wastewater. One method that refers to the removal of organic pollutants in wastewater is photocatalytic activity.

Semiconducting photocatalyst TiO_2 has been widely used because of its good performance as a nanoparticle powder [1-2]. However, TiO_2 can only absorb ultraviolet (UV) light due to its wide band gap (3.2 eV) and high photogenerated electron-hole pairs [3-5]. To overcome this problem, perovskite oxide has been used as a good photocatalyst in the visible light region because of its environment friendliness, high electron mobility, non-toxicity, and high chemical stability [6-7]. The perovskite oxide that has been used to degrade organic pollutants contained in wastewater is $LaMnO_3$ nanoparticles. However, $LaMnO_3$ nanoparticles have showed some limitations in their practical application such as separating materials from liquid. Adding content of Fe_3O_4 to the $LaMnO_3$ overcame the separation problem. The presence of Fe_3O_4 in $LaMnO_3$ shows an easily-composited photo-generated electron-hole pairs and a small surface area that may reduce catalytic efficiency. Therefore, to overcome several problems associated with this material, many efforts have been introduced to improve their photocatalytic efficiencies, such as growing $LaMnO_3/Fe_3O_4$ on the layers of carbon-based materials.

Graphene is a type of ultra-thin two dimensional two layered of carbon-based material that has attracted considerable attention because of its excellent electronic conductivity, high specific surface area, high stability and large surface to volume ratio [8]. Thus compounding different types of graphene (GN-1 and GN-2) in a semiconductor photocatalyst has the potential to enhance photocatalytic performances. The present work investigates the addition of different types of graphene (GN-1 and GN-2) in $LaMnO_3/Fe_3O_4$ nanocomposites affects methylene blue (MB) degradation as a model for an organic pollutant under a visible light source.

2. Experimental details

2.1. Chemicals

All the chemicals used in this work are of analytical grade. The chemicals used in the synthesis process of the $LaMnO_3/Fe_3O_4$ with the addition of different types of graphene are lanthanum (III) chloride heptahydrate ($LaCl_3.7H_2O$, Merck, 99%), manganese (II) chloride tetrahydrate ($MnCl_2.4H_2O$, Merck, 99%), sodium hydroxide (NaOH, Merck, 99%)), iron (II) sulphate heptahydrate ($FeSO_4.7H_2O$, Merck, 99%), acetic acid (20%), ethanol and ethylene glycol. They are commercially available from Merck and were used in this study without further purification. Furthermore, different types of graphene (NGP and graphene) were available from Angstrom materials.

2.2. Catalyst preparation

$LaMnO_3$ and Fe_3O_4 nanoparticles were synthesized using the co-precipitation [9] and the sol gel [10] method, respectively as described in our previous study. Briefly, $LaMnO_3$ nanoparticles were dissolved into 100 ml distilled water and Fe_3O_4 nanoparticles were dissolved into 50 ml ethanol. Both solutions were mixed until blended. Then, the mixed solutions were ultrasonicated for 2 hour and allowed to stand at a room temperature for a few minute to get thermal equilibrium. After that, the mixed solutions were centrifuged to separate materials from residual reaction. The obtained materials are allowed to stand at a room temperature to have an aging process overnight. The $LaMnO_3/Fe_3O_4$ nanocomposites were obtained after being dried at a temperature 100 °C under vacuum conditions for an hour. The molar ratio of $LaMnO_3/Fe_3O_4$ that will be used to synthesize the sample on different types of graphene is 1:0.5.

The NGP and graphene amount was added to $LaMnO_3/Fe_3O_4$ in a constant 5 wt%. The $LaMnO_3/Fe_3O_4$ with the addition of NGP and graphene were also synthesized using the co-precipitation method. Briefly, 100 mg of NGP were dissolved in 80 ml water and 40 ml ethanol using an ultrasonic treatment for two hours. Next, 2 g of $LaMnO_3/Fe_3O_4$ were added to the NGP solution. Then, the mixed solution was stirred for two hours to achieve a homogeneous solution. The suspension was heated at 120°C for three hours to affect the deposition of the $LaMnO_3/Fe_3O_4$ on the graphene sheets. Then, the solution was centrifuged to separate the materials from the residual reaction. It also performed the same way to synthesize $LaMnO_3/Fe_3O_4$ with the addition of graphene. The $LaMnO_3/Fe_3O_4$/GN-1 and $LaMnO_3/Fe_3O_4$/GN-2 composites were obtained after being dried under vacuum conditions over night. The $LaMnO_3/Fe_3O_4$ with the addition NGP was labeled $LaMnO_3/Fe_3O_4$/GN-1 and the $LaMnO_3/Fe_3O_4$ with the addition of graphene was labeled $LaMnO_3/Fe_3O_4$/GN-2.

2.3. Characterization

The synthesized catalysts were characterized employing standar analytical techniques such as XRD. A crystal structure analysis was carried out on a Rigaku Miniflex 600 with Cu-Kα (λ = 1.54060) radiation operated in 40 kV and 20 mA. Thermogravimetric analysis (TGA) was conducted to estimate the amount of GN in the samples.

IOP Conf. Series: Journal of Physics: Conf. Series **820** (2017) 012021 doi:10.1088/1742-6596/820/1/012021

2.4. Catalytic activity

The catalytic activity was examined by observing the degradation rate of MB in the presence of $LaMnO_3/Fe_3O_4$/GN-1 and $LaMnO_3/Fe_3O_4$/GN-2. The degradation of MB was tested in an alkaline condition (pH 13). A 40 W Xe lamp light was used as a visible light source. The suspensions were placed in a dark condition for 30 minutes to achieve absorption and desorption equilibrium. Then, the suspensions irradiated under visible light for two hours. The suspension was analyzed at regular 15 minute using a Hitachi UH5300 UV-visible spectrophotometer. The MB degradation rate was also analyzed by adding some scavengers, sodium sulfate as an electron scavenger, di-ammonium oxalate as a holes scavenger and tert-butyl alcohol as a hydroxyl radical scavenger. The catalyst's stability was also checked four times.

3. Results and Discussion

The XRD patterns of $LaMnO_3/Fe_3O_4$ with different types of graphene (GN-1 and GN-2) have been analyzed. Figure 1.a shows the XRD pattern of graphene in different types (GN-1 and GN-2), $LaMnO_3/Fe_3O_4$, $LaMnO_3/Fe_3O_4$/GN-1 and $LaMnO_3/Fe_3O_4$/GN-2. The observed diffraction peaks at (2θ = 22.12°, 22.96°, 25.06°, 31.08°, 32.21°, 33.2°, 38.94°, 40.05°, 45.4°, 47.07°, 56.19°, 57.6°, 58.56°, 67.68°, and 75.68°) shows the orthorhombic structure of $LaMnO_3$ nanoparticles[11-12] and (2θ= 30.36°, 35.76°, 43.47°, 53.94°, 57.51°, and 63.74°) shows the cubic spinel of Fe_3O_4 nanoparticles

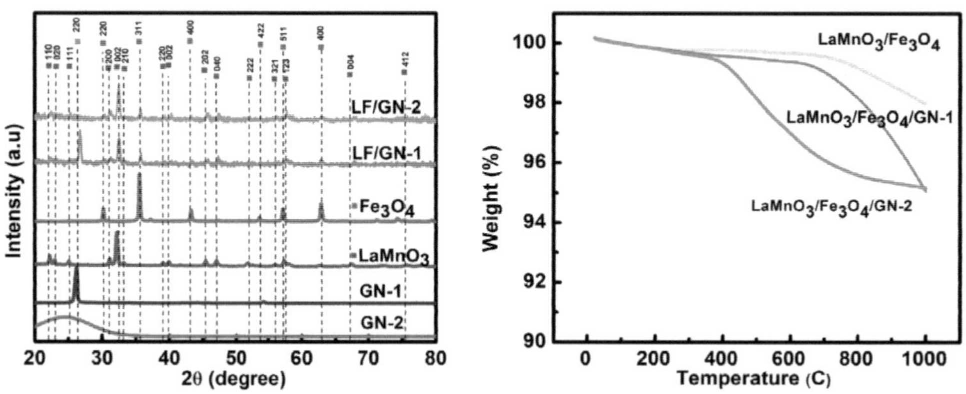

(a) (b)

Figure 1. (a) XRD pattern; (b) TGA curves of $LaMnO_3/Fe_3O_4$, $LaMnO_3/Fe_3O_4$/GN-1 and $LaMnO_3/Fe_3O_4$/GN-2.

TABLE 1. The lattice parameter and grain size <D> of $LaMnO_3/Fe_3O_4$ and $LaMnO_3/Fe_3O_4$/GN composites.

Sample	Lattice Parameter			Lattice Parameter	<D> $LaMnO_3$	<D> Fe_3O_4
	a (Å)	b (Å)	c (Å)	a=b=c (Å)	(nm)	(nm)
$LaMnO_3/Fe_3O_4$	5.731	7.705	5.538	8.367	17	25
$LaMnO_3/Fe_3O_4$/GN-1	5.518	7.669	5.713	8.340	32	17
$LaMnO_3/Fe_3O_4$/GN-2	5.512	7.658	5.722	8.339	49	15

[13]. After NGP was introduced to the $LaMnO_3/Fe_3O_4$ for photocatalysis, the XRD pattern showed a peak at $2\theta = 26.4°$ and its diffraction peaks reveal the presence of graphite structure. The graphite structure of NGP in $LaMnO_3/Fe_3O_4$/GN-1 composites was confirmed. Furthermore, the peaks of graphene in $LaMnO_3/Fe_3O_4$/GN-2 cannot be identified by XRD measurements because of its broad peak. There were no peaks of graphene observed in the $LaMnO_3/Fe_3O_4$/GN-2 composite because of the small grapahene content (5wt%). The crystallite size of the samples which was calculated by the scherrer equation [14-15] and the unit cell parameter was tabulated in Table 1.

To estimate the amount graphene (GN-1 and GN-2) in the $LaMnO_3/Fe_3O_4$/GN-1 and $LaMnO_3/Fe_3O_4$/GN-2 composites, thermogravimetric analysis (TGA) was carried out in an air flux. Figure 1.b shows the TGA curve of $LaMnO_3/Fe_3O_4$ nanocomposite and $LaMnO_3/Fe_3O_4$ with the addition of different types of graphene (GN-1 and GN-2). The results indicate that $LaMnO_3/Fe_3O_4$ has

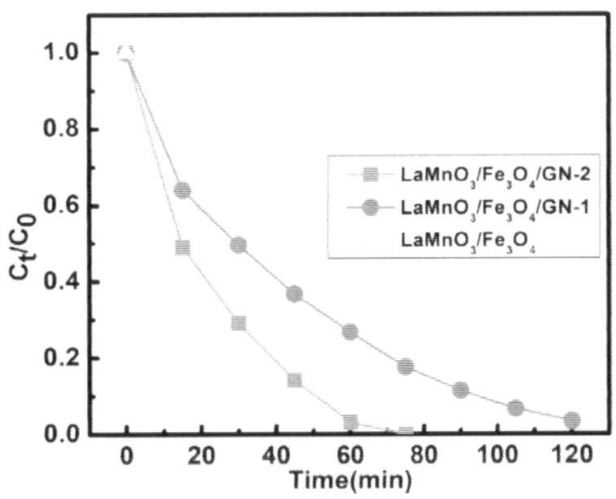

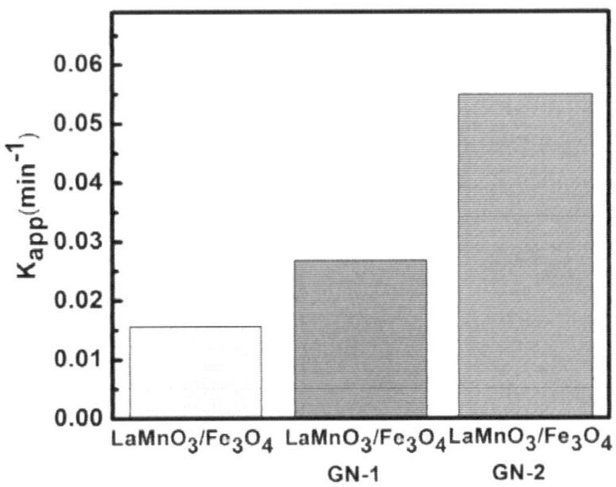

(a) (b)

Figure 2. (a) The photocatalytic activity of $LaMnO_3/Fe_3O_4$ nanocomposites and $LaMnO_3/Fe_3O_4$ with different types of graphene; and (b) kinetics of photocatalytic degradation by $LaMnO_3/Fe_3O_4$ and $LaMnO_3/Fe_3O_4$ with different types of graphene (GN-1 and GN-2).

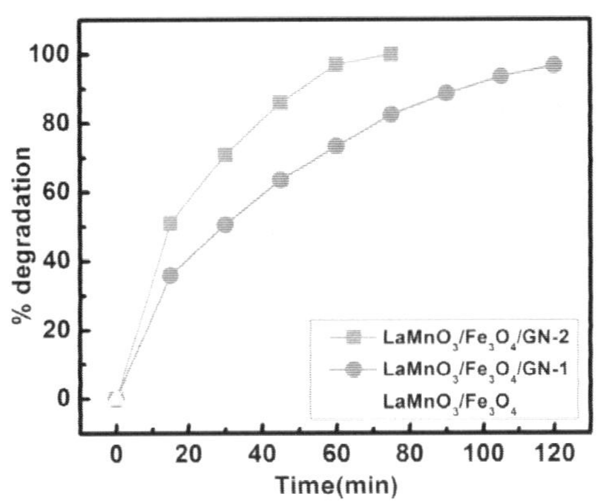

Figure 3. The % degradation in photocatalytic activity of $LaMnO_3/Fe_3O_4$ nanocomposites and $LaMnO_3/Fe_3O_4$ with different types of graphene.

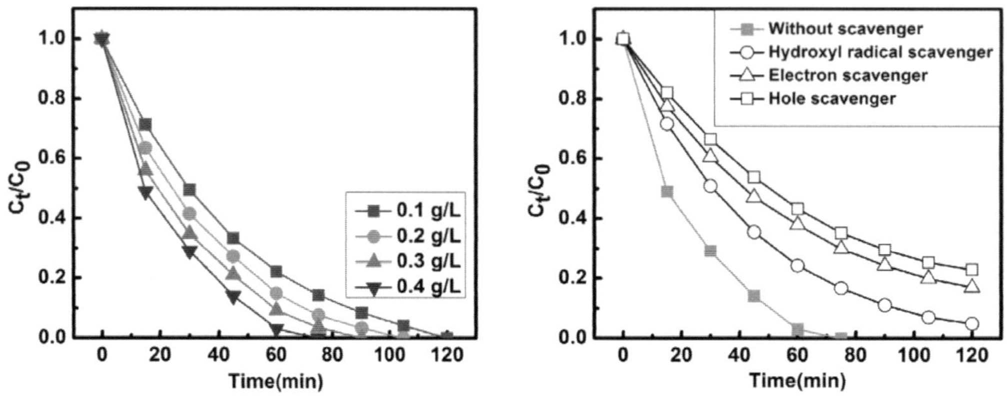

(a) **(b)**

Figure 4. (a) The effect of catalyst dosage on the degradation of MB using $LaMnO_3/Fe_3O_4$/GN-2 composite; and (b) The effect of scavengers on photocatalytic activity.

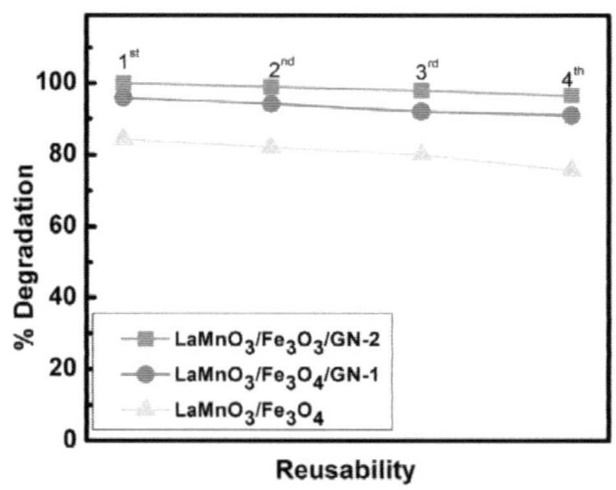

Figure 5. The stability of catalyst using $LaMnO_3/Fe_3O_4$, $LaMnO_3/Fe_3O_4$/GN-1 and $LaMnO_3/Fe_3O_4$/GN-2.

the lowest weight loss percentage and that it is the more stable sample. However, $LaMnO_3/Fe_3O_4$ with the addition of nanographene platelets and graphene has the weight loss for about 5%. The differences in weight loss between $LaMnO_3/Fe_3O_4$, $LaMnO_3/Fe_3O_4$/GN-1 and $LaMnO_3/Fe_3O_4$/GN-2 with the addition of different types of graphene (GN-1 and GN-2) after the oxidation could be directly translated to the amount of NGP and graphene in the $LaMnO_3/Fe_3O_4$/GN-1 and $LaMnO_3/Fe_3O_4$/GN-2 [16].

Figure 2.a shows the photocatalytic activity of $LaMnO_3/Fe_3O_4$ and $LaMnO_3/Fe_3O_4$ with the addition of different types of graphene (GN-1 and GN-2). As can be seen from the figure, $LaMnO_3/Fe_3O_4$ with the addition of NGP and graphene has better photocatalytic activity than that $LaMnO_3/Fe_3O_4$ nanocomposite. $LaMnO_3/Fe_3O_4$/GN-2 also has the highest photocatalytic performances among other samples. Compared with $LaMnO_3/Fe_3O_4$ and $LaMnO_3/Fe_3O_4$/GN-1, the addition of graphene expands the $LaMnO_3$ light response range, high separation efficiency of electron-hole recombination and it also shows that graphene can act as an electron acceptor and transporter in the $LaMnO_3/Fe_3O_4$/GN-2 composite [17]. In the $LaMnO_3/Fe_3O_4$/GN-2, the excited electrons can transfer from the conduction band to graphene. Thus, graphene can act as an acceptor of the generated electrons. Both the graphene's electrons accepting and transporting can achieve a higher photocatalytic performances for $LaMnO_3/Fe_3O_4$/GN-2. Furthermore, the large surface contact between graphene and $LaMnO_3/Fe_3O_4$ was also able to increase its photocatalytic performance.The first-order kinetics of photocatalytic degradation for $LaMnO_3/Fe_3O_4$, $LaMnO_3/Fe_3O_4$/GN-1 and $LaMnO_3/Fe_3O_4$/GN-2 are shown in figure 2.b. The figure plotted the function of ln (C_t/C_0) vs. irradiation time (t) and the photodegradation of the MB can be formulated using the first-order kinetic equation. Ln (C_t/C_0) = -k.t is used to fit experimental data, where C_0 is the concentration of MB at t=0 and C_t is the concentration of MB at (t), respectively, and k is the first order rate constant. The figure also showed that $LaMnO_3/Fe_3O_4$ with the addition of graphene has the maximum values of the apparent rate. The absorption peak of MB was almost eliminated using $LaMnO_3/Fe_3O_4$/GN-2 after irradiating for about 75 minutes. As shown in figure 3, graphene to $LaMnO_3/Fe_3O_4$ made the decoloration rate of the dye 99.96 %.

The effect of catalyst dosage in $LaMnO_3/Fe_3O_4$/GN-2 is shown in Figure 4.a. increasing the amount of catalyst dosage to approximately 0.1g/L-0.4g/L can increase, the photocatalytic activity in the MB degradation as a model for organic pollutants. This is is due to the increase of active species on the photocatalytic activity, which increases a group of (•OH) to degrade MB as a model for organic pollutants. The main active species in the photocatalytic activity of $LaMnO_3/Fe_3O_4$/GN-2 could be determined by adding some scavengers as shown in Figure 4.b. The addition of tert-butyl alcohol as a hydroxyl radical scavenger, di-amonium oxalate as a holes scavenger and sodium sulfate (Na_2SO_4) as an electron scavengers was able to inhibit the degradation rate. This demonstrates that the main active species in the photocatalytic activity of $LaMnO_3/Fe_3O_4$ was holes> electron > hydroxyl radical (•OH). The samples' stability was also checked in four cycle processes by using the highest dosage concentration (0.4 g/L) as shown in Figure 5. The results show that, the samples have good stability and reusability even after the four cycle processes.

4. Conclusion
$LaMnO_3/Fe_3O_4$ with the addition of different types of graphene has been successfully synthesized using the co-precipitation method. The photocatalytic activity of $LaMnO_3/Fe_3O_4$ nanocomposites was evaluated under visible light irradiation. The results show that $LaMnO_3/Fe_3O_4$/GN-2 demonstrates superior photocatalytic activity to $LaMnO_3/Fe_3O_4$-GN-1 because the addition of graphene in $LaMnO_3/Fe_3O_4$/GN-2 expands the $LaMnO_3$ light response range, high separation efficiency of electron-hole recombination and it also shows that graphene can act as an electron acceptor and transporter in the $LaMnO_3/Fe_3O_4$/GN-2 composite. Furthermore, adding a dosage concentration up to 0.4 g/L in the photocatalytic activity increases MB degradation. The samples have good stability and reusability even after the four cycle processes after the additional of NGP and graphene. Furthermore, the main active species with the addition of some scavenger were holes >electrons >hydroxyl radical (•OH).

5. References
[1] Zhong J B, Li J Z, Lu Y, He X Y, Zheng J, Hu W, Shen Y C 2012 *Appl. Surf. Sci.* **258** 4929
[2] Gao K, Li S D 2012 *Appl. Surf. Sci.* **258** 6460
[3] Tang P S, Tong Y, Chen H F, Cao F, Pan G X 2013 *Curr. Appl. Phys.* **13** 340.
[4] McEvoy J G, Cui W Q, Zhang Z S 2014 *Appl. Catal., B.* **144** 702
[5] Fakhouri H, Pulpytel J, Smith W, Zolfaghari A, Mortaheb H R, Meshkini F, Jafari R, Sutter E, Khonsari F A 2014 *Appl. Catal., B.* **144** 12
[6] J. Zhang, Q. Ping, M. Niu, H. Shi and N, Li 2013 *Appl clay sci* . **83** 12–16
[7] G. S. Lotey and N. K. Verma 2014 *Mater sci semicond process.* **21** 206–21
[8] Elif Das, Selmiye alkan Gursel, Lale isikel Sanli, Ayse Bayrakceken Yurtcan 2016 *int. J. Hydrogen Energy.* **41** 9755-9761

[9] N. Afifah and R. Saleh 2016 *Mater Sci Forum.* **864** 99-105.
[10] Taufik, I. K. Susanto, and R. Saleh 2015 *Mater Sci Forum.* **827** 37-42.
[11] A. Galal, N. F. Atta and S. M. Ali, *Electrochim. Acta.* 2011 **56** 5722
[12] C. Zhang, Y. Guo, Y. Guo, G. Lu, A. Boreave , L. Retailleau, A. Baylet and A. Giroir-Fendler 2014 *Appl. Catal., B.* **48–149** 490–498
[13] L.M. Rossi, F.P. Silva, L.L.R. Vono, P.K. Kiyohara, E.L. Duarte, R. itri, R. Landers and G. Machado 2007 *Green Chem.* **9** 379-385
[14] R.Y. Hong, SZ. Zhang, G.Q. Di, H.Z. Li, Y. Zheng, J. Ding, D,G. Wei, 2008 *Mater Ress Bull.* **43** 2457-2468
[15] A. Hamrouni, H. Lachheb, A. Houas 2013 *Mater Sci Eng.* **B 178** 1371-1379,
[16] Y. shi et al 2013 *J. power sources.* **244** 684-689, 2013.
[17] J. Hu, J. Ma, L. wang, H. Huang 2014 *J. Alloys Compd.* **583** 539-545.

Photodegradation of Methylene Blue by $LaFeO_3$/ZnO Nanocomposites under Visible and UV Light Irradiation

Nur Afifah[1,2] and Rosari Saleh[1,2,a]
[1]Departemen Fisika, Fakultas MIPA, Universitas Indonesia, Kampus UI Depok, 16424 Depok, Indonesia
[2]Integrated Laboratory of Energy and Environment, Fakultas MIPA, Universitas Indonesia, Kampus UI Depok, 16424 Depok, Indonesia
[a]Corresponding's author: rosari.saleh@gmail.com, rosari.saleh@ui.ac.id

Abstract. In this study, $LaFeO_3$/ZnO nanocomposites with different molar ratios of ZnO (0.1, 0.3, 0.5, and 1) have been prepared using the sol-gel method. The structural and elemental analyses of the samples were measured by X-ray diffraction (XRD) and X-ray fluorescence (XRF) spectroscopy. XRD results showed that the synthesized samples exhibited an orthorhombic structure of $LaFeO_3$ and a hexagonal wurtzite structure of ZnO. However, the ZnO hexagonal wurtzite structure could not be detected below the ZnO molar ratio of 0.5. The presence of ZnO on the nanocomposites was confirmed using XRF measurements. The results revealed that the XRF spectra of the samples consisted of La, Fe, and Zn atoms. The photodegradation of methylene blue (MB) using $LaFeO_3$/ZnO nanocomposites was then monitored under visible and UV light irradiation. The photodegradation results showed that the presence of $LaFeO_3$/ZnO nanocomposites could degrade the MB solution. The highest MB degradation was found in the presence of ZnO at a molar ratio of 0.5. All of the $LaFeO_3$/ZnO nanocomposites showed better photocatalytic performance than did $LaFeO_3$ nanoparticles alone. Furthermore, several parameters, such as the effect of catalyst dose, initial concentration of MB, and various radical scavengers, were also measured to obtain the maximum condition and the main active species involved in the photocatalytic process.

1. Introduction

Photocatalysis is the process of destroying the complex chemical bonds of organic or inorganic pollutants to transform them into harmless chemicals using a photocatalyst and light as an irradiation source [1]. Among several photocatalysts, ZnO has shown great potential due to its high catalytic activity, high electron mobility, high photosensitivity, chemical stability, low price and non-toxicity [2-4]. However, ZnO has a low response in the visible light region, and high electron and hole-pair recombinations due to its wide band gap of about 3.2 eV and low quantum efficiency, which hinders its use in practical applications [5-6]. However, coupling ZnO with narrow band gap catalysts is a promising solution to this problem by extending the photoresponse of ZnO into the visible light region and maximizing the utilization of sunlight as a renewable energy source [7].

A catalyst that has been widely used in the visible light region is perovskite, $LaFeO_3$. $LaFeO_3$ has attracted considerable attention due to its structural stability, suitable gaps in the visible light region, and unique optoelectronic properties [8-9]. Perovskite materials as catalysts also show high efficiency, high thermal stability, nontoxicity, and unique physicochemical properties [10]. The photocatalytic activity of perovskite materials such as $LaFeO_3$ [11] and $LaMnO_3$ [12] nanoparticles has been reported in our previous work. However, it photocatalytic performance didn't showed the complete degradation of MB solution due to its high electron and hole recombination. In order to fully utilize sunlight, $LaFeO_3$ can be combined with ZnO to improve its photocatalytic performance by inhibiting the electron and holes recombination. Moreover, combining $LaFeO_3$ as visible light catalyst and ZnO as UV light catalyst in this study is expected to extend the catalyst response in both visible and UV light irradiation. The $LaFeO_3$/ZnO nanocomposites can also be applied as the promising catalyst in the sunlight as a renewable energy source. Furthermore, several parameters that play important roles in the photocatalytic process were studied and discussed.

2. Experimental details

Catalyst preparation: $LaCl_3{\cdot}7H_2O$, $FeCl_2{\cdot}4H_2O$, $ZnSO_4{\cdot}7H_2O$, and NaOH were purchased from Merck to be used as raw materials without further purification. The $LaFeO_3$/ZnO nanocomposites with different molar ratios of ZnO (0.1, 0.3, 0.5, and 1) were obtained by the sol-gel method. Briefly, solution A was prepared by dissolving 0.00125 mol $ZnSO_4{\cdot}7H_2O$ into 31 ml of distilled water under a magnetic stirrer. Then, 0.0025 NaOH was added to 62 ml of distilled water and dissolved into solution A until a pH value of 12 was reached. After that, the mixture was stirred and heated to a temperature of 80 °C. Meanwhile, solution B was obtained by dissolving the $LaFeO_3$ nanoparticles (prepared by following our previous work [11]) into 31 ml of distilled water. Solution B was then dissolved into solution A under the continuous agitation of the magnetic stirrer at 80 °C for 2 hours. Simple centrifugation was used to separate the precipitate and the solution. Finally, the precipitate was allowed to stand overnight at room temperature, then dried at 120 °C for one hour under vacuum conditions.

Characterization: The $LaFeO_3$/ZnO nanocomposite crystal structures were characterized by X-ray diffraction (XRD) measurements. A monochromatic Cu-K_α (λ = 1.54060 Å), which operated at 30 kV and 20 mA in the range of 10°–90°, was used as the radiation source. The elemental analyses of the samples and the concentration (C) of each element were analyzed using X-ray *fluorescence (XRF) spectroscopy*.

Photocatalytic experiment: The photodegradation of MB in the presence of $LaFeO_3$/ZnO nanocomposites as catalysts was performed under visible and UV light irradiation. First, an amount of catalyst was added to MB solution under a magnetic stirrer. Then, the solution was then kept in dark for 30 min to reach absorption and desorption equilibrium. After that, the solution was treated under light irradiation using a 40W Xe lamp or a 40W UV lamp as the visible and UV light sources, respectively. Finally, to obtain the percentage degraded MB, the 10 ml of MB solution was collected at different time intervals and then analyzed using a UV-visible spectrophotometer. In order to obtain the maximum condition in the photocatalytic experiment, the effects of catalyst dosage and initial concentrations of MB were investigated. Moreover, the effects of various radical scavengers were also determined to obtain the main active species involved in the photocatalytic process. Sodium sulfate, di-ammonium oxalate, and tert-butyl alcohol were used as electrons, holes, and hydroxyl radical scavengers, respectively.

3. Results and Discussion

The XRD pattern of the ZnO nanoparticles, $LaFeO_3$ nanoparticles, and $LaFeO_3$/ZnO nanocomposites with different molar ratios of ZnO are shown in Figure 1a. The XRD lines of the ZnO nanoparticles were observed at 31.73° (100), 34.35° (002), 36.22° (101), 47.52° (102), 56.51° (110), 62.82° (103), 66.42° (200), 67.89° (112), and 69.04° (201), which are well indexed by the hexagonal wurtzite structure of ZnO [13-14]. Meanwhile, the XRD lines of $LaFeO_3$ nanoparticles were observed at 22.6° (101), 32.2° (121), 39.8° (220), 46.2° (202), 57.5° (240), 67.4° (242), and 76.7° (204), which are well indexed by the orthorhombic structure of $LaFeO_3$ [15-16]. Moreover, all diffraction peaks of $LaFeO_3$/ZnO nanocomposites were in agreement with those of the standard patterns for orthorhombic $LaFeO_3$ and hexagonal wurtzite ZnO. However, at the low ZnO molar ratio, the hexagonal wurtzite structure of ZnO could not be detected, possibly due to the lower ZnO content in the nanocomposites. The grain size <D> of the prepared samples was calculated using the Debye-Scherrer equation [17]: $\langle D\rangle = 0.9\lambda/\beta\cos\theta$, where λ is the Cu-K_α wavelength, β is the full width of half-maximum, and θ is the Bragg angle. The

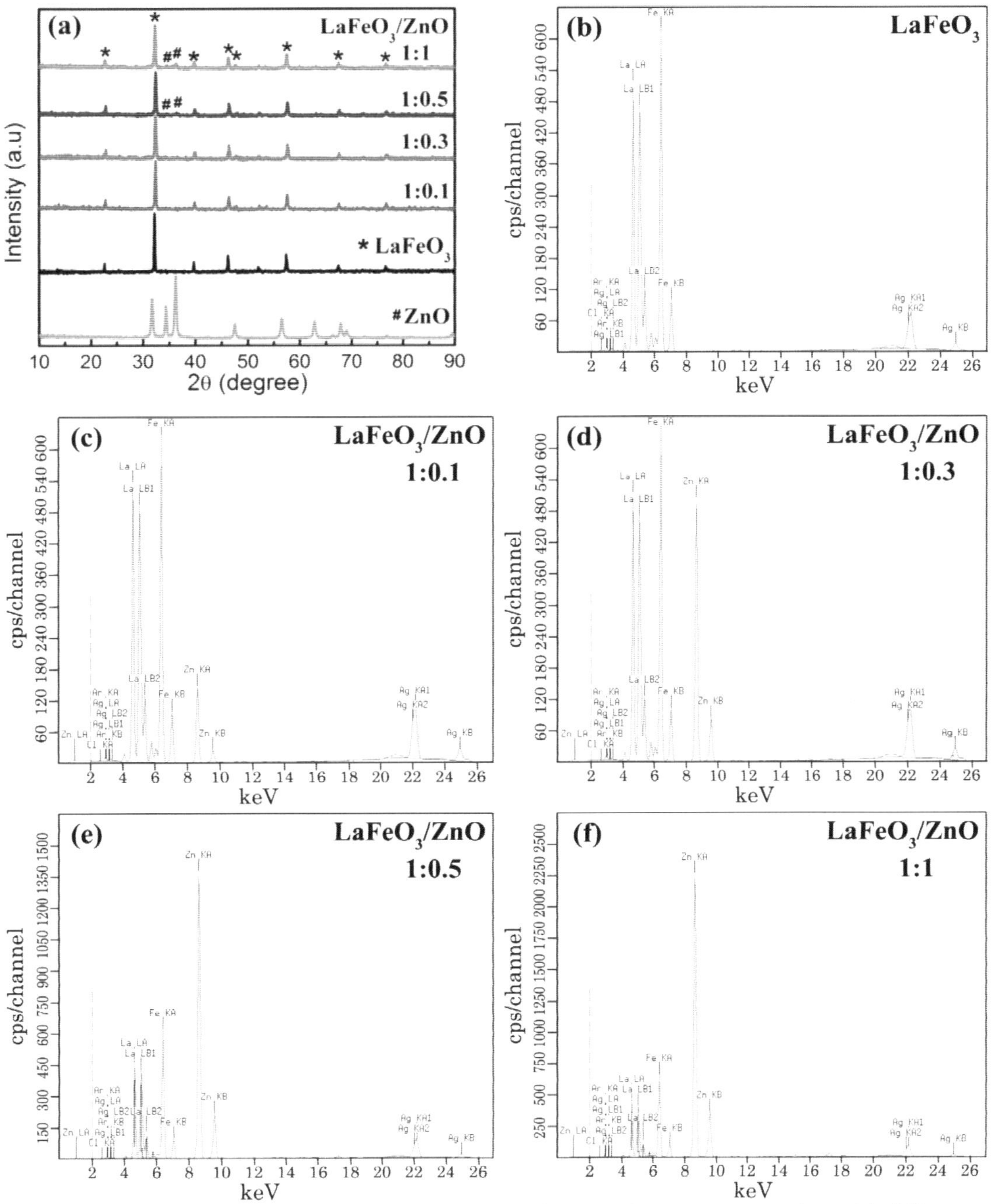

Figure 1. (a) XRD pattern; and (b-f) XRF spectra of $LaFeO_3$/ZnO nanocomposites with different molar ratios of ZnO.

(121) peak of $LaFeO_3$ and (101) peak of ZnO were used to estimate the crystallite size, and the values are tabulated in Table 1. The lattice parameter calculated by the Rietveld refinement method using the MAUD program is also tabulated in Table 1.

Table 1. The lattice parameter and grain size <D> of $LaFeO_3$/ZnO nanocomposites.

Sample	Molar Ratio	Lattice Parameter $LaFeO_3$			Lattice Parameter ZnO		<D> $LaFeO_3$ (nm)	<D> ZnO (nm)
		a (Å)	b (Å)	c (Å)	a=b (Å)	c (Å)		
$LaFeO_3$/ZnO	1:0	5.5436	7.8456	5.5528	-	-	58	-
	1:0.1	5.5205	7.8263	5.5503	3.2088	5.2368	37	30
	1:0.3	5.5948	7.8771	5.5640	3.2180	5.2344	31	26
	1:0.5	5.4970	7.8220	5.5492	3.2305	5.2484	28	24
	1.1	5.5402	7.8407	5.5537	3.2356	5.2497	25	21

Table 2. The chemical composition of $LaFeO_3$/ZnO nanocomposites.

Sample	Molar Ratio	La Concentration (%)	Fe Concentration (%)	Zn Concentration (%)	Cl Concentration (%)
$LaFeO_3$/ZnO	1:0	71.621	26.834	0	1.545
	1:0.1	71.186	26.184	2.436	0.195
	1:0.3	66.395	25.451	7.923	0.232
	1:0.5	57.714	21.653	20.435	0.198
	1.1	52.040	20.045	27.759	0.156

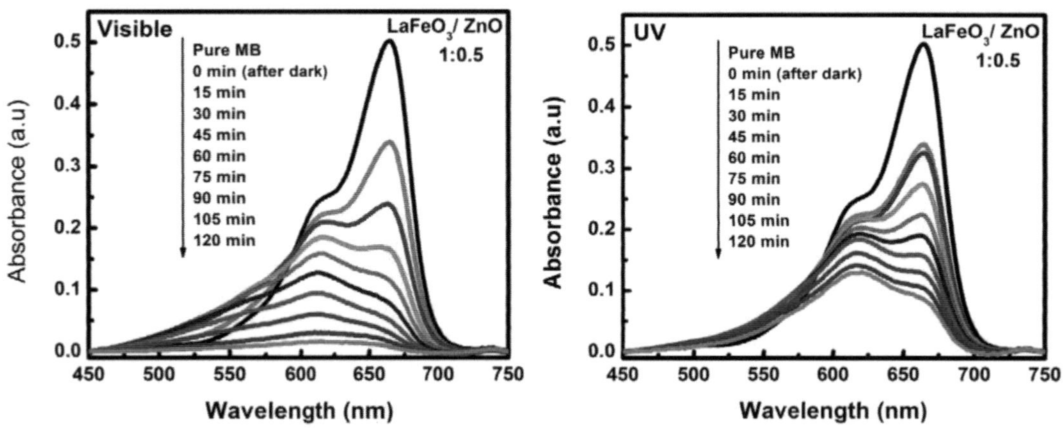

Figure 2. The UV-Visible spectra of MB solution at different time intervals.

The elemental analyses of the prepared samples were performed using XRF spectroscopy. Figure 1(b-f) shows the XRF spectra of the $LaFeO_3$/ZnO nanocomposites with different molar ratios of ZnO. As a comparison, the XRF spectra of the $LaFeO_3$ nanoparticles are also shown. The XRF spectra of the $LaFeO_3$ nanoparticles and $LaFeO_3$/ZnO nanocomposites consisted of La and Fe atoms, while the Zn peak was also detected in the XRF spectra of the nanocomposites. The presence of Zn atoms in the samples confirmed the formation of $LaFeO_3$/ZnO in the nanocomposite form. The appearance of Ag, Ar, and Cl peaks in the spectra were confirmed by the X-ray tube anode element of the XRF tools. The chemical composition of each element was also determined by XRF measurements. The concentration (C) of each element is tabulated in Table 2. As seen in the table, the concentration of elemental Zn increased with the increasing ZnO molar ratio.

To evaluate the photocatalytic activity of the prepared sample, MB degradation was monitored. During the visible and UV light irradiation in the presence of $LaFeO_3$/ZnO nanocomposites as catalysts,

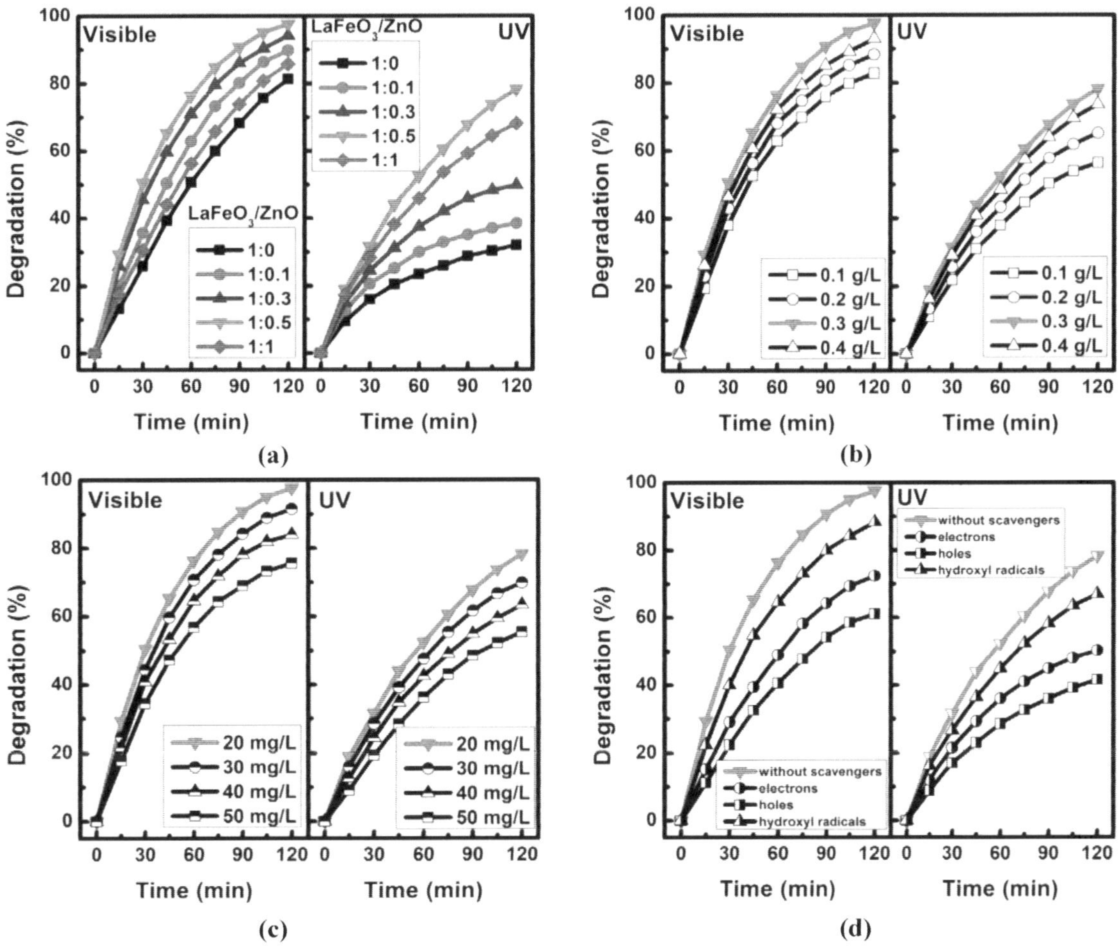

Figure 3. The percentage of degraded MB in the presence of different $LaFeO_3/ZnO$ molar ratios (a). The effect of catalyst dosage (b), initial concentration of MB (c), and various radical scavengers (c) on the percentage of degraded MB.

the MB color gradually changed to colorless, which was further confirmed by monitoring the UV-visible spectra (450–750 nm) of the MB solution at different time intervals, as shown in Figure 2. The UV-visible absorbance spectra was then used to study the photocatalytic efficiency in terms of percentage of degraded MB with the following equation [17]: Degradation (%) = $((C_0-C_t)/C_0) \times 100$, where C_0 is the MB initial concentration (after dark) and C_t is the MB concentration at each irradiated time interval (min). The percentage of degraded MB under visible and UV light irradiation in the presence of different $LaFeO_3/ZnO$ molar ratios is shown in Figure 3a. The percentage of degraded MB in the presence of $LaFeO_3$ nanoparticles is also shown as a comparison. The obtained results revealed that the incorporation of ZnO in the nanocomposite form could increase the degradation efficiency of $LaFeO_3$ nanoparticles. The incorporation of ZnO into the nanocomposite could serve as an electron trapper, which would inhibit electron and hole recombinations and enhance the photocatalytic performance [13, 18]. The $LaFeO_3/ZnO$ nanocomposites at a molar ratio of 1:0.5 showed the best degradation efficiency both under visible and UV light irradiation.

To obtain the maximum condition, several parameters were adjusted in the photocatalytic experiments. Figure 3b shows the effect of catalyst dosage on the percentage of degraded MB using

$LaFeO_3/ZnO$ nanocomposites at a molar ratio of 1:0.5 under visible and UV light irradiation. The initial MB concentration was held constant at 20 mg/L. As seen in the figure, the degradation efficiency increased with the increasing catalyst dosage from 0.1 g/L to 0.3 g/L; thereafter, at the catalyst dosage of 0.4 g/L, the degradation efficiency decreased. At lower maximum catalyst dosages, low active sites were available, while at higher maximum catalyst dosages, agglomerated particles, high turbidity, and scattering effects might cause lower degradation efficiency [13, 19]. Furthermore, the effects of MB initial concentration on the photocatalytic performance of $LaFeO_3/ZnO$ nanocomposites were also evaluated under visible and UV light irradiation. The $LaFeO_3/ZnO$ nanocomposites at a molar ratio of 1:0.5 and the catalyst with maximum dosage concentration of 0.3 g/L were used as the dependent parameters, while the initial concentration of MB varied from 20 mg/L to 50 mg/L. The results presented in Figure 3c show that the lower initial concentration of MB exhibited the best photocatalytic performance. At higher initial concentrations of MB, the penetration of light that could travel to the active site of the catalyst was blocked, resulting in a decrease in the total number of active sites that played roles in the degradation process [20].

It has been reported that photocatalytic activity depends on the number of active species that play important roles in the photocatalytic process [18, 21]. During irradiation, excited electrons in the conduction band of the catalyst react with the oxygen molecules (O_2) to form superoxide radicals ($\bullet O_2$). In contrast, holes left behind in the valence band of the catalyst will react with water (H_2O) or hydroxyl (OH^-) molecules to form hydroxyl radicals (•OH). These active radicals are known as strong reducers and oxidizers, which can destroy the complex chemical bonds of dyes such as MB [21]. For this reason, a control experiment with the addition of a scavenger such as electrons, holes, and hydroxyl radicals would be interesting to monitor. Sodium sulfate, di-ammonium oxalate, and tert-butyl alcohol were used as electrons, holes, and hydroxyl radical scavengers, respectively, in this study. After a 2-hour irradiation time, the presence of hydroxyl radical scavengers (as shown in Figure 3d) decreased the degradation of $LaFeO_3/ZnO$ nanocomposites without additional scavengers. Moreover, the presence of electron scavengers further decreased the degradation efficiency. The lowest degradation was obtained in the presence of di-ammonium oxalate as a hole scavenger, indicating that holes are the main active species in the photodegradation of MB.

4. Conclusion

The photodegradation of MB using $LaFeO_3/ZnO$ nanocomposites with different molar ratios of ZnO was successfully monitored under visible and UV light irradiation. The presence of ZnO in the orthorhombic and hexagonal wurtzite structures of the $LaFeO_3/ZnO$ nanocomposites improved the photocatalytic performance of the $LaFeO_3$ nanoparticles. The concentration of 0.3 g/L of $LaFeO_3/ZnO$ nanocomposite at a molar ratio of 1:0.5 exhibited the best photocatalytic performance for degrading a 20 mg/L MB solution. The main active species in the photocatalytic process is holes followed by eletrons and hydroxyl radical scavengers.

References

[1] Sakellis I, Giamini S, Moschos I, Chandrinou C, Travlos , Kim C-Y, Lee J-H, Kim J-G and Boukos N 2014 *Energ. Procedia* **60** 37
[2] Mekasuwandumrong O, Pawinrat P, Praserthdam P and Panpranot J 2010 *Chem. Eng. J.* **164** 77
[3] Xia Y, Zhang W, Zhang Y, Yu X and Chen F 2014 *Mater. Lett.* **131** 178
[4] Fu X, Xie M, Luan P and Jing L 2014 *ACS Appl. Mater. Interfaces* **6** 18550
[5] Guo X, Zhu H and Li Q 2014 *Appl. Catal. B.* **160–161** 408
[6] Awasthi G P, Adhikari S P, Ko S, Kim H J, Park C H and Kim C S 2016 *J. Alloy. Compd.* **682** 208
[7] He Z, Xia Y, Tang B, Jiang X and Su J 2016 *Mater. Lett.* **184** 148
[8] Shen H, Xue T, Wang Y, Cao G, Lu Y and Fang G 2016 *Mater. Res. Bull.* **84** 15
[9] Tang P, Tong Y, Chen H, Cao F and Pan G 2013 *Curr. Appl. Phys.* **13** 340
[10] Li S, Zhao Y, Wang C, Li D and Gao K 2016 *Mater. Lett.* **170** 122

[11] Afifah N and Saleh R 2016 *J. Phys. Conf. Ser.* **710** 012030
[12] Afifah N and Saleh R 2016 *Mater. Sci. Forum* **864** 99
[13] Derikvandi H and Nezamzadeh-Ejhieh A 2017 *J. Hazard. Mater.* **321** 629
[14] Afifah N, Djaja N F and Saleh R 2015 *Mater. Sci. Forum* **827** 19
[15] Phokha S, Pinitsoontorn S, Maensiri S and Rujirawat S 2014 *J. Sol-Gel Sci. Technol.* **71** 333
[16] Abazari R and Sanati S 2013 *Superlattice. Microst.* **64** 148
[17] Kaur J and Singhal S 2014 *Ceram. Int.* **40** 741
[18] Yang Y, Que W, Zhang X, Xing Y, Yin X and Du Y 2016 *J. Hazard. Mater.* **317** 430
[19] Nezamzadeh-Ejhieh A and Ghanbari-Mobarakeh Z 2015 *J. Ind. Eng. Chem.* **21** 668
[20] Saleh R and Djaja N F 2014 *Superlattice. Microst.* **74** 217
[21] Cai A, Wang X, Qi Y and Ma Z 2017 *Appl. Surf. Sci.* **391** 484

Construction of solvable potential partner of Generalized Hylleraas potential in one-dimensional Schrodinger system

A Suparmi, C Cari, Beta Nur Pratiwi and Dewanta Arya Nugraha

Physics Department, Graduate Program, Sebelas Maret University, INDONESIA

Email: soeparmi@staff.uns.ac.id

Abstract. The higher order potential partner of generalized Hylleraas potential was constructed by using Supersymmetry potential partner equation. The superpotential of shape invariant system was determined from the ground state wave function by applying Supersymmetry operators' properties. The ground state wave function and the ground state energy were determined from the solution of Schrodinger equation with generalized Hyllerras potential using Nikiforov-Uvarov method. The higher order potential partner was combination of the generalized Hylleraas potential, ground state energy, and super potential.. The energy eigenvalue and energy eigen function of potential partner which was solvable potential were also determined using NU method.

1. Introduction

Supersymmetric Quantum Mechanics (SUSY QM) which was developed as a simplest model of SUSY field theory was proposed by Witten.[1] SUSY QM since then has developed as a powerful technique used to solve one dimensional Schrodinger equation without solving the second order differential equation directly.[2,3,4] By applying the properties of SUSY QM, SUSY QM operators, and the ground state wave function of the original potential, the isospectral potential partner which had higher order was constructed.[5,6]

One dimensional Schrodinger equation for some potentials are exactly solved by some methods such as AIM [7], SUSY QM [2,3], Factorization method, Nikiforov-Uvarov (NU) method [8-10], Romanovsky polynomials [11], and others. NU method is one of those methods that has more general application and has been used widely since NU differential equation has been formulated with generalized parametric of generalized hypergeometric like equation [12].

Hylleraas potential is an interesting short range potential that could be used to describe atomic physics, nuclear physics such as an interaction of nucleon – nucleon, particle physics such as in meson-meson and various branches of nuclear physics [13-14]. For wider application, Hylleraas potential has been generalized into the form given as

$$V(x) = A\frac{a+e^{\alpha x}}{-(1/\mathrm{p})+e^{\alpha x}} - B\frac{b+e^{\alpha x}}{-(1/\mathrm{p})+e^{\alpha x}} \tag{1}$$

with a, b, and p are the Hylleraas parameters, A and B are proportional to potential depth, and α is proportional to the range of potential.

By using NU method the energy eigen value and the energy eigen function was obtained from the solution of Schrodinger equation with energy potential expressed in equation (1). By knowing the ground state wave function of the system then the construction of potential partner of generalized

Hylleraas potential equation (1) will be performed by using properties of SUSY quantum mechanics algebra.

The proposed of the research is to construct new potential which is potential partner of generalized Hylleraas potential by manipulating the potential partner equations with the ground state wave function of generalized Hylleraas potential. This paper is organized as follows. Brief introduction is presented in section 1, brief review of NU method and the properties of SUSY QM, and reduction of Schrodinger equation for Hylleraas potential into generalized hypergeometric like equation with generalized parametric are presented in section 2. Result and discussion are presented in section 3 and brief conclusion is in section 4.

2. Method

2.1. Review of NU method

The one-dimensional solvable Schrodinger equation is reduced into hypergeometric or confluent hypergeometric type differential equation by suitable variable transformation. The hypergeometric type differential equation, which is solved using Nikiforov-Uvarov method [8], is presented as

$$\frac{\partial^2\psi(s)}{\partial s^2}+\frac{\bar{\tau}(s)}{\sigma(s)}\frac{\partial\psi(s)}{\partial s}+\frac{\bar{\sigma}(s)}{\sigma^2(s)}\psi(s)=0 \tag{2}$$

where $\sigma(s)$ and $\bar{\sigma}(s)$ are mostly second order polynomials, and $\bar{\tau}(s)$ is first order polynomial. Equation (2) is solved using variable separation method by setting

$$\psi(s)=\phi(s)\,y(s) \tag{3}$$

then equation (2) becomes hypergeometric type equation

$$\sigma\frac{\partial^2 y(s)}{\partial s^2}+\tau(s)\frac{\partial y(s)}{\partial s}+\lambda y(s)=0 \tag{4}$$

and $\phi(s)$ a logarithmic derivative that fullfil the condition that $\frac{\phi'}{\phi}=\frac{\pi}{\sigma}$ with $\pi(s)$, λ, and τ are

$$\pi=\left(\frac{\sigma'-\bar{\tau}}{2}\right)\pm\sqrt{\left(\frac{\sigma'-\bar{\tau}}{2}\right)^2-\bar{\sigma}+k\sigma} \tag{5}$$

$$\tau=\bar{\tau}+2\pi\,,\quad \lambda=k+\pi'\text{ and}\qquad \lambda=\lambda_n=-n\tau'-\frac{n(n-1)}{2}\sigma'',\quad n=0,1,2,\ldots \tag{6}$$

The value of k in equation (5) is found from the condition that the expression under the square root of equation (5) must be square of mostly first-degree polynomial and thus the discriminate of the quadratic expression is zero. The new energy eigenvalue from equation (4) is obtained with the condition that $\tau'<0$. The solution of the second part of the wave function, $y_n(s)$, which is connected to Rodrigues relation [10], is given as

$$y_n(s)=\frac{C_n}{w(s)}\frac{d^n}{ds^n}\left(\sigma^n(s)w(s)\right) \tag{7}$$

where C_n is normalization constant, and the weight function $w(s)$ must satisfy the condition

$$\frac{\partial(\sigma w)}{\partial s}=\tau(s)w(s) \tag{8}$$

If in the parametric generalization, the hypergeometric type equation in equation (2) is written as [12]

$$\frac{\partial^2\psi(s)}{\partial s^2}+\frac{(c_1-c_2 s)}{s(1-c_3 s)}\frac{\partial\psi(s)}{\partial s}+\frac{(-\varepsilon_1 s^2+\varepsilon_2 s-\varepsilon_3)}{s^2(1-c_3 s)^2}\psi(s)=0 \tag{9}$$

By comparing equation (2) and equation (9) then the energy value equation and eigen function obtained from equation (9) are respectively given as

$$c_2 n-(2n+1)c_5+(2n+1)\left(\sqrt{c_9}+c_3\sqrt{c_8}\right)+n(n-1)c_3+c_7+2c_3c_8+2\sqrt{c_8c_9}=0 \tag{10}$$

and

$$\psi(s)=N_{n1}s^{c_{12}}\left(1-c_3s\right)^{-c_{12}-(c_{13}/c_3)}P_n^{(c_{10}-1,\,(c_{11}/c_3)-c_{10}-1)}\left(1-2c_3s\right) \tag{11}$$

with

$$\begin{aligned} &c_4=\frac{1}{2}(1-c_1),\ c_5=\frac{1}{2}(c_2-2c_3),\ c_6=c_5^2+\varepsilon_1,\ c_7=2c_4c_5-\varepsilon_2,\ c_8=c_4^2+\varepsilon_3\\ &c_9=c_3c_7+c_3^2c_8+c_6,\ c_{10}=c_1+2c_4+2\sqrt{c_8},\ c_{12}=c_4+\sqrt{c_8}\\ &c_{11}=c_2-2c_5+2\left(\sqrt{c_9}+c_3\sqrt{c_8}\right), c_{13}=c_5-\left(\sqrt{c_9}+c_3\sqrt{c_8}\right)\end{aligned} \tag{12}$$

The ground state energy and wave function are obtained from equations (10) and (11) for $n=0$

2.2. Brief review of SUSY QM

As proposed by Witten [1]the pair of SUSY Hamiltonian (H_{ss}) for two charge operators that are commute with H_{ss} is given by

$$H_{ss}=\begin{pmatrix} -\frac{d^2}{dx^2}+\frac{d\varphi(x)}{dx}+\varphi^2(x) & 0\\ 0 & -\frac{d^2}{dx^2}-\frac{d\varphi(x)}{dx}+\varphi^2(x)\end{pmatrix}=\begin{pmatrix} H_+ & 0\\ 0 & H_-\end{pmatrix} \tag{13}$$

with partner Hamiltonian $H_-=H_1$ and $H_+=H_2$, and potential partners $V_-=V_1$ and $V_+=V_2$ are

$$V_-(x)=V_1=\varphi^2(x)-\varphi'(x)\ \text{and}\ V_+(x)=V_2=\varphi^2(x)+\varphi'(x) \tag{14}$$

The relation between the effective potential V_{ef} and the first potential partner is

$$V_{ef}(x)=V_-(x;a_0)+E_0=V_1(x;a_0)+E_0 \tag{15}$$

where E_0 is the ground state energy for effective potential V_{ef} . By setting the new SUSY operators, lowering operator A and raising operator A^+ as

$$A^+=-\frac{d}{dx}+\varphi(x)\ \text{and}\ A=\frac{d}{dx}+\varphi(x) \tag{16}$$

then the SUSY Hamiltonians in equation (13) becomes

$$H_-(x)=H_1=A^+A,\ \text{and}\ H_+(x)=H_2=AA^+ \tag{17}$$

and so

$$A\psi_0^{(-)}=A\psi_0=0 \tag{18}$$

The potential partner V_2 of Hylleraas potential is obtained from equations (14) and (18) given as

$$V_2(x)=V_1-2\frac{d^2}{dx^2}\ln\psi_0 \tag{19}$$

Here ψ_0 is the ground state wave function of Schrodinger with potential of V_1 whose its potential partner is V_2.

3. Result and discussion

The Schrodinger equation with generalized Hylleraas potential which expressed in equation (1) is given as

$$-\frac{\hbar^2}{2m}\frac{\partial^2\psi(x)}{\partial x^2}+\left(A\frac{a+e^{\alpha x}}{-(1/p)+e^{\alpha x}}-B\frac{b+e^{\alpha x}}{-(1/p)+e^{\alpha x}}\right)\psi(x)=E\psi(x) \tag{20}$$

By setting $\hbar=1, 2m=1$, and $e^{\alpha x}=s$ in equation (20) then we have

$$\frac{\partial^2\psi}{\partial s^2}+\frac{(1-ps)}{s(1-ps)}\frac{\partial\psi}{\partial s}+\frac{\left(-p(p_1-p_2-pE)s^2+(a_1p+p_1+b_1p-p_2-2Ep)s-(-a_1+b_1+E)\right)}{\alpha^2s^2(1-ps)^2}\psi(s)=0 \quad (21)$$

with $$a_1=Apa,\ b_1=Bpb,\ p_1=Ap,\ p_2=Bp \quad (22)$$

By comparing equations (9) and (21) we obtain

$$c_1=1,\ c_2=c_3=p,\ \varepsilon_1=\frac{p(p_1-p_2-pE)}{\alpha^2},\varepsilon_2=(\frac{(a_1+b_1)p+p_1-p_2-2Ep}{\alpha^2})$$
$$\varepsilon_3=\frac{(-a_1+b_1+E)}{\alpha^2} \quad (23)$$

By manipulating equations (12) and (23) we get

$$c_4=0,\ c_5=-\frac{p}{2},\ c_6=p^2\left(\frac{1}{4}-\frac{E}{\alpha^2}\right)+\frac{p(p_1-p_2)}{\alpha^2},\ c_7=-(\frac{(a_1+b_1-2\mathrm{E})p+p_1-p_2}{\alpha^2}),$$
$$c_8=\frac{(-a_1+b_1+E)}{\alpha^2},\quad c_9=p^2\left(\frac{(-2a_1+2E)}{\alpha^2}+\frac{1}{4}\right),\quad c_{10}=1+2\sqrt{\frac{(-a_1+b_1+E)}{\alpha^2}}$$
$$c_{11}=p+p+2\left(p\sqrt{\left(\frac{(-2a_1+2E)}{\alpha^2}+\frac{1}{4}\right)}+p\sqrt{\frac{(-a_1+b_1+E)}{\alpha^2}}\right), \quad (24)$$
$$c_{12}=\sqrt{\frac{(-a_1+b_1+E)}{\alpha^2}},\quad c_{13}=-\frac{p}{2}-\left(p\sqrt{\left(\frac{(-2a_1+2E)}{\alpha^2}+\frac{1}{4}\right)}+p\sqrt{\frac{(-a_1+b_1+E)}{\alpha^2}}\right)$$

The ground state wave function generated from equations (11) and (24) is given by

$$\psi_0(\mathrm{x})=N_0s^{c_{12}}\left(1-c_3s\right)^{-c_{14}}=N_0e^{c_{12}\alpha x}\left(1-c_3e^{\alpha x}\right)^{-c_{14}} \quad (25)$$

with $$c_{14}=c_{12}+\frac{c_{13}}{c_3}=-\frac{1}{2}-\sqrt{\left(\frac{(-2a_1+2E)}{\alpha^2}+\frac{1}{4}\right)} \quad (26)$$

In equation (26) $E=E_0$ is the ground state energy of generalized Hylleraas potential and is obtained numerically from equation (28). By using equations (10) and (21-24) we obtain the energy eigen value equation given as

$$pn+(2n+1)\frac{p}{2}+(2n+1)\left(\sqrt{(-2a_1-2\mathrm{E}+\frac{1}{4})p^2}+p\sqrt{(-a_1+b_1+E)}\right)+n(n-1)p$$
$$-((a_1+b_1-2\mathrm{E})p+p_1-p_2)+2p(-a_1+b_1+E)+2\sqrt{(-a_1+b_1+E)(-2a_1-2\mathrm{E}+\frac{1}{4})p^2}=0 \quad (27)$$

By setting $n=0$ in equation (27) we get the ground state energy eigen value equation which is given as

$$+\frac{p}{2}+p\left(\sqrt{(-2a_1-2\mathrm{E}_0+\frac{1}{4})}+\sqrt{(-a_1+b_1+E_0)}\right)+$$
$$((-3a_1+b_1+4\mathrm{E}_0)p-p(\mathrm{A}-B))+2p\sqrt{(-a_1+b_1+E_0)(-2a_1-2\mathrm{E}_0+\frac{1}{4})}=0 \quad (28)$$

The ground state energy is calculated numerically from equation (28). The corresponding wave function generated from equations (11) and (24) is given by

$$\psi(s)=N_{n1}s^{c_{12}}\left(1-c_3s\right)^{-c_{12}-(c_{13}/c_3)}P_n^{(c_{10}-1,\,(c_{11}/c_3)-c_{10}-1)}\left(1-2c_3s\right) \quad (29)$$

3.1. Construction of the potential partner of Hylleraas potential

By using equations (15), (19), and (25) we get the potential partner V_2 of generalized Hylleraas effective potential given as

$$V_2(x) = V_1 - 2\frac{d^2}{dx^2}\ln\psi_0 = \left(-\frac{a_1 + p_1 s}{1-ps} + B\frac{b_1 + p_2 s}{1-ps}\right) - E_0 - 2\left(\frac{c_{14} p\alpha^2 s}{(1-ps)^2}\right) \tag{30}$$

with $c_{14} = -\frac{1}{2} - \sqrt{\left(\frac{(-2a_1 + 2E_0)}{\alpha^2} + \frac{1}{4}\right)}$. It is seen in equation (30) that the potential partner shows higher order function potential compared to the generalized Hylleraas potential in equation (1). It will also be shown that the potential partner in equation (30) is solvable. By writing the Schrodinger equation for potential partner in equation (30) as

$$\begin{aligned}&\frac{\partial^2\psi}{\partial s^2} + \frac{(1-ps)}{s(1-ps)}\frac{\partial\psi}{\partial s} - \frac{\left(p(\mathrm{p}_1 - p_2 - pE_{2n} + p\,\mathrm{E}_0)s^2\right)}{\alpha^2 s^2 (1-ps)^2}\psi \\ &+\left[\frac{(a_1 p + p_1 + b_1 p - p_2 - 2E_{2n}p - 2c_{14}p\alpha^2 + 2p\,\mathrm{E}_0)s - (-a_1 + b_1 + E_{2n} + E_0)}{\alpha^2 s^2 (1-ps)^2}\right]\psi = 0\end{aligned} \tag{31}$$

then we can se that equation (31) is similar with equation (21), therefore equation (31) can also be solved using NU method. So the one dimensional Schrodinger equation with potential partner V_2 which is expressed in equation (31) are exactly solvable by NU method with parameter given as

$$\begin{aligned}&c_1{}' = 1,\ c_2{}' = c_3{}' = p,\ \varepsilon_1{}' = \frac{p(\mathrm{p}_1 - p_2 - pE_{2n} + pE_0)}{\alpha^2},\ \varepsilon_3{}' = \frac{(-a_1 + b_1 + E + E_0)}{\alpha^2} \\ &\varepsilon_2{}' = (\frac{(a_1 + b_1)p + p_1 - p_2 - 2E_{2n}p - 2c_{14}p\alpha^2 + 2p\,\mathrm{E}_0}{\alpha^2}),\end{aligned} \tag{32}$$

By manipulating equations in equation (32) and together with equation (12) we generate

$$\begin{aligned}&c_4{}' = 0,\ c_5{}' = -\frac{p}{2},\ c_6{}' = \frac{p^2}{4} + \frac{p(\mathrm{p}_1 - p_2 - pE + pE_0)}{\alpha^2}, \\ &c_7{}' = -(\frac{(a_1 + b_1)p + p_1 - p_2 - 2Ep - 2c_{14}p\alpha^2 + 2p\,\mathrm{E}_0}{\alpha^2}),\ c_8{}' = \frac{(-a_1 + b_1 + E + E_0)}{\alpha^2} \\ &c_9{}' = (\frac{2p^2(E_{2n} + c_{14}\alpha^2 - 2a_1)}{\alpha^2}) + \frac{p^2}{4},\ c_{10}{}' = 1 + 2\sqrt{\frac{(-a_1 + b_1 + E + E_0)}{\alpha^2}}, \\ &c_{11}{}' = 2p + 2p\left(\sqrt{(\frac{2(E_{2n} + c_{14}\alpha^2 - 2a_1)}{\alpha^2}) + \frac{1}{4}}\right) + p\sqrt{\frac{(-a_1 + b_1 + E_{2n} + E_0)}{\alpha^2}}, \\ &c_{12}{}' = \sqrt{\frac{(-a_1 + b_1 + E + E_0)}{\alpha^2}},\ c_{13}{}' = -\frac{p}{2} - p\left(\sqrt{(\frac{2(E + c_{14}\alpha^2 - 2a_1)}{\alpha^2}) + \frac{1}{4}} + \sqrt{\frac{(-a_1 + b_1 + E + E_0)}{\alpha^2}}\right)\end{aligned} \tag{33}$$

By using equations (10) and (11) and together with equations in equation (33) we get the energy eigen value equation and energy eigen function for potential partner given as

$$(2n^2+2n+1)\frac{p}{2}+(2n+1)\left(\sqrt{\left(\frac{2p^2(E+c_{14}\alpha^2-2a_1)}{\alpha^2}\right)+\frac{p^2}{4}}+p\sqrt{\frac{(-a_1+b_1+E+E_0)}{\alpha^2}}\right)$$
$$+p\frac{(-3a_1+b_1+4E-A+B+2c_{14}\alpha^2)}{\alpha^2} \tag{34}$$
$$+2\sqrt{\left[\left(\frac{2p^2(E+c_{14}\alpha^2-2a_1)}{\alpha^2}\right)+\frac{p^2}{4}\right]\frac{(-a_1+b_1+E+E_0)}{\alpha^2}}=0$$

and

$$\psi(s)=N_{n1}s^{c_{12}'}\left(1-c_3's\right)^{-c_{12}'-(c_{13}'/c_3')}P_n^{(c_{10}'-1,\,(c_{11}'/c_3')-c_{10}'-1)}\left(1-2c_3's\right) \tag{35}$$

with all c's parameters are listed in equation (33).

It was shown that from exactly solvable potential in quantum system could be generated another potential that was also solvable potential in quantum system. This new solvable potential is simply generated by using the properties of SUSY QM. Due to the differences in parameters between equation (21) and (31) then it is expected that the potential partner has different energy with the original ones.This potential partner has different energy spectra compared to the energy spectra of the original potential. The result of this work is in agreement with the result of the research about the simple construction of potential partner for Hulthen potential in Ref. (6)

4. Conclusion

The new exactly solvable potential in quantum system was generated from generalized Hylleraas potential using SUSY QM. By applying the properties of a pair of potential partner V_1 and V_2 and SUSY operators, potential partner V_2 was generated from another potential partner V_1 and with known ground state wave function of Hylleraas potential. Both ground state energy and wave functions of generalized Hylleraas potential were determined by using parametric generalization NU method. The energy eigenvalue equation and wave function of potential partner V_2 was also analytically obtained using NU method. The potential partner has different energy spectra compared to the energy spectra of the original potential and so the wave function. This potential partner was considered to be new potential that could be applied to describe the interaction model in atomic system. It would be worthy to explore more potential partner of solvable potential system which could be used to describe the behavior of particles in atomic system.

Acknowledgement
This research was partly supported by Sebelas Maret University Higher Education Project Grant with contract no. 632/UN27.21/LT/2016

References

[1] Witten E 1981 *Nucl.Phys.B***188** 513
[2] Sukumar C V 1985 *J. Phys. A***18** L57
[3] Cooper F and Freedman B 1983 *Ann. Phys***146** 262
[4] Suparmi and Cari 2014 *J. Math. Fund. Sci* **46** 205
[5] Bhoi J and Laha U 2016 *Brazilian J. Phys***46** 129
[6] Bhoi J and Laha U 2016 *Lat. Am. J. Phys. Educ.* **10** 1
[7] Soylu A, Bayrak O and Boztosun I 2008 *Journal of Physics A: Mathematical and Theoretical* **41** 065308
[8] Nikiforov A F and Uvarov V B 1998 Special Functions of Mathematical Physics (Birkhauser, Basel)
[9] Suparmi A, Cari C and YulianiH 2013*Adv. Phys. Theo. And Appl.* **16**
[10] Ikhdair S M and Sever R 2009 *Ann. Phys***18** 747

[11] Suparmi A, Cari C, Handhika, Yanuarief C and Marini H 2012 *IOSR J. Appl. Phys* **2** 43
[12] Tezcan C and Server R 2009 *Int. J. Theor. Phys* **48** 337
[13] Ita B I, Ikeuba A I and Ikot A N 2014 *Commun. Theor. Phys*. **61** 149
[14] Hylleraas E A 1938 *J. Chem. Phys*. **3** 595
[15] Varshni Y P 1957 *Rev. Mod. Phys*. **29** 664

Skull outer contour extraction image using compass scanning

Pauladie Susanto[1,2], Eko Mulyanto Yuniarno[1], Mauridhi Hery Purnomo[1]

[1] Departement of Electrical Engineering, Institut Teknologi Sepuluh Nopember Surabaya, Indonesia

[2] Departement of Computer Engineering, Institut Bisnis dan Informatika Stikom Surabaya, Indonesia

pauladie.s@gmail.com

Abstract. Skull outer contour extraction is an important step for craniofacial reconstruction. It useful for performing surface reconstruction. Skull outer contour obtained by three stages: thresholding, defining region of interest, and skull outer contour extraction. Compass scanning is used to this extraction. Scanning is performed using eight compass directions, four scanning are done from the four edges of ROI area, and the other four from corner of ROI area. This proposed method successfully extracts skull outer contour for four sample slices: upper temple, eye, nose, and mouth. Scanning from four corners of ROI area gives better result than four edges of ROI area, but using all eight scannings will give the best extraction performance.

Keywords: skull outer contour extraction, craniofacial reconstruction, compass scanning.

1. Introduction

Two common modalities used in neuroimaging are the MRI and the CT [1]. Due to MRI's inherent weakness concerning hard tissue imaging; few works have been published so far on skull segmentation in newborns using MRI [2][3]. In all the proposed methods, the extracted skull models suffer from low accuracy and absence of any information about the location of fontanels.[1]

Unlike MRI, CT scanning has some limitations such as the need to expose the subject to ionizing radiation and the lack of good contrast for soft tissues. Nevertheless, since it provides the best contrast for bone definition, CT is a preferred modality in cranial studies.[1]

Some works has done using medical images. Galdames[4] using Simplex Mesh and Histogram Analysis Skull Stripping (SMHASS) to perform skull stripping methods that are designed to eliminate the non-brain tissue in magnetic resonance (MR) brain images. They get better accuration compared to Brain Extraction Tool (BET), Brain Surface Extractor (BSE), and Hybrid Watershed Algorithm (HWA). Tilotta[5] uses curve evolution method to perform contour extraction and makes new public database which is compose of 85 CT-scans of healthy subjecy aged 20 to 65 years old. Foi[6] has presented a fully automatic method to compute the OFD and BDP measurements from fetal ultrasound images. The method is based on fitting an ellipse modelling the head contour of the fetus to the provided ultrasound images by minimizing a cost function with respect to the parameters of the ellipse. Dogdas[7] presents a new technique for segmentation of skull and scalp in T1-weighted magnetic resonance images (MRIs) of the human head. The method uses mathematical morphological operations to generate realistic models of the skull, scalp, and brain.

Craniofacial reconstruction is a useful tool in the identification of skeletal remains when there is a lack of other forensic evidense to suggest an identity. The aim of craniofacial reconstruction is to recreate a likeness of the real face using relationship between the soft tissues and the underlying skull.[8] With the development of computer science and medical imaging, many computer-aided methods have been proposed for craniofacial reconstruction.[9][10][11]

In craniofacial reconstruction, there is an important stage to do, it is extraction of the outer contour of the skull. To build a 3D model of the head, skull outer contour extraction must be performed first. Deng[8] shows two phases of his method framework, first is extraction of the relationship between skull and face regions, and it consists of six steps: 3D CT reconstruction, global coordinate adjustment, dense point registration, segmentation, local coordinate adjustment, and PCA and PLSR. The second phase is craniofacial reconstruction for a skull, and it consists of seven steps: global coordinate adjustment, dense point registration, segmentation, locale coordinate adjustment, facial regions reconstruction, local coordinate readjustment, and fusion. This proposed method is a sub-work of 3D CT reconstruction in first phase.

2. Method

According to a study [12], skull outer contour extraction is one of segmentation method. Segmentation being classified into three categories: structural techniques, stochastic techniques, and hybrid techniques. The classification is done based on the approach used for segmentation. Structural techniques utilize some information about the structure of the region in segmenting it. Stochastic techniques are the ones that are applied on discrete voxels without any consideration for the structure of the region. The final category is the hybrid methods which include those techniques which posses characteristics of both structural and stochastic techniques.

Structural techniques tries to find structural property from region that will be segmented. Intersection surface is one of that property and done by an edge beetween 2 slices. The outer contour from each sllice will become an important information to form those edges.

Skull outer contour extraction has three stages: thresholding, finding region of interest (ROI), and compass scanning. Fig 1 shows the stages of this work. This research is using head CT image as data, we already set some rules to meet the data, specially on stage ROI. We may try to use this extraction method on other part of body with some rule adjustments.

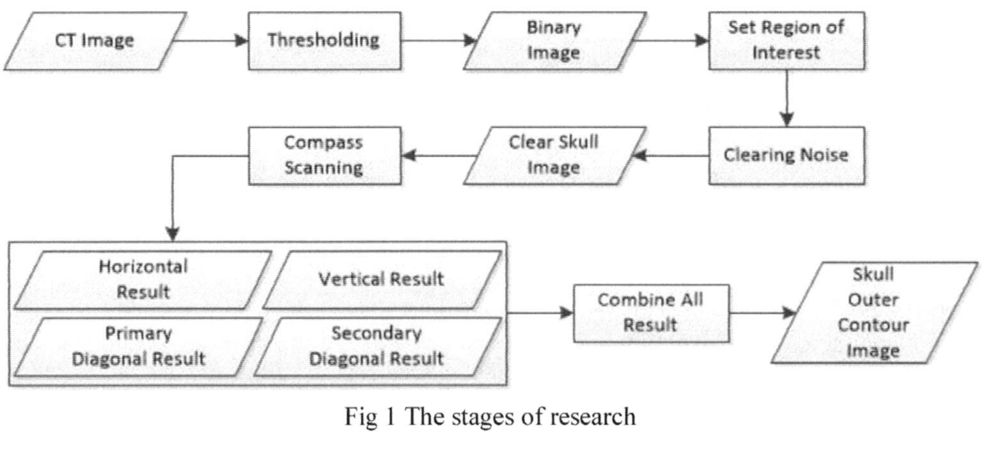

Fig 1 The stages of research

2.1. Thresholding

In thresholding, one assumes that the foreground can be characterized by its brightness [13]. One selects a value θ, $\min_x(f(x)) \leq \theta \leq \max_x(f(x))$, and sets foreground voxels, accordingly.

$$g(x) = \begin{cases} 1 & if\ f(x) \geq \theta \\ 0 & else \end{cases} \quad (1)$$

In this research, we set the value of θ manualy to satisfy with the data characteristic. Its value is 0.518.

2.2. Finding ROI

This stage tries to keep the important region and erase the other region. Our focus is skull outer countour, we will keep everything inside the skull and remove everything outside the skull. We use ellipse to select the important region. The most important information in the CT image is slice of skull and its form is ellipse as well. The position and the size of ellipse are set manually. After we adjust ellipse and agree with its position and size, we can make the ellipse become the valid ROI.

The result of this ellipse is a binary mask. The area inside the ellipse will have logical value 1 and the outside will have logicall value 0. We operate this mask with binary image produced by thresholding stage with logical operator AND. AND operator will make region inside ellipse area still remain like what they are, but changes area outside ellipse to logical value 0. After this stage we successfully clear some objects outside the ellipse.

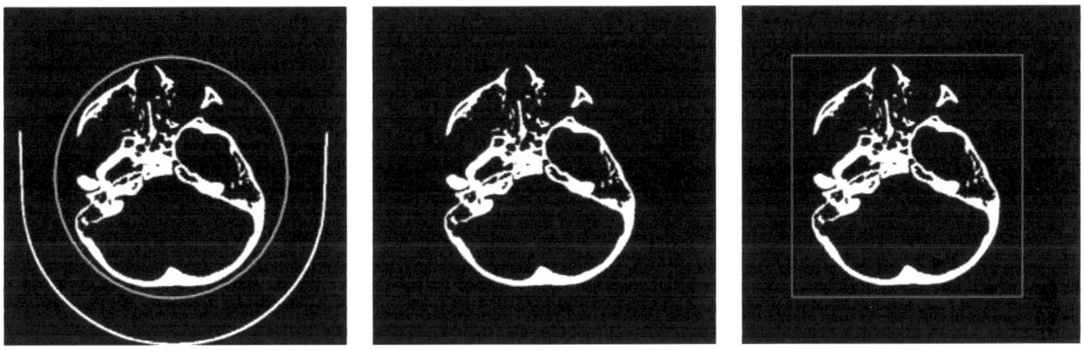

(a) (b) (c)

Fig 2 Region of interest stage; (a) Blue ellipse, (b) After clearing, (c) Minimize the computational area

The other advantage of this ellipse is its position. We can use its position to minimize the computational area. We wish we can save a little computational time. Fig 2 show the result of stage ROI, Fig 1a shows the adjustable blue ellipse, we can move or resize it. Fig 1b show the result after being operated with the mask. Fig 1c show the position of the ellipse after being converted to a rectangle.

2.3. Compass scanning

After stage ROI, we get clear binary image with smaller area, and we run compass scanning within this area. The concepts of compass scanning is simple, de do the scanning activity according to eight direction of wind. We can have 4 scaning groups, horizontal, vertical, and two diagonals. Horizontal scanning covers west and east, vertical scanning covers north and south, primary diagonal covers north west and south east, and secondary diagonal covers north east and south west. Scanning on a certain direction, actually the same as running multiple line using linier equation.

Horizontal from left and right	$y = c$, with: $ymin \leq c \leq ymax$	(2)
Vertical from top and bottom	$x = c$, with: $xmin \leq c \leq xmax$	(3)
Primary Diagonal	$y = x + c$, with: $ymin \leq c \leq ymax$	(4)

form top left and bottom right	$x = y + c$, with: $xmin \leq c \leq xmax$	(5)
Secondary Diagonal	$y = -x + c$, with: $ymin \leq c \leq ymax$	(6)
from top right and bottom left	$x = -y + c$, with: $xmin \leq c \leq xmax$	(7)

Each scanning through a certain line will search the first white pixel and stop, save the pixel, then continue to next line. That white pixel represent the outer contour of the skull. We have four scanning groups, as the result, we will have four results as well. We count the number of white pixel to determine which scanning is more efective.

3. Result and Discussion

We have done some experiment using those stages to four samples area, upper temple, eye, nose, and mouth. Each location has ten samples. Fig 3 shows result image on eye location. Fig 4 shows result image on upper temple location.

From the result image we can say that when we working with hard tissue (Fig 4), this method can give us excellent result, but when there is much soft tissue in it we can have some false detection and that pixel become noise (Fig 3). The noise can be produced by some holes on outer contour of skull. When the scanning process is working and there is a hole on its path, the scanning will continue to working until it find the first white pixel. This white pixel will be a false pixel, because its location no longer serves as outer contour of skull.

Now we discuss the other result we have, the counter of white pixel for each scanning group. We record all counter for each experiments. Table 1 shows the record from nose locations. Table 2 shows the record from mouth location.

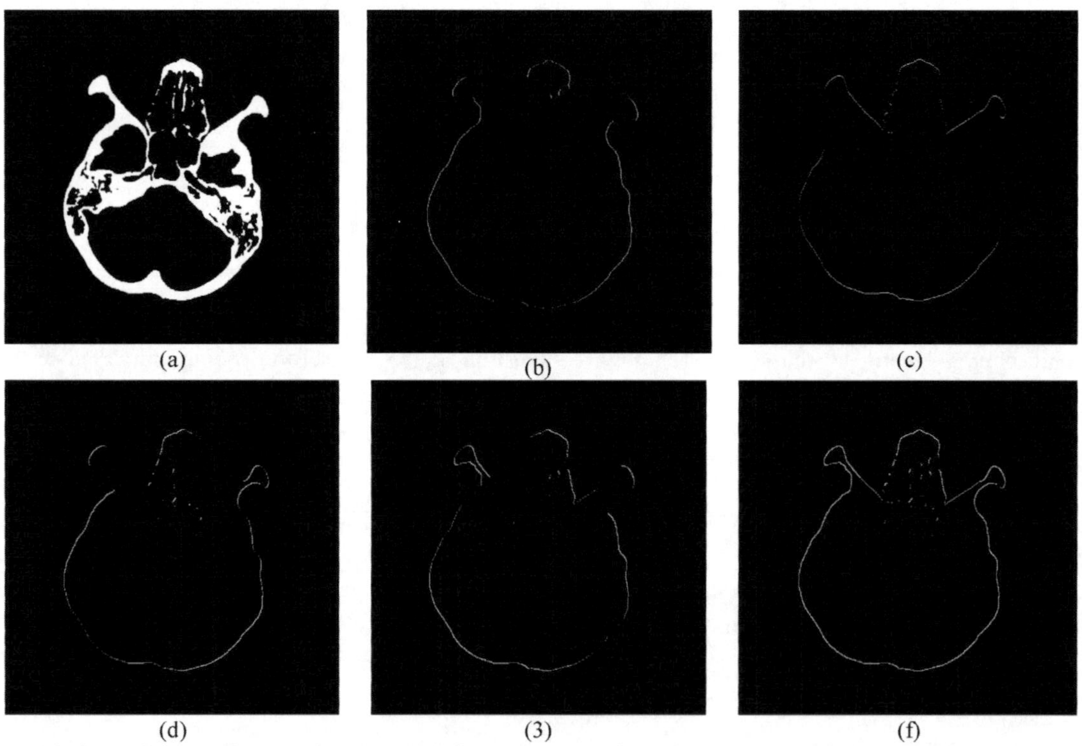

Fig 3 Result image from eye location; (a) ROI result, (b) Horizontal result, (c) Vertical result, (d) Primary diagonal result, (e) Secondary diagonal result, (f) Combine result

IOP Conf. Series: Journal of Physics: Conf. Series **820** (2017) 012024 doi:10.1088/1742-6596/820/1/012024

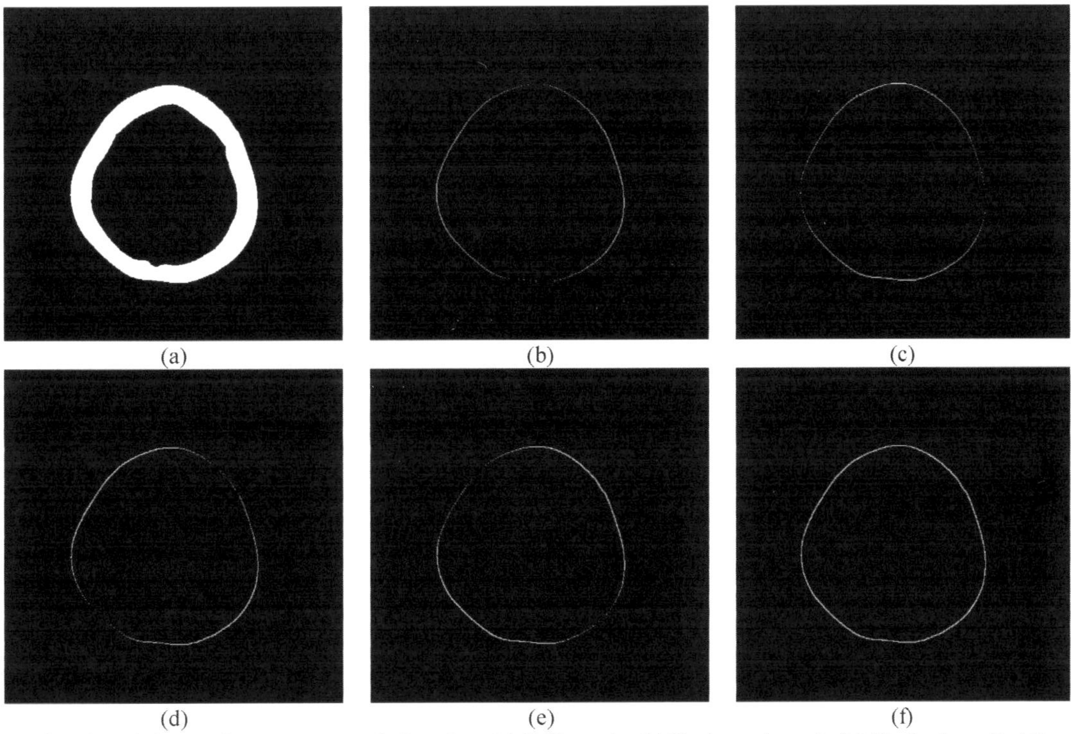

Fig 4 Result image from upper temple location; (a) ROI result, (b) Horizontal result, (c) Vertical result, (d) Primary diagonal result, (e) Secondary diagonal result, (f) Combine result

Table 1 Counter record for nose location

Hor	Ver	H+V	Dia1	Dia2	D1+2	ALL
488	528	844	776	786	1225	1448
520	524	869	767	783	1239	1464
530	520	869	749	780	1216	1444
536	514	871	737	776	1193	1430
539	504	859	728	774	1184	1424
540	488	839	706	770	1181	1412
540	486	831	698	766	1173	1415
547	486	827	696	762	1181	1419
555	486	819	696	756	1189	1420
574	480	813	696	741	1181	1405

Table 2 Counter record for mouth position

Hor	Ver	H+V	Dia1	Dia2	D1+2	ALL
446	377	621	682	655	1088	1193
446	374	620	680	656	1085	1183
434	368	605	675	658	1075	1159
439	364	607	667	659	1081	1157
432	360	597	666	660	1098	1156
429	354	592	670	664	1104	1169
427	350	589	673	664	1102	1173
421	344	581	673	664	1102	1172
408	338	572	670	660	1099	1162
390	334	559	654	656	1070	1134

From the table we can see that D1+2 has better performance than H+V. We can have decision only to depend on D1+2 to scan the image, but it still has a disadvantage, it can not cover all outer contour. As a prove, we compare D1+2 and ALL. D1+2 is combining result from primary and secondary diagonal scanning, and ALL is combining reslut from D1+2 and H+V. We can see that ALL always greater than D1+2. It means that although D1+2 has better performance, it doesn't mean that it can cover all skull outer contour. H+V can have many pixels that D1+2 doesn't have. There is a possibility, if we only use D1+2 we may have some holes leaving by H+V. According to our discussion above, a hole can bring noise, and more holes can bring more noise as well.

We already discuss this research's results. We have two bottom line, first is holes on skull outer contour can bring noise. As a further works, we have atleast two options to fix this, first after finding ROI we can use some particularly methods to close the hole, then perform the scanning. Second option is we perform all proposed stages, then remove those noises using some particularly methods. Whatever option we choose, it need a lot of effort to do. But those are challenges to enhance this research.

Second bottom line is about scanning, we have some scanning groups, those group is not suppose to eliminate each other, but to take over other scanning weakness. Using all scanning groups together will give us better result.

References

[1] Ghadimi S Moghaddam HA Grebe R Wallois F 2016 Skull Segmentation and Reconstruction From Newborn CT Images Using Coupled Level Sets *IEEE Journal Of Biomedical And Health Informatics, Vol. 20, No. 2. 563-573*

[2] Daliri M Moghaddam HA Ghadimi S Momeni M Harirchi F Giti M 2010 Skull Segmentation in 3D Neonatal MRI using Hybrid Hopfield Neural Network. *Proc. Conf. IEEE Eng. Med. Biol. Soc., Buenos Aires, Argentina. 4060–4063*

[3] Mahapatra D 2012 Skull stripping of neonatal brain MRI: Using prior shape information with graph cuts *J. Digit Imag., vol. 25, no. 6. 802–814*

[4] Galdames FJ Jaillet F Perez CA 2011 An Accurate Skull Stripping Method Based on Simplex Meshes and Histogram Analysis in Magnetic Resonance Images *Rapport de recherche RR-LIRIS-019*

[5] Tilotta F Richard F Glaunes J Berar M Gey Sverdeille S Rozenholc Y Gaudy JF 2008 *PREPRINT MAP5 11*

[6] Foi A Maggioni M Pepe A Tohka J - Head Contour Extraction From Fetal Ultrasound Images By Difference Of Gaussians Revolved Along Elliptical Paths

[7] Dogdas B Shattuck DW Leahy RM 2005 Segmentation of Skull and Scalp in 3-D Human MRI Using Mathematical Morphology *Human Brain Mapping 26:273-285*

[8] Deng Q Zhou M Wu Z Shui W Ji Y Wang X Liu CYJ Huang Y Jiang H 2016 A Regional Method For Craniofacial Reconstruction Based On Coordinate Adjustment And A New Fusion Strategy *Forensic Science Internasional 259 19-31*

[9] Claes P Vandermeulen D De Greef S Willems G Clement JG Suetens P 2010 Computerized Craniofacial Reconstruction: Conceptual Framework and Review *Forensic Sci. Int. 201 (1-3) 138-145*

[10] Wilkinson C 2005 Computerized Forensic Facial Reconstruction: A review of Current System *Forensic Sci. Int.Med. Pathol 1 (3) 173-177*

[11] De Gref S Claes P Mollemans W Vandermeulen D Suetens P willems G 2005 Computer-Assisted Facial Reconstruction: Recent Development and Trends *Rev. Belge. Med. Dent. 60 (3) 237-249*

[12] Lakare S 2000 3D Segmentation Techniques for Medical Volumes

[13] Wirjadi O 2007Survey of 3D Image Segmentation Methods *Berichte des Fraunhofer ITWM, Nr. 123*

The Research on Informal Learning Model of College Students Based on SNS and Case Study

Peng Lu[1,a], Xiao Cong[2,b], Fangyan Bi[3,c], Dongdai Zhou[3,d]
1. Department of Media Technology and Communication, Northeast Electric Power University, Jilin, Jilin, 132012, P.R.China;
2. College of Science, Northeast Electric Power University, Jilin, Jilin, 132012, P.R.China;
3. Engineering and Research Center of E-learning, Northeast Normal University, Changchun, Jilin, 130117, P.R.China

E-mail: [a]lup595@nenu.edu.cn; [b]peng.lu2008@gmail.com; [c]peng1984_2004@sina.com; [d]pengphoenix@qq.com

Abstract. With the rapid development of network technology, informal learning based on online become the main way for college students to learn a variety of subject knowledge. The favor to the SNS community of students and the characteristics of SNS itself provide a good opportunity for the informal learning of college students. This research first analyzes the related research of the informal learning and SNS, next, discusses the characteristics of informal learning and theoretical basis. Then, it proposed an informal learning model of college students based on SNS according to the support role of SNS to the informal learning of students. Finally, according to the theoretical model and the principles proposed in this study, using the Elgg and related tools which is the open source SNS program to achieve the informal learning community. This research is trying to overcome issues such as the lack of social realism, interactivity, resource transfer mode in the current network informal learning communities, so as to provide a new way of informal learning for college students.

1. Introduction

According to the Pareto Principle, people's knowledge acquired in life about 80% comes from informal learning, only about 20% of the knowledge from formal learning of formal education institutions such as schools, while inputs and concerns to informal learning of people are less than 20% [1]. Informal learning can be intentional, conscious, or unintentional, unconscious, however, the informal learning which intentional and conscious can have a positive impact on learning efficiency and learning quality according to learning characteristics. College students have a strong desire for knowledge, strong learning and innovation ability, but also become the main force of informal learning.

Social relationships in the online learning has become an important resource in informal learning, people through the establishment of a stable social network to achieve the sharing of network resources with others. The Social Networking Services (SNS) focus on social authenticity, that breaking the network community of the virtual, so that interpersonal relationships with more authenticity. SNS integration of a variety of applications as a Web2.0 concept and technology, with interactive, cooperative, initiative, communication and other characteristics. Learners can acquire knowledge from established personal social networks and facilitate interaction by introducing SNS into informal learning, and form a stable and sustainable social network that promotes learners communicate and sharing of resources to achieve the positive effects of informal learning. SNS can effectively promote the formation and development of social networks among learners. By using the information push mechanism provided by SNS, it is possible to perfect the network learning information transmission mode and to create a flexible and convenient network learning interactive way, others, it is possible to construct a flexible and diverse network learning environment and to divide the groups according to the theme, to meet the learners' common learning interests and needs [2]. The characteristics and functions of SNS make it have great development potential in the informal learning of students.

2. Related research

Informal learning as the main mode of learning in people's daily life and work, was introduced by Malcolm Knowles in the 1950s, in which he defined the informal learning as a form of learning relative to formal learning[4]. Since then, more and more researchers have paid attention to the informal learning.

(1) Research on informal learning in network environment. Research focused on social software to support informal learning, informal learning models, informal learning resources and resource integration, suggestions and strategies to promote informal learning, and informal learning communities and so on. Elina Jokisalo and Antoni Riu studied informal learning in the Web 2.0 era, pointing out that people can learn through a variety of ways [5]. Bull, G., Thompson, A., et al. proposed the use of informal learning to assist students in formal school learning for the function of social software, and the use of relevant social media connection formal learning and informal learning theoretical model [6]. Ferretti, S., Mirri, S. and others attempts to Web2.0 technology used in informal learning practice and proposed a Web2.0 technology-based learning tools for users to edit educational resources, and through collaborative work analysis of multimedia content[7]. Carmen Holotescu studied the use of Cirip.eu in informal learning, which is a microblogging platform developed specifically for education. It can provide a variety of functions to facilitate the sharing and delivery of open educational resource [8].

(2) Research on informal learning based on SNS. Research focused on the advantages and problems that the social networking sites to education, the use of open-source SNS social networking platform that developed for education to establish education-based social networking sites and so on. Jones, B. & Conceicao, S. C. O. proposed that SNS provides a platform for learners to participate. Through the SNS website, learners can establish new interpersonal relationships and promote personality development[9]. M.D. Roblyer et al. investigated the use of Facebook in college education and explored how to use SNS to facilitate student learning[10]. Sacide Güzin Mazman and others argue that social networking tools have the interactive, collaborative, active participation, information and resource sharing features, and applying social networks in education and instructional contexts has great potential [11]. John D. Ophus and others discuss the SNS website in the university curriculum application, and students hope take the Facebook and other social networking systems as a communications platform [12]. Kevin P. Brady builds a personalized educational social networking site with Ning, enabling learners to communicate and collaborate on a global platform[13].

It is a good chance for college students of informal learning and how to use SNS to construct the informal learning community and get better results of informal learning is the research focus. The research mainly from the following aspects based on SNS informal learning. (1) Construct the informal learning model of students based on SNS. Analyzes the main elements of the informal learning model based on SNS, on this basis, the theoretical model of informal learning is constructed. (2) A case study of informal learning community based on SNS. According to the theoretical model proposed in this paper, the open-source SNS program Elgg are used to develop the informal learning community.

3. Related concepts and theories

3.1. SNS

SNS is an internet services for social network. It extends people's social network in real life to the virtual network, and enlarges the social scope so that the users can establish a wider network of social relations. Boyd and Ellison defined SNS as a Web-based service, (1) allow individuals to establish a public or semi-public profile within a privileged system; (2) connect with other users and share the connection with them; (3) to browse or share connections made by other users within the system, and establish contact with others[14]. Another common explanation of SNS is the social networking site, which based on the idea of social network relationship system, such as Facebook, Myspace and other social networking sites. SNS also has an interpretation refers to the social network software, which is the next generation of personal network-based software that used distributed technology[15].

In this study, SNS is defined as *social network service*, which refers to Internet application services that help people to build social networks. It also refers to *social software* or *social networking sites*. For example, Blog, microblogging, podcasts, BBS, Wiki, Podcast, RSS, social networking sites, etc. The features of SNS: 1) User-centric. Users are both information publishers and creators. It to achieve the information exchange and sharing; 2) Provides a real network environment. It has a real social function to meet the real social needs of users to achieve virtual to authenticity; 3) SNS provides an open platform. SNS integrated a variety of technical applications to achieve a variety of forms of aggregation.

3.2. Informal Learning

The concept of informal learning is proposed in relation to formal learning. Danlelle Colard and Jens Bjnavold argue that formal learning refers to learning taking place in an organized and structured context, as learner's perspective it is intended and planned[16]. However, the concept of informal learning has not been uniformly defined, some representative points of view are: Marslck and Watkins argue that informal learning is usually conscious learning, but not rigorous organizational learning activities and it is closely related to life and work situations with learners themselves. The initiative of informal learning rests in the hands of learners which can happen in public institutions [1]. Merrill argues that informal learning is often purposeful, but not tightly organized. Informal learning may take place in the learning institutions, and the learning initiative is in the hands of the learner[17]. Table 1 provides a comparison of formal and informal learning from multiple perspectives.

Table 1. Comparison of formal and informal learning

	Formal learning	**Informal learning**
Place	Educational or Training Institutions place	Work, Family or Daily activities place

Flexibility	Structured learning objectives, learning time and learning support services	Flexible learning objectives, learning time and learning support services
Certification	Certification	Normally no authentication
Mode	Intentional	Most of the time it is not intentional
Content	Content fixed, pre-arranged	Content is not fixed, diversification of resources
Knowledge	Mainly Explicit Knowledge	Mainly Tacit Knowledge
Culture	The same cultural environment	Culture environment is quite different

In this study, the informal learning is defined as learning that occurs in everyday activities related to work, family, or leisure. It is unorganized or structured in terms of goals, time, and learning support services, but the results of learning are usually not supported by some authorities' confirmation or certification. Informal learning may be unintentional learning and may be unrecognizable, and it is also possible to learn intentionally that in order to promote their own knowledge or skills, so learners actively learn. At this time, the learners have strong motivation to study and learn more flexible time, and they can also seek diversified learning services themselves.

3.3. Related theories

(1) *Six-degree separation theory*. Six-degree separation theory proposed by the Harvard University psychology professor of Stanley Milgram. This theory explains the *weak connection* which is prevalent in society and it can play a very important role in interpersonal communication. It provides a good theoretical basis for the development of many social networking sites, and SNS site also proved the powerful power of this theory. The character of SNS is that the user's interpersonal network from the reality of social life to the network. Each individual's social circle is constantly enlarged and finally become a large network.

(2) *Connectivism theory*. Connectivism theory was first proposed by Canadian scholar George Simmons. The theory is that learning is the process of forming connections and creating networks. This process takes place in a complex and chaotic environment. Learning can take place not only within the learner, but also in the community, organization and equipment. The task of internal cognition processing is offloaded to the knowledge network[18]. Learning becomes an interrelated process, and each learner constitutes a knowledge network. Learner needs to learn new knowledge constantly and integrate it into the internal network.

(3) *Social constructivism theory*. The early Soviet famous psychologist Vygotsky's social constructivist learning is defined as learning is a unique high-level psychological structure and skills of people, and this function is not generated from the internal spontaneous, but only from people's collaboration activities and interpersonal relationships[19]. Informal learning is realized through the interaction with others. Learners promote informal learning in the process of communication and construction, and make informal learning develop in the direction of meaning construction.

(4) *Long tail theory*. The Long Tail concept first proposed by Chris Anderson[20]. It has evolved from the economic field, but has a close relationship with informal learning. At present, formal learning in schools and other formal educational institutions is the *head* of the long tail curve, while the informal learning methods such as network-based autonomous learning and collaborative learning are the *long tail* of learning. Informal learning is more of the problem in the context of non-conforming real-world situations. Different from formal learning, learners can acquire more useful knowledge and transform

them into their own invisible knowledge system when they spend less effort in informal learning.

4. The Construction of Informal Learning Model for Students Based on SNS

Based on the analysis of the important supportive role of SNS on informal learning in concept and service, this paper constructs the informal learning theory model of students based on SNS by combining the needs of informal learning community.

4.1. Informal Learning Environment Model for College Students Based on SNS

Based on the theory of Six-degree Separation, Connectivism theory, Social Constructivism, Long Tail theory, this paper analyzes the constituent elements of informal learning model based on SNS and first constructs the informal learning environment model of students is shown in Figure1.

Figure 1. SNS-based informal learning environment model for college students.

The informal learning environment model of college students based on SNS is constructed on the basis of learners, learning resources, learning service, social interaction and learning emotion. The model is divided into five layers. From the inside to the outside is learning social network layer, learning service layer, learning application layer, learning resource layer, learning mode innovation layer. In addition, learning emotions throughout the environment model. Because the students in the network informal learning in the cognitive and emotional is integrated, and strengthening the emotional participation of students is cornerstone to improve the effectiveness of learning the. The following is a description of the informal learning environment model for college students at different layers.

(1) *Learning social network layer*. Learning social network layer is the core of the model, and it refers to the various links that students in the SNS-based informal learning community. Construction of social network not only facilitate the acquisition of learning resources for students, but also connected the students together who at different times and regions. Learning social network plays an important role in cultivating students' collective wisdom and enhancing their creativity.

(2) *Learning service layer*. Learning service layer is the support factor of SNS-based informal learning community. It mainly provides some auxiliary tools for informal learning of students, and provides hardware and software conditions for establishing learning community. SNS provides a variety of learning services for informal learning of students which provide methodological and technical support. The basic learning services are Blog, RSS, IM, E-mail, Wiki, Group, Podcast, etc.

(3) *Learning application layer*. Learning application layer which based on the learning service layer is to build applications that students need for learning, and students can choose and gather the necessary services to play its overall role in order to form a variety of learning applications, such as content subscriptions, search engines, personal publishing, and resource sharing, etc. These learning applications meet the diverse learning needs of students and facilitate the sharing of knowledge in informal learning communities.

(4) *Learning resource layer*. On the basis of various learning services, students could share learning resources and knowledge, and will share their own audio, video, works with others, others can also give comments and modify comments. Students not only gain knowledge in their own community, but also access the wider learning resources through the community's search engine to the Internet.

(5) *Learning method Innovation layer*. In the SNS-based informal learning community, college students are knowledge sharers, creators and publishers. It can effectively improve the effect of autonomous learning of students, and promote active cooperative learning and improve the effect of research learning. SNS is not only a service provided to informal learning, but also a learning style and atmosphere. Students in the organization of learning resources, learning services and learning applications, while also organizing their own social relations.

(6) *Learning Emotion*. The achievement of students' learning goals and the guarantee of learning effects require the maintenance of community emotion and certain motivation. The informal learning community can meet the needs of students in emotional maintenance, communication and interaction.

4.2. Informal Learning Process Model of College Students Based on SNS

According to the theory of connectivism, social constructivism, from the point of view of students' informal learning process and integrate the learning service of SNS effectively, in order to construct the process model of SNS-based informal learning, as shown in figure 2.

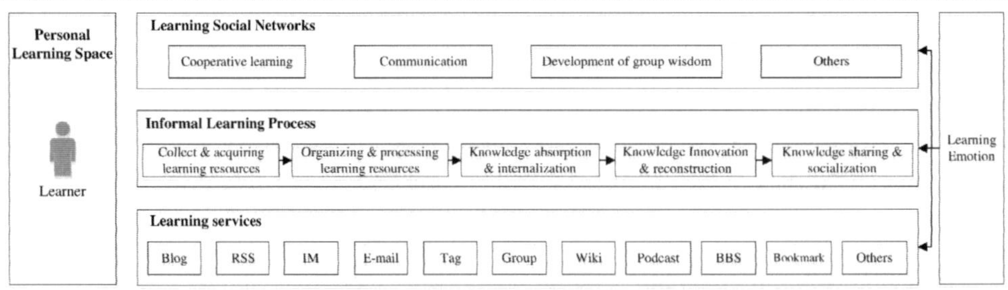

Figure 2. The process model of SNS-based informal learning.

(1) *Collect and acquiring learning resources*. Collecting and acquiring learning resources is the starting point of the informal learning process of students. Students can use the services provided by SNS to access learning resources, such as subscribe resources through RSS, as improve the efficiency

of informal learning from the learning community or group to share the resources released by others and the collaboration to obtain valuable learning resources, etc.

(2) *Organizing and processing learning resources.* Students identify learning resources to their own learning content from the network, and need to have the ability to identify information. When the students through various means to obtain learning resources, they need to organize learning resources, processing, storage and organization orderly, in order to facilitate search and update in the future.

(3) *Knowledge absorption and internalization.* The absorption and internalization of knowledge is the key process in the informal learning of students. Students will be extract and absorption knowledge from the outside to internalization of their own knowledge network. There are two ways, one is the students to the meaningful extraction and absorption of resources and internalization; the other is in the unconscious state to get the knowledge to complete the internalization of knowledge.

(4) *Knowledge Innovation and reconstruction.* The innovation and reconstruction of knowledge is the most important link in the process of students' informal learning. The goal of the students to absorb and internalize knowledge is to complete the innovation and reconstruction of knowledge. The learning service provided by SNS can effectively realize the innovation and reconstruction of knowledge and connect them to learning social network which can provide more opportunities from communication and learning for students to form their own unique knowledge system.

(5) *Knowledge sharing and socialization.* Knowledge sharing and socialization are the prerequisite for the informal learning of students. The learning resources of acquisition, collation and processing, and the internalization and innovation of knowledge are all based on the sharing and socialization of knowledge. Knowledge sharing and socialization require students to re-processing their own knowledge and through SNS can effectively achieve the preservation and dissemination and knowledge flow between students to achieve knowledge sharing and socialization.

5. Realization of College Students' Informal Learning Community Based on SNS

5.1. The Design of Informal Learning Community Based on SNS

Based on the analysis of the specific needs of the student to the informal learning community, and according to the informal learning model proposed in this study, the design of the informal learning community of functional modules as shown in Figure 3.

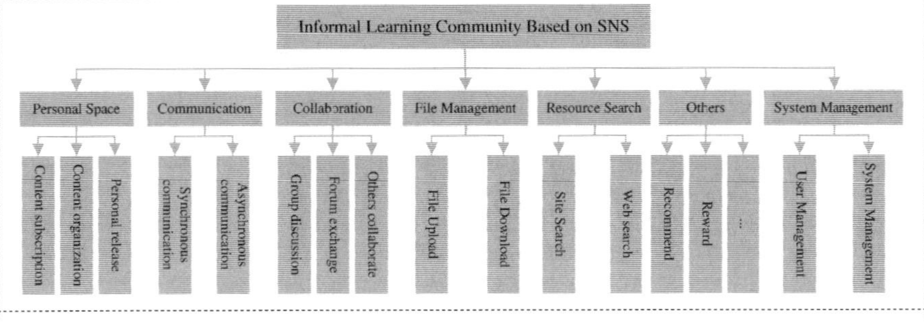

Figure 3. The Functional Structure of Informal Learning Community Based on SNS.

The informal learning community mainly includes personal space modules, include personal diary, learning materials, network resources, etc.; Communication module, including synchronous or asynchronous communication, online discussion, etc.; Group collaboration module, such as the establishment of forums, groups, etc.; File management module, include the PPT, Word, pictures, audio/video and other learning materials to upload and download; Resource search module, for search and access to learning resources, support for internal and external search, etc.; Other modules, including some useful features, such as initiating voting, learning resource recommendation, incentive mechanism; System management module, include the user login, user rights, related settings, etc.

5.2. The Development of Informal Learning Community Based on SNS

5.2.1. Develop Technology

The informal learning community in this study uses the open-source SNS program Elgg to build and secondary development. This open approach can reduce the difficult of development and development cycle is short. In addition, developers can take full control and have a greater flexibility, such as add new features. Elgg is a free open source social networking program designed specifically for education, and it achieve social networking around the blog. From the social point of view, Elgg is a social platform to interests as the core. It integrates a variety of applications and have a good scalability is very strong and also can install a lot of plug-ins.

5.2.2. Main function of the community

(1) system management module. This module is management and maintenance to the site, pages, activities, other users, plug-ins of the administrator, and can also track learners. As shown in Figure 4.

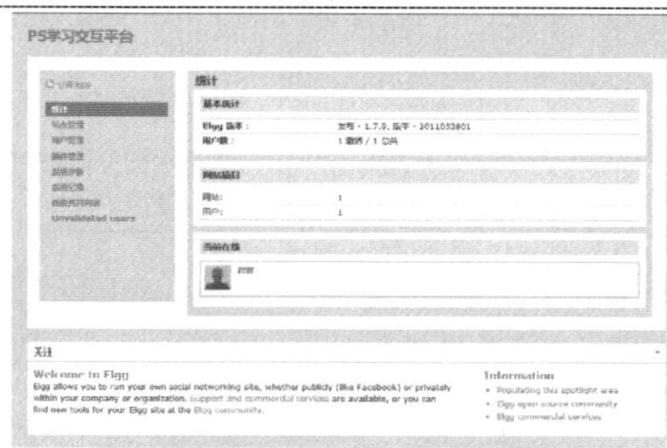

Figure 4. System Management.

(2) Communication module. If there are problems between the members can send site message. You can receive the message sent by others as long as online, as shown in Figure 5.

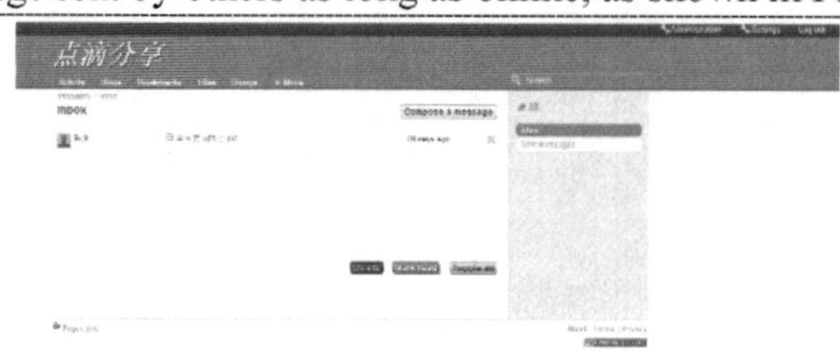

Figure 5. Site Information Management.

(3) Group collaboration module. Learners can create their own groups, and also can join other groups interested. Learners who enter a group initiate a topic and invite others to join the group forums and present different opinions. In the group, any participating learner can clearly see the state of other learners in the platform, as shown in Figure 6.

Figure 6. Group collaboration.

6. Conclusion

This study summarizes the related research of SNS and informal learning, and the concept of informal learning is defined and proposes an informal learning model for college students based on SNS. On this basis, the use of Elgg tools which is an open source SNS program for empirical research and established the informal learning community of students. Mainly include the following aspects in detail: (1) It discusses the supportive role of SNS in informal learning from concept and service. (2) In-depth study of the main elements of constructing the informal learning model based on SNS, including learners, resources, emotion, service and social interaction. On this basis, the theoretical model of informal learning is constructed and elaborated in detail. (3) Implementation of the practice according to the theoretical model and construction principle Proposed in this paper, and use open-source SNS program Elgg and related tools for development the informal learning community. There are some shortcomings in the study, in the follow-up study will expand the scope of the informal learning community and will continue to improve in the future study, such as expand the scope of the informal learning community and constantly improve the model, and hope to provide reference for further research.

Acknowledgments

The work was supported by the Jilin Province Educational Science "12th Five-Year Plan " (GH150295) and Northeast Electric Power University of Soft Science and Humanities Social Sciences special funding (201415)

References

[1] Jay Cross. Informal Learning the other 80%. http://www.internettime.com/Learning/The20%Other%2080%25.htm.
[2] Xiao Guangde, Gao Danyang. Study on Constructing Web-based Learning Environment by Using SNS Website Function. China Educational Technology, (04) (2010)121-124.
[3] Cheng gang, Xu jin, Yu Shengquan. The Latest Development of Learning Resource Standards and a Perspective of Learning Resource. Journal of Distance Education, (4) (2009)6-12.
[4] Gerald A. Straka. Informal learning: genealogy, concepts, antagonisms and questions. Univ. of Bremen, (2004)5-6.

[5] Elina Jokisalo, Antoni Riu. Learning in the era of Web2.0. eLearning Papers, (2009).
[6] Glen Bull, Ann Thompson, Mike Searson, Joe Garofalo, John Park, Carl Young, John Lee. Connecting Informal and Formal Learning Experiences in the Age of Participatory Media. Contemporary Issues in Technology and Teacher Education, (8)(2010)100-107.
[7] Stefano Ferretti, Silvia Mirri, Ludovico Antonio Muratori, Marco Roccetti, Paola Salomoni. E-learning 2.0: you are We-LCoME. Proceedings of the 2008 international cross-disciplinary conference on Web accessibility, (2008)116-125.
[8] Carmen Holotescu, Gabriela Grosseck. Learning to microblog and microblogging to learn. A case study on learning scenarios in a microblogging context. The 6th International Scientific Conference eLearning and Software for Education. Bucharest, (2010)15-16.
[9] Barbara E. Jones, Simone. C. O. Conceicao. Can social networking tools foster informal learning?. Paper presented at the 24th Annual conference on distance teaching and learning, (2008)1-5.
[10] M.D. Roblyer, Michelle McDaniel, Marsena Webb, James Herman, James Vince Witty. Findings on Face book in higher education: A comparison of college faculty and student uses and perceptions of social networking sites. Internet and Higher Education, 13(3) (2010)134-140.
[11] Sacide Güzin Mazman, Yasemin Koçak Usluel. Modeling educational usage of Facebook. Computers & Education, 55(2) (2010)444-453.
[12] John D. Ophus, Jason T. Abbitt. Exploring the Potential Perceptions of Social Networking Systems in University Courses. Journal of Online Learning and Teaching, 5(4) (2009)639-648.
[13] Kevin P. Brady, Lori B. Holcomb, Bethany V. Smith. The Use of Alternative Social Networking Sites in Higher Educational Settings: A Case Study of the E-Learning Benefits of Ning in Education. Journal of Interactive Online Learning, 9(2) (2010)151-170.
[14] Danah M. Boyd, Nieole B. Ellison. Social Network Sites: Definition, History, and Scholarship. Journal of Computer-mediated Communication, 13(l) (2007) 210-230.
[15] Baidu Encyclopedia. SNS entry.http://baike.baidu.com/view/8258.htm.
[16] Danlelle Colardyn, Jens Bjornavold. Validation of Formal, Non-formal and Informal Learning. European journal of Education, (1) (2004)68-88.
[17] Feng Li. The Construction of Social Software-based Informal Learning Environment Model. Southwest University, (2009).
[18] Wang Yuoumei, Zhu zhiting. From Connectionism to Connectivism, New Orientation of Learning Theory. China Educational Technology, (3) (2006)5-8.
[19] Sang Xinmin. Learning sciences and learning technologies: The Cultivation of College Students' Learning Ability in the Information Age. Beijing: Higher Education Press, (2004).
[20] Chris Anderson.The long tail. Beijing: China CITIC Press, (2006).

Development of Fe_3O_4/ZrO_2 Composite Powered by Nanographene Platelets (NGP) for Degradation of Water Pollutants via Photo- and Sonocatalysis

Yogi Kristianto[1,2], Ardiansyah Taufik[1,2], and Rosari Saleh[1,2*]
[1] Departemen Fisika, Fakultas MIPA, Universitas Indonesia, Kampus UI Depok, Depok 16424, Indonesia

[2] Integrated Laboratory of Energy and Environment, Fakultas MIPA, Universitas Indonesia, Kampus UI Depok, Depok 16424, Indonesia

[*]Corresponding author: rosari.saleh@ui.ac.id, rosari.saleh@gmail.com

Abstract. In this study, a series of Fe_3O_4/ZrO_2/nanographene platelets (NGP) composite, with various weight percent (wt%) of NGP (5%, 10% and 15%), were prepared successfully using ultrasonic-assisted followed by simple hydrothermal method. Their physicochemical properties were fairly characterized by X-ray diffraction, fourier transform infrared and thermal gravimetric analysis. Furthermore, their catalytic activities were investigated toward anionic congo red (CR) and cationic methylene blue (MB) as models of organic pollutant under ultraviolet (UV) and ultrasonic (US) irradiation, respectively. The experimental results showed that the incorporation of NGP in Fe_3O_4/ZrO_2 composite improved its efficiency in degrading CR and MB and became maximum at 10wt% of NGP. In addition, the role of active radicals involved in catalytic activities were discussed.

1. Introduction

Recently, semiconductor-based materials have been attracted great attention due to its application in various fields, such as hydrogen production, water purification and reduction of heavy metals [1-3]. Various types of semiconductor metal oxides such as TiO_2, ZnO, SnO_2 and ZrO_2 have been studied for their catalytic activity [4-7]. Among various metal oxides, zirconium dioxide (ZrO_2) play an important role due to several unique properties, such as excellent physical and chemical properties, low cost, high stability, chemical inertness, lack of toxicity and high specific surface area [8-10] that make it versatile for several applications, especially in water purification. Peng et al. [7] and Poungchan et al. [8] reported the photocatalytic degradation of organic pollutants in wastewater by ZrO_2. The reports demonstrated that ZrO_2 could degrade organic pollutants. Unfortunately, applications of ZrO_2 in water purification is hindered because of high recombination of photoinduced electron-hole pairs [11], which causes low catalytic performance, and difficult to recover from wastewater solution for recycle process, which causes low efficiency when applied in large-scale.

To solve mentioned problems many researchers used magnetite and carbon-based material graphene as support materials to make composites [8,12]. Magnetite material, such as Fe_3O_4, introduced into the composite in order to provide magnetic properties so it would make the composite become more easily recovered from the solution by external magnetic field [13]. Furthermore, the Fe_3O_4 is conductive which could improve separation of electron-hole pairs [14]. The carbon-based material graphene has superior properties that make it greatly note, such as large specific surface area and good mobility of charge carriers, as well as good electrical and thermal conductivity [15-17]. Based on these properties, the role of carbon-based material graphene in the composite is an electron transport channel, which could inhibit electron-hole recombination, enhance separation of photogenerated electron-hole pairs and increase the number of active sites.

Over the last decade, photo- and sonocatalytic using semiconductor metal oxide as photocatalyst has caused attention in the treatment of organic pollutant. Because of many advantages, such as eco-friendly, did not cause any secondary pollutant, high efficiency and simple operation, they are considered as an efficient technique for degrading organic pollutants [18,19].

Based on these issues, in this paper, we attempt to synthesize Fe_3O_4/ZrO_2 composite through ultrasonic-assisted method, and then combined with NGP by simple hydrothermal methods as a superior photocatalyst for degradation of water pollutants. The ability of Fe_3O_4/ZrO_2/NGP with various wt% of NGP photocatalyst in degrading water pollutants was tested via photo- and sonocatalytic degradation of MB and CR under UV light and ultrasonic irradiation. In addition, influence of contact time, pH of solution, catalyst stability and effect of reactive species on the catalytic activites were also investigated and the results were discussed.

2. Experimental

Materials: All the chemical reagents were of an analytical grade and were used without further purification. Iron(II) sulfate heptahydrate ($FeSO_4.7H_2O$, 99%), zirconium(IV) chloride ($ZrCl_4$) and sodium hydroxide (NaOH) were purchased from Merck. NGP was purchased from Angstron Materials.

Preparation of samples: The Fe_3O_4 and ZrO_2 nanoparticles were synthesized using the same method as described previously [20,21]. The Fe_3O_4/ZrO_2 composites were synthesized using the ultrasonic-assisted method. Briefly, ZrO_2 nanoparticles were mixed with Fe_3O_4 in distilled water and ethanol mixture. Then, the mixed solution was given ultrasonic for 2 hours followed by centrifugation to obtain the product on its precipitation. The obtained product was dried in vacuum condition at 80°C for 12 hours to obtain Fe_3O_4/ZrO_2 composites, which labelled as FZ. The molar ratio of Fe_3O_4 to ZrO_2 is set 0.5:1. The FZ/NGP composites were synthesized using the simple hydrothermal method. Briefly, the NGP powder was dispersed in distilled water and an ethanol solution via ultrasonic treatment for 2 hours, which was followed by addition of the prepared FZ composites into NGP solution. The mixed solution was stirred for 1 hour. Then, the suspension was heated at 120°C for 3 hours to effect deposition on to the graphene sheets. The product was collected through centrifugation and was dried in a vacuum at 70°C for 12 hours in order for FZ/NGP composites to be obtained.

Characterization: X-ray Diffraction (XRD) performed on a Rigaku Miniflex 600 diffractometer with Cu K-α radiation (λ =1,5406 Å) was used to obtain the crystal structure of samples. The infrared absorption spectra of samples were recorded using a fourier transform infrared (FT-IR). The thermal stability of samples was analyzed using a thermal gravimetric analysis (TGA).

Photo-, and sonocatalytic activity: MB and CR were used as models organic pollutant. The photo- and sonocatalytic activity of samples was investigated via degradation of MB and CR in aqueous solution with the initial concentration of 20 mg/L at room temperature with UV light / ultrasonic, respectively. Firstly, the catalyst was dissolved into MB and CR solution, separately. The solution was continuously stirred in the dark for 30 minutes to ensure that an equilibrium was reached between the adsorption and desorption dyes molecules on the catalyst surface. Subsequently, the solution was exposed with UV light / ultrasonic, respectively. A small amount of the sample water was collected through centrifugation every 15 min to measure the MB concentration with UV-vis spectrophotometer. Two lamps of 40 W UV-C were used as UV light source and commercially ultrasonic bath which operated at fixed frequency and power of 40 kHz and 150 W was used as ultrasonic source.

Scavenger experiment: In order to determine the species involved in photo- and sonocatalytic activities, different radical scavengers were added to the solutions of MB and CR, separately. Di-ammonium oxalate, sodium sulfate, and tert-butyl alcohol (TBA) were used as trapping agents for holes, electrons, and hydroxyl radical, respectively. The same measurements as described above were used.

3. Result and Discussion

Figure 1 (a) shows the XRD spectra of the FZ/NGP with various NGP wt%. For comparison, the XRD spectra of Fe_3O_4, ZrO_2 and NGP were also plotted. Peaks at 2θ = 30.14°, 35.49°, 43.28°, 53,76°, 57.20° and 62.83 were attributed to the cubic spinel phase of the Fe_3O_4, and peaks at 2θ = 30.18°, 35.03°, 50.36°, and 59.94° were attributed to the tetragonal phase of the ZrO_2. In addition, graphitic-like structure of the NGP was detected at 2θ = 26.49°, indicating the presence of NGP in the FZ composites. Furthermore, it can be seen also that the intensity of the NGP, increases as increasing wt% of NGP, indicating there is more NGP content in the FZ/NGP composite. No undesired peaks were detected by XRD, suggesting that the as-prepared samples have good purity.

Evaluation of the thermal stability of FZ and FZ/NGP composites with various NGP wt% were carried out by means of TGA under argon atmosphere, from ambient temperature until 1000°C, and the results are displayed in fig. 1 (b). The TGA plot showed the thermal degradation of the samples passed through one weight loss stages for FZ and two stages for FZ/NGP. The first weight loss, which took place below 150°C was attributed to the removal of adsorbed water [22]. The second weight loss in the region above 670°C corresponded to the combustion process of NGP in the FZ/NGP composite [23]. The percentage loss of the samples in region above 670°C could be identified as the total amount of NGP in the sample.

Figure 2 (a) represents FT-IR spectra of NGP, FZ and FZ/10% NGP. The broad band around 3400 cm^{-1} and small band around 1400 cm^{-1} correspond to the physical adsorbed water on the sample [24]. The characteristic features in the FT-IR spectra of NGP are the absorption bands corresponding to C=C and carboxyl groups vibration in range 1500 - 1630 cm^{-1} [25]. The strong bands at 587 cm^{-1} were assigned to Fe–O in the Fe_3O_4 phase [26]. The absorption band below 500 cm^{-1} was attributed to the symmetric streching vibrations of Zr–O–Zr [27].

The photo- and sonocatalytic of MB and CR were conducted to evaluate the ability of FZ and FZ/NGP with various NGP wt%. As we can see in fig. 2 (b) and fig. 3 (a), the incorporation of NGP could improve catalytic performance of FZ composite for degrading MB and CR, respectively. It is due to the superior properties of carbon-based material, such as large specific surface area and could inhibit recombination rate of electron-hole pairs, leading to increased catalytic performance [15-17].

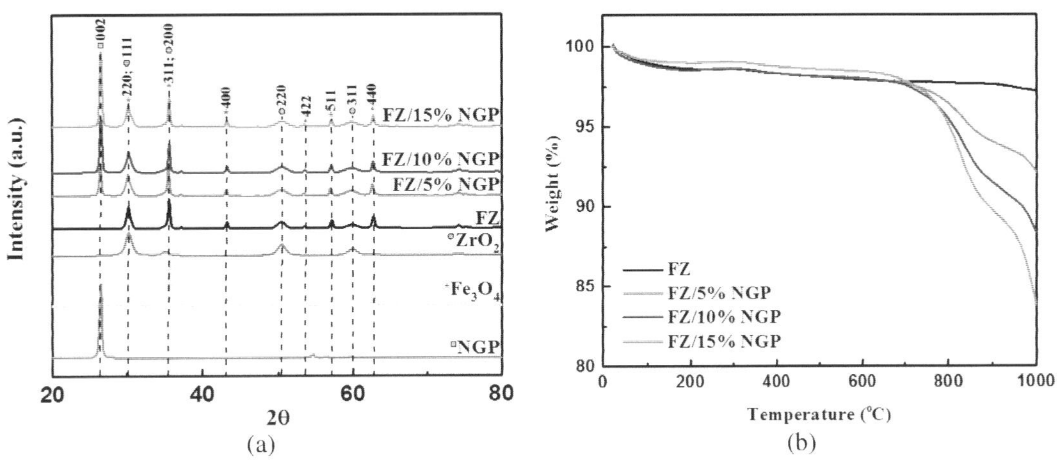

Figure 1. (a) XRD patterns of all prepared samples. (b) TGA plot of FZ and FZ/NGP with various wt% of NGP.

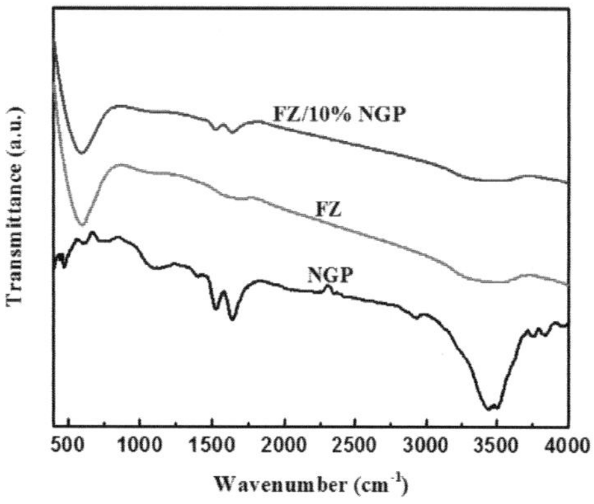

(a)

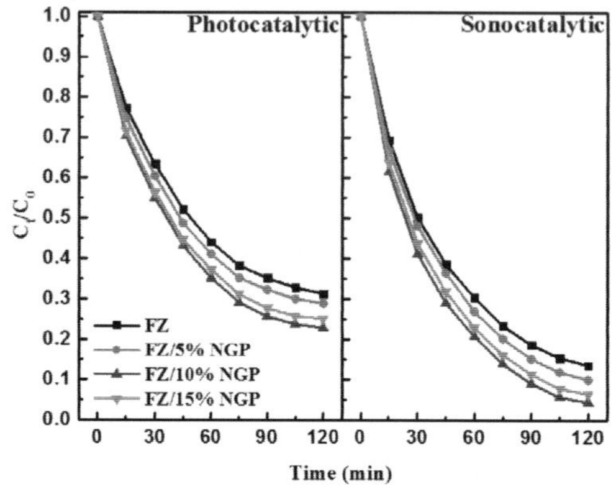

(b)

Figure 2. (a) FT-IR spectra of NGP, FZ and FZ/10% NGP. (b) Catalytic activities of FZ and FZ/NG various NGP wt% in degrading MB.

The catalytic performance of the sample, increases as increasing NGP wt% until 10%, but at 15% it decreases. It might be due to excessive NGP content can causes agglomeration of NGP, causes a stacking and decrease of interlayer distance, leading to low electron conductivity and the light-harvesting competition between catalyst and NGP was intensified, leading to decrease in catalytic performance [28]. From the figure, it can be seen also that sonocatalytic could presents better catalytic performance than photocatalytic.

As an important variable in catalytic reactions, the effect of solution pH on the degradation reaction

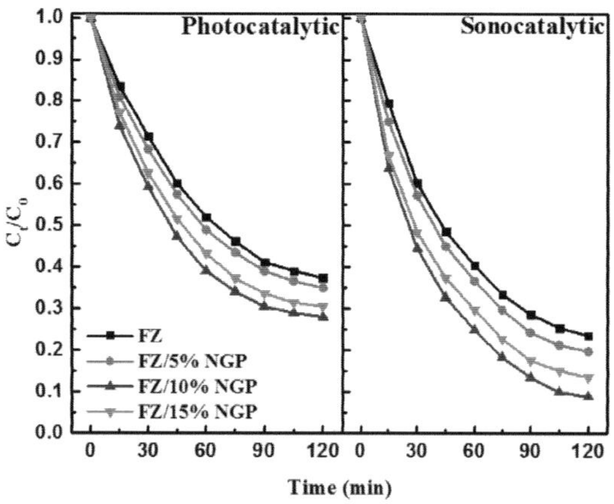

(a)

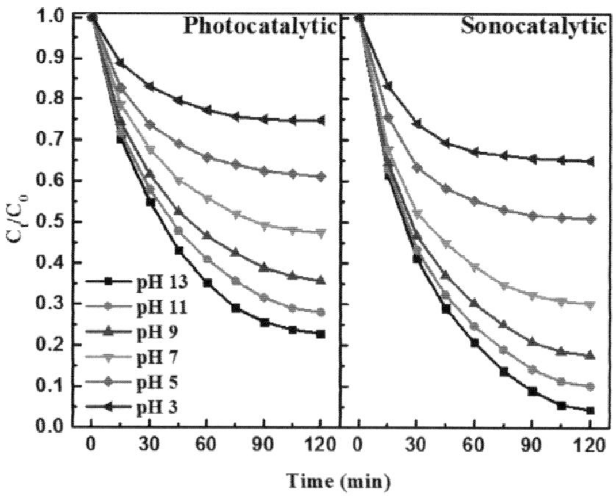

(b)

Figure 3. (a) Catalytic activities of FZ and FZ/NGP with various NGP wt% in degrading CR. (b) Effect of pH solution of FZ/10% NGP in degrading MB.

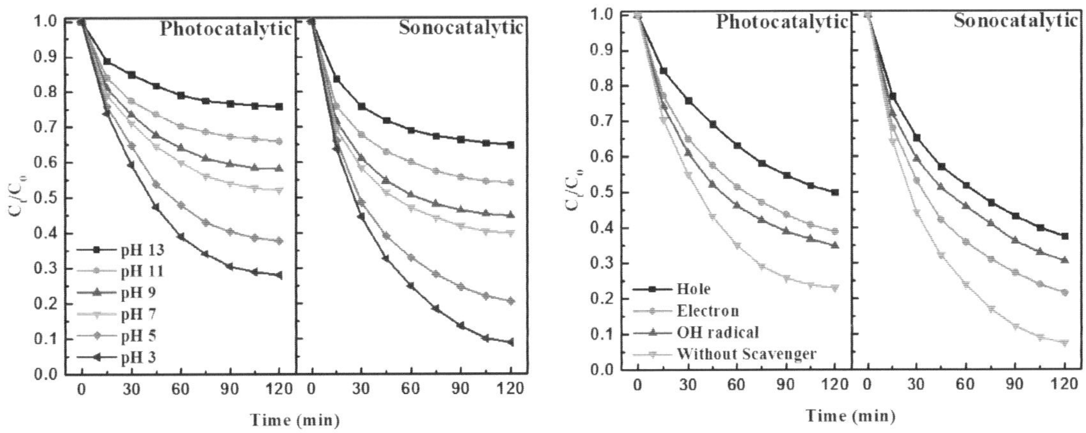

(a) (b)

Figure 4. (a) Effect of pH solution of FZ/10% NGP in degrading CR. (b) Effect of main active species of FZ/10% NGP in degrading MB (pH 13).

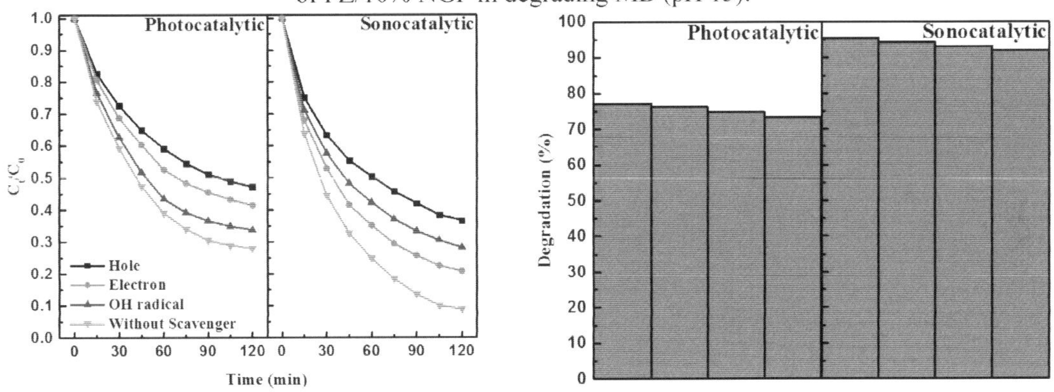

(a) (b)

Figure 5. (a) Effect of main active species of FZ/10% NGP in degrading CR (pH 3). (b) Reusability of FZ/10% NGP in degrading MB (pH 13).

was investigated by maintaining pH of solution at 3, 5, 7, 9, 11 and 13. The results of pH effect on degradation reaction are plotted in fig. 3 (b) and fig. 4 (a). At higher pH, electrostatic attractions between the negative catalyst surface and MB cations lead to strong adsorption and enhancing degradation rate. Meanwhile, in anionic CR, there would be electrostatic repulsions between catalyst and CR, resulting in negligible adsorption and lower degradation rate. In the case of lower pH, the catalyst surface will be positively charged and hence the MB experience electrostatic repulsions, while the CR electrostatic attractions [29].

In order to investigate the role of the reactive radicals that involve in photo- and sonocatalytic activity, three typical chemicals, Di-ammonium oxalate (a quencher of hole), sodium sulfate (a quencher of electron), and TBA (a quencher of hydroxyl radical) were adopted as the scavengers at the optimum condition of both dyes, respectively. The result was shown in fig. 4 (b) and fig. 5 (a). The catalytic performance of MB and CR were obviously suppressed in the presence of Di-ammonium oxalate. In addition, sodium sulfate and TBA slight impeded the degradation of both dyes. It can be concluded that holes are the main active species for the degradation of MB and CR.

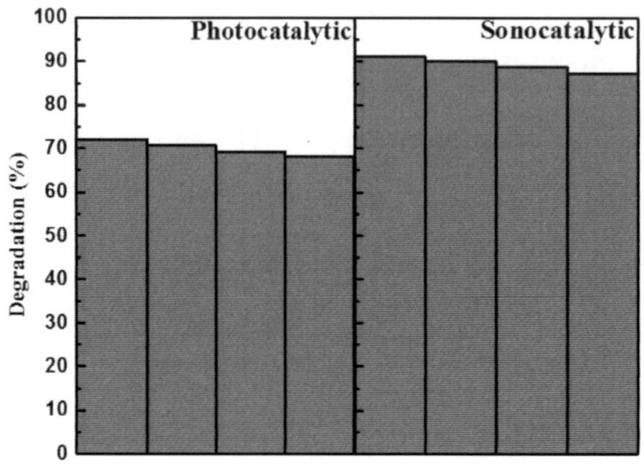

Figure 6. Reusability of FZ/10% NGP in degrading CR (pH 3).

Easily recoverable and good reusability are superior advantages for their practical application in large scale. Therefore, it is quite necessary to investigate the reusability of the catalyst. Four cycles of MB and CR catalytic activities were performed with the same FZ/10% NGP to investigate its reusability, also at optimum condition. At the end of each run, the catalysts were separated by external magnetic field and reused in the next run for fresh MB and CR, respectively. Results are shown in the fig. 5 (b) and fig. 6. It is clear that the catalyst has good reusability and retains its activity the successive degradation reaction.

4. Conclusions

In summary, we reported a series of Fe_3O_4/ZrO_2/NGP composites with NGP wt% as photocatalyst for for degradation of water pollutants via photo- and sonocatalytic that characterized by XRD, TGA and FT-IR. Catalytic activities of prepared samples were investigated by degradation of MB and CR in aqueous solution. Compared to photocatalytic, sonocatalytic could presents better catalytic performance. The catalytic performance greatly increases with weight percent of NGP up to 15% and became maximum at 10%. Based on the effects of solution pH, cationic MB could be effectively degraded at alkaline condition, while anionic CR at acid condition. In addition, from scavenger experiment, it was concluded that holes have vital role in photo- and sonocatalytic degradation of both dyes. In the last, the composites have good reusability for successive four runs.

5. References

[1] K. Yu, C. Zhang, Y. Chang, Y. Feng, Z. Yang, T. Yang, L. L. Lou and S. Liu 2017 *Applied Catalysis B: Environmental* **200** 514-5200
[2] F. Chen, D. Li, B. Luo, M. Chen and W. Shi 2017 *Journal of Alloys and Compounds* **694** 193-200.
[3] I. Ghiloufi, J. E. Ghoul, A. Modwi and L. E. Mir 2016 *Materials Science in Semiconductor Processing* **42** 102-106.
[4] H. Aydin, K. Alireza, K. Semra, K. Canan and G. Peyman 2016 *Ultrasonics Sonochemistry* dx.doi.org/10.1016/j.ultsonch.2016.09.027
[5] K. Thenozhi and A.S.S. Stanly 2016 *Journal of Molecular Structure* **1125** 358-365
[6] W. Junxia, L. Can, L. Xinchuan, W. Yongqian, Z. Zhengxin and M. Dawei 2017 *Materials and Design* **115** 103-111
[7] Z.A. Sahar and S.N. Masoud 2016 *Journal of Molecular Liquids* **216** 545-551
[8] G. Poungchan, B. Ksapabutr and M. Panapoy 2016 *Materials and Design* **89** 137-145
[9] H.R. Pouretedal, Z. Tofangsazi and M.H. Keshavarz 2012 *Journal of Alloys and Compounds*

513 359-364
[10] H.P. Peng, R.P. Liang, L. Zhang, J.D. Qiu 2011 *Electrochimica Acta* **56** 4231-4236
[11] W.A. William, O.B.O. Samuel, K.S. Sudheesh, S.A. Eric and P.G. Pooami 2016 *Progress in Natural Science: Materials International* **26** 354-361
[12] J. Xiaohua, D. Rongrong, L. Dandan, H. Song, W. Xiangyang and S. Haojie 2017 *Applied Surface Science* **392** 268-276
[13] X.F. Bian, K.Q. Hong, L.Q. Liu and M.X. Xu 2013 *Applied Surface Science* **280** 349-353
[14] W. Wu, C.Z. Jiang and V.A.L. Roy 2015 *Nanoscale* **7** 38-58
[15] M. Aleksandrzak, P. Adamski, W. Kulkulka, B. Zielinska and E. Mijowska 2015 *Applied Surface Science* **331** 193-199
[16] S.M. Torres, L.M.P. Martinez, J.L. Figueiredo, J.L. Faria and A.M. Silva 2013 *Applied Surface Science* **275** 361-368
[17] B.F. Machado and P. Serp 2012 *Catalyst Science Technology* **2** 54-75
[18] A. Rizwan, A. Zaki, U.K. Asad, R.M. Naila, A. Muhammad and K. Joenghwan 2016 *Journal of Environmental Chemical Engineering* **4** 4143-4164
[19] A. Khataee, A. Karimi, S.A. Oskoui, R.D. C. Soltani, Y. Hanifepour, B. Soltani and S.W. Joo 2015 *Ultrasonics Sonochemistry* **22** 371-381
[20] A. Taufik, I.K. Susanto and R. Saleh 2015 *Materials Science Forum* **827** 37-42
[21] Y. Kristianto, A. Taufik and R. Saleh 2016 *Journal of Physics: Conference Series* **776** 012040
[22] K.M. Rahulan, G. Vinitha, L.D. Stephen and C. C. Kanakam 2013 *Ceramics International* **39** 5281-5286
[23] A. Taufik and R. Saleh, *in The 3rd International Conference on Advanced Materials Science and Technology-2015, AIP Conf. Proc. 1725*, edited by Sutikno et al. (American Institute of Physics, Melvile, NY, 2016), 020089.
[24] M. S. Gohari and A. H. Yangjeh 2015 *Cemarics International* **141** 1467-1476
[25] A. R. Unnithan, H. P. Chan and C. S. Kim 2016 *Composite Part B: Engineering* **90** 503-511
[26] H. Zhang, F. Huang, D. L. Liu and P. shi 2015 *Chinese Chemical Letters* **26** 1137-1143
[27] P. Bansal, G. R. Chaudhary and S. K. Mehta 2015 *Chemical Engineering Journal* **280** 475-485
[28] T. W. Kim, M. Park, H. Y. Kim and S. J. Park 2016 *Journal of Solid State Chemistry* **239** 91-89
[29] B. Pratibha, R. C. Ganga and S. K. Mehta 2015 *Chemical Engineering Journal* **280** 475-485

IOP Conf. Series: Journal of Physics: Conf. Series **820** (2017) 012027 doi:10.1088/1742-6596/820/1/012027

Comparison of Photocatalytic Performance of Different Types of Graphene in Fe_3O_4/SnO_2 Composites

Valentinus Paramarta[1,2], Ardiansyah Taufik[1,2], Rosari Saleh[1,2,*]
[1]Departemen Fisika, Fakultas MIPA-Universitas Indonesia, 16424 Depok, Indonesia
[2]Integrated Laboratory of Energy and Environment, Fakultas MIPA-Universitas Indonesia, 16424 Depok, Indonesia

E-mail: rosari.saleh@gmail.com

Abstract. We have reported the role of annealing temperature Fe_3O_4/SnO_2 nanocomposites as a photocatalyst for remove methylene blue (MB) dye from aqueous solution. However, how to enhanced the degradation performance of Fe_3O_4/SnO_2 nanocomposites is important to its photocatalytic application. Therefore, in this work Fe_3O_4/SnO_2 nanocomposites was combined with two different types of graphene materials (NGP and grahene) to improve the photocatalytic performance for remove methylene blue (MB) dye. Fe_3O_4/SnO_2/NGP and Fe_3O_4/SnO_2/graphene have been successfully synthesized by co-precipitation method. The influence of two types graphene on Fe_3O_4/SnO_2 nanocomposites properties were systematically investigated by means of X-ray diffraction (XRD), Fourier transform infrared (FTIR) spectroscopy and Thermal gravimetric analysis (TGA). Degradation of methylene Blue (MB) in aqueous solution was studied in detail under photocatalytic process. Effect of catalyst dosage (0.1-0.4 g/L) and scavengers on dye degradation were carried out to check the efficiency of photocatalyst. The results indicated Fe_3O_4/SnO_2/graphene displayed higher photocatalytic activity than other catalyst. The reusability tests have also been done to ensure the stability of the used photocatalyst.

1. Introduction

In recent years heterogeneous photocatalysis has attracted much attention to developed sustainable technologies for energy production and storage, green chemical synthesis, and water treatments [1-2]. Semiconductor nanoparticles are useful as an effective technique to eliminate pollutants from air and wastewater due to their unique optical and electronic properties [3]. Among various semiconductor nanoparticles, such as SiC, TiO_2, ZnO, Fe_2O_3, WO_3, SnO_2 nanoparticles have attracted considerable attention because of their high physicochemical stability, low cost, lack of toxicity, easy production, and good photoactivity [4-6]. Because of their several advantages, SnO_2 potentially to be ideal catalyst. However, as a photocatalyst, SnO_2 has several fundamental issue for photocatalytic application such as: (1) the difficulty process of splitting the catalyst material from dye waste after photocatalytic process [7]; (2) The recombination rate of electron and hole pairs are too rapid [8].

To overcome the problem, one strategy is to make magnetically separable photocatalyst for recovery and reuse of SnO_2 nanoparticle [9]. Combining SnO_2 nanoparticle with magnetic material such as Fe_3O_4 was used to enhance separation properties of the photocatalyst from the treated water [10]. Furthermore, hybridizing SnO_2 nanoparticle with graphene materials is a good method to enhance the photocatalytic activity of SnO_2 for degrade organic dyes by enhancing the electron mobility and also inhibit the recombination of photogenerated electron and hole pairs. Due to graphene properties such as excellent electronic conductivity, mechanical properties and high surface area [11]. Therefore, it is believed that, combining SnO_2, Fe_3O_4 and graphene materials would exhibit excellent photocatalytic performance with highly effectively recovery by magnetic separation technique.

In this study, Fe_3O_4/SnO_2 nanoparticle will be combined with two different types of graphene materials: 1) nanographene platelets (NGP); 2) graphene that are applied as photocatalyst to degrade methylene blue (MB) dye. To hybridize with graphene materials, Fe_3O_4/SnO_2/NGP and Fe_3O_4/SnO_2/graphene nanocomposites were prepared using co-precipitation method. Photocatalyst powders are characterized by using X-ray diffraction (XRD), Fourier transform infrared (FTIR) spectroscopy and Thermal gravimetric analysis (TGA). In the present study, the photocatalytic activity of the Fe_3O_4/SnO_2 nanocomposites with two different types of graphene materials were analyzed using methylene blue as organic pollutant model in aqueous solution under UV light irradiation. As expected, the result shows Fe_3O_4/SnO_2 combined with graphene materials exhibit better photocatalytic performance than Fe_3O_4/SnO_2 for degradation of methylene blue (MB).

2. Experimental

2.1. Chemicals
All reagents used were analytical grade and were used without further purification. Anhydrous tin chloride ($SnCl_2$), iron (II) sulfate heptahydrate ($FeSO_4 \cdot 7H_2O$), sodium hydroxide (NaOH), methylene blue (MB), ethanol, ethylene glycol (EG) were purchased from Merck (Kenilworth, NJ, USA). Graphene and nanographene platelets (NGP) were bought from Angstron Material.

2.2. Preparation of SnO_2 nanoparticle and Fe_3O_4/SnO_2 nanocomposites
The Fe_3O_4 nanoparticles were synthesized using the same method used in our previous study [12]. The SnO_2 nanoparticles were synthesized using a modification of the method reported by Yue Li and co-workers [13]. First, $SnCl_2$ was dissolved in a mixture of ethanol and aqueous solution, which was then added into the NaOH solutions using magnetic stirring. Then, the mixed solutions were heated at 180 °C for 3 h, then chilled to room temperature. The precipitate was obtained by centrifugation and washed using aqueous solutions and ethanol several times. The precipitated particles were dried under vacuum at 80°C, and SnO_2 particles were obtained by calcination for 3 h at 700 °C.

The Fe_3O_4/SnO_2 nanocomposites were synthesized using the sol-gel method. First, SnO_2 nanoparticles were mixed with Fe_3O_4 in a mixture of ethanol and aqueous solution. Each mixture was ultrasonicated for 2 h and then centrifuged to obtain the precipitate. The resulting product was then dried under vacuum at 80 °C for 12 h to obtain Fe_3O_4/SnO_2 nanocomposites with the molar ratio 1:2.

2.3. Preparation of Fe_3O_4/SnO_2/NGP and Fe_3O_4/SnO_2/graphene composites
Fe_3O_4/SnO_2/graphene is synthesized using the hydrothermal method. First, graphene is dissolved into DI water and ethanol through ultrasonic treatment for 2 hours, then Fe_3O_4/SnO_2 nanocomposites is poured into the solution and stirred magnetically. The mixed solution is then heated at 120 °C for 3 hours. The result of solution is then centrifuged and dried at 70 °C under vacuum condition. The same method is also applied to synthesize Fe_3O_4/SnO_2/NGP.

2.4. Characterization
The sample was characterized by XRD using a Rigaku Miniflex 600 (Rigaku, Tokyo, Japan) with a Cu K-α radiation source (λ =1.5406 Å), the spectrum of infrared absorption from the samples is obtained using Shimadzu FTIR spectrophotometer in the range of 400-4000 cm^{-1} and thermal gravimetric analysis (TGA) of the sample is obtain using Rigaku TG8121 with the temperature from 27°C until 900°C.

2.5. Photocatalytic experiments

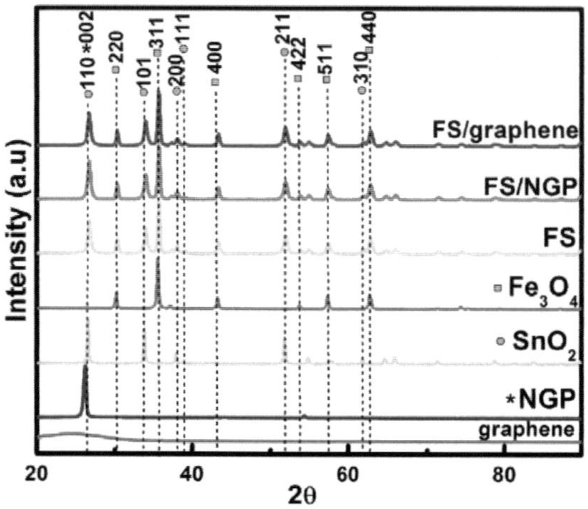

FIGURE 1. XRD pattern of Fe_3O_4/SnO_2 nanocomposite, Fe_3O_4/SnO_2/NGP and Fe_3O_4/SnO_2/graphene composites.

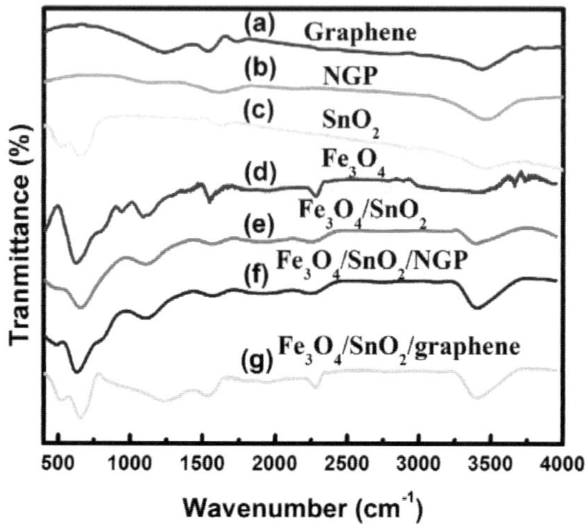

FIGURE 2. FTIR spectra of (e) Fe_3O_4/SnO_2 nanocomposite, (f) Fe_3O_4/SnO_2/NGP and (g) Fe_3O_4/SnO_2/graphene composites.

The photocatalytic experiment was done by mixing the sample into 100mL methylene blue (MB) solutions used as pollutant model with concentrations 20mg/L respectively, and pH solutions adjusted NaOH. The solution was allowed to stand in stirring state for adsorption and desorption balance. The photocatalytic test, the solution was given UV light irradiation with power 40W and wavelength of 320-400 nm. The solution was given UV light irradiation for 2 hours. Every 15 minutes span the concentration of MB solution was analyzed using UV-vis spectrometer. The maximum degradation were calculated using the following equation:

$$\text{Maximum degradation (\%)} = \frac{(C_t - C_0)}{C_0} x\, 100\% \qquad (1)$$

where C_0 is the initial concentration of MB dye solution (mg/L), C_t is the concentration of dye at certain irradiation time (mg/L).

2.6. Scavenger experiments

To determine the most influential species in sonocatalytic activity, it was given the difference radical-scavanger in methylene blue (MB) solution. Ammonium oxalate, sodium sulfate, and tert-butyl alcohol (TBA) were used each for scavenger hole, electron and hydroxyl radical. The measurement done in the same way as described above.

3. Result

The X-ray diffraction (XRD) patterns of the Fe_3O_4/SnO_2 nanocomposite, Fe_3O_4/SnO_2/NGP and Fe_3O_4/SnO_2/graphene composites which labeled as FS, FS/NGP and FS/graphene are shown in Fig. 1, respectively. The diffraction pattern of Fe_3O_4 sample observed at the value 2θ = 30.14°, 35.49°, 43.28°, 57.20°, 62.83° and 74.8° which shows each area (220), (311), (400), (511), (440) and (533) of the cubic spinel structure. Diffraction pattern of SnO_2 nanoparticle sample observed at value 2θ = 26,5°, 33.8°, 38°, 39°, 51.8°, 54.8°, 58°, 62°, 64.7°, 65.8°, 71.2°, 78.2°, 81.2°, 83.7° which shows the existence of areas (110), (101), (200), (111), (211), (220), (002), (310), (112), (301), (202), (321), (400), (222) of tetragonal structure of SnO_2 nanoparticle. The XRD pattern of Fe_3O_4/SnO_2 nanocomposites shows the existence of cubic spinel phase of Fe_3O_4, followed by the addition of tetragonal SnO_2 nanocomposites. The peak at 2θ of 26.0° indicates (002) plane of the NGP structure. For graphene, there is only a broad peak at around 2θ of 25°, which indicates the structure of

IOP Conf. Series: Journal of Physics: Conf. Series **820** (2017) 012027 doi:10.1088/1742-6596/820/1/012027

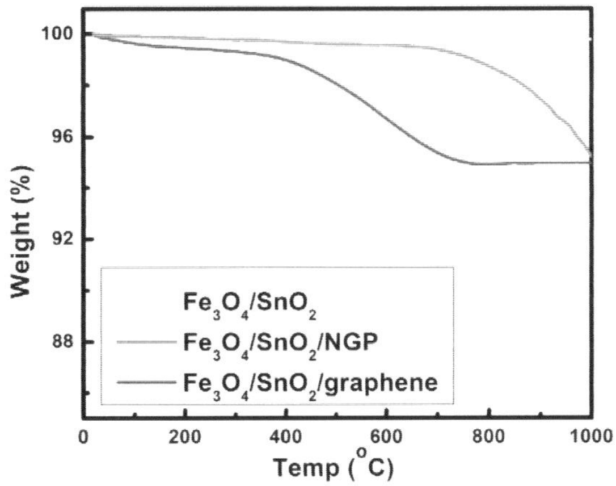

FIGURE 3. TGA curve of Fe_3O_4/SnO_2 nanocomposite, Fe_3O_4/SnO_2/NGP and Fe_3O_4/SnO_2/graphene composites.

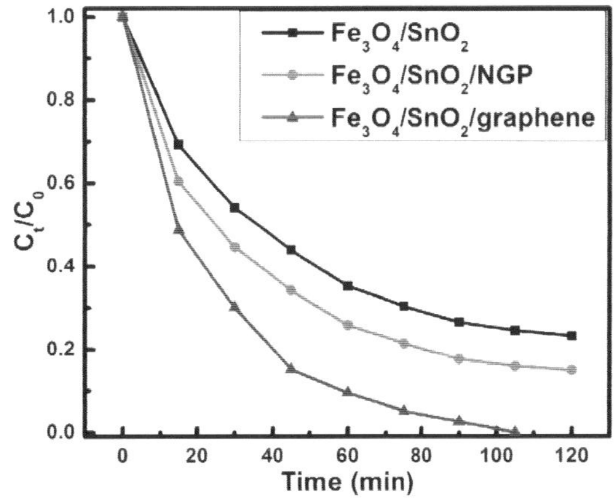

FIGURE 4. Photocatalytic degradation of Fe_3O_4/SnO_2 nanocomposite, Fe_3O_4/SnO_2/NGP and Fe_3O_4/SnO_2/graphene composites.

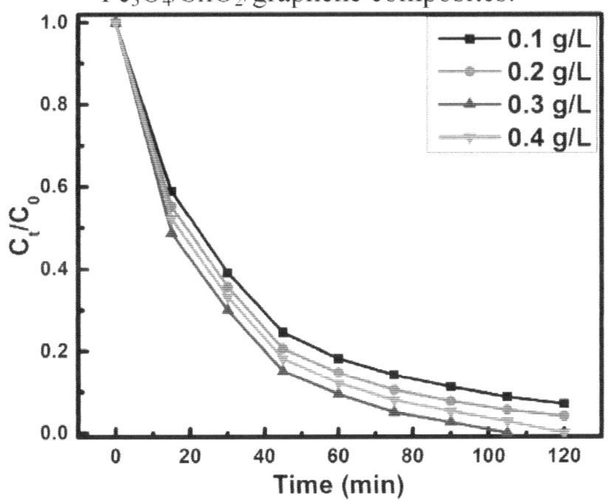

FIGURE 5. Effect of dosage on photocatalytic activity of Fe_3O_4/SnO_2/graphene composites.

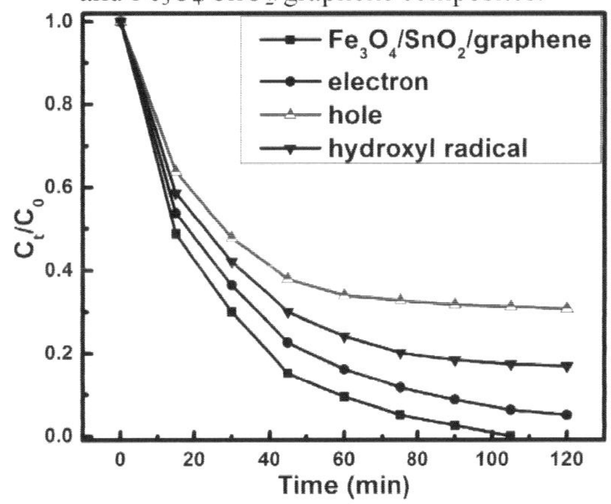

FIGURE 6. Effect of scavengers on photocatalytic activity of Fe_3O_4/SnO_2/graphene composites.

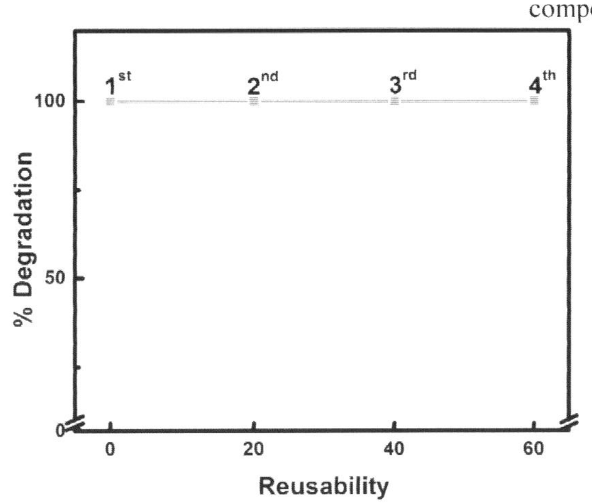

FIGURE 7. Reusability of Fe_3O_4/SnO_2/graphene composites.

graphene. The diffraction peaks of SnO_2, NGP and graphene at Fe_3O_4/SnO_2/NGP and Fe_3O_4/SnO_2/graphene composites are overlapped at 2θ of around 26°. As it is clear from this figure, there is no any impurity in the nanoparticle sample.

The FTIR spectrum of Fe_3O_4/SnO_2 nanocomposite, Fe_3O_4/SnO_2/NGP and Fe_3O_4/SnO_2/graphene composites are presented in Fig. 2. Fig. 2f shows the FTIR spectra for Fe_3O_4/SnO_2/NGP composites, while the FTIR spectra for Fe_3O_4/SnO_2/graphene composites is shown in Fig. 2g. The absorption of SnO_2 nanoparticle appears in the range of 540-660 cm^{-1} which indicates O-Sn-O and Sn-O stretching vibration modes. The Fe_3O_4 show the characteristic band of Fe-O stretching vibration at 594 cm^{-1} [7]. The absorptions in range of 3400 cm^{-1} and 1634 cm^{-1} indicate the O-H group stretching and bending vibration modes originating from water molecule [14] while absorption in the range of 1220 cm^{-1} and 1527 cm^{-1} indicates the C-OH and C-O stretching vibration modes [15,16].

Figure 3 shows the weight loss curve for Fe_3O_4/SnO_2/NGP and Fe_3O_4/SnO_2/graphene composites obtained from TGA measurement. As can be seen, Fe_3O_4/SnO_2 nanoparticle do not exhibit the significant percentage loss in the result of TGA measurement. Which indicate that Fe_3O_4/SnO_2 nanoparticle without any addition of graphene is stable until 1000°C. The weight loss started around 100°C due to the volatilization of adsorbed water [17]. The weight loss stage from 400°C to 600°C which indicates the combustion effect from NGP and graphene material and thereafter no further decomposition takes place thereafter [18]. Furthermore, the weight loss of graphene at TGA measurement can confirm the existence of the NGP and graphene in Fe_3O_4/SnO_2/NGP and Fe_3O_4/SnO_2/graphene composite samples.

The photocatalytic activities of prepared Fe_3O_4/SnO_2, Fe_3O_4/SnO_2/NGP and Fe_3O_4/SnO_2/graphene catalysts were investigated under UV light irradiation in aqueous solution of methylene blue (MB) dye (Fig 4). It can be seen that photocatalytic activities of Fe_3O_4/SnO_2/graphene composite are obviously better than the Fe_3O_4/SnO_2/NGP composite, especially better than Fe_3O_4/SnO_2 nanocomposite. More than 95% of MB has been degraded over them within 90 min. However, only 76.83% and 85.1% of MB could be decomposed over Fe_3O_4/SnO_2 and Fe_3O_4/SnO_2/NGP after 120 min, respectively.

Fig. 5 shows the effect of Fe_3O_4/SnO_2/graphene dosage, from 0.1 g/L to 0.4 g/L, on the decolorization efficiency of methylene blue at pH 13. As shown in Fig. 5, maximum degradation at Fe_3O_4/SnO_2/graphene dosages of 0.1 g/L, 0.2 g/L, 0.3 g/L and 0.4 g/L were 92.97%, 95.93% after 120 min, 97.4% after 90 min and 97% after 105 min, respectively. The decreasing maximum degradation at dosage 0.4 g/L it could be due to as the amount of Fe_3O_4/SnO_2/graphene increasing, the scattering of UV by Fe_3O_4/SnO_2/graphene also increased, because of shielding effect [19]. The above observation indicates the optimal Fe_3O_4/SnO_2/graphene dosage was 0.3 g/L.

To know the most involved reactive species MB degradation to the all samples, the addition scavenger was done. In Fig. 6 shows the result of MB degradation in the photocatalytic process with scavenger addition influence. The scavenger addition serves to bind the reactive species to not take role in the photocatalytic process. The result obtained shows that the addition of hole scavenger shows the lowest degradation capacity. This indicates that the most influential reactive species for the samples is hole.

The stability of Fe_3O_4/SnO_2/graphene composite was investigated by a 4-run cycling test under the same condition. For each run, the photocatalyst was recycled, cleaned, and dried from its solutions using magnetic external. The photodegradation efficiency of Fe_3O_4/SnO_2/graphene composite shows no apparent decrease after the 4 reuse cycles, indicating Fe_3O_4/SnO_2/graphene composite is stable (Fig. 7).

4. Conclusion

The hybridizing Fe_3O_4/SnO_2 with two different types of graphene material (Fe_3O_4/SnO_2/NGP and Fe_3O_4/SnO_2/graphene) have been successfully synthesized via co-precipitation method. Fe_3O_4/SnO_2/graphene compsite show the best photocatalytic activity under UV light irradiation. More than 95% of MB has been degraded over them within 90 min. The optimum catalyst dosage for Fe_3O_4/SnO_2/graphene composite sample is 0.3 g/L. Scavengers experiment shows that the most influential reactive species for the samples is hole. Sample showed good cycling capacity result after reuse three times.

5. Reference

[1] Abdessalem Hamrouni, Noomen Moussa, Agatino Di Paola, Leonardo Palmisano, Ammar Houas, Francesco Parrino 2015 *Journal of Photochemistry and Photobiology A: Chemistry* **309** 47-54
[2] Hadj Benhebal, Messaoud Chaib, Angelique Leonard, Stephanie D. Lambert, Michel Crine 2011 *Journal of Molecular Structure* **1004** 222-226
[3] Nisha Bayal, P. Jeevanandam 2013 *Materials Research Bulletin* **48** 3790-3799
[4] Sung Phil Kim, Myong Yong Choi, Hyun Chul Choi 2015 *Applied Surface Science* **357** 302-308
[5] Huanhao Xiao, Fengyu Qu, Ahmad Umar, Xiang Wu 2016 *Materials Research Bulletin* **74** 284-290
[6] N. Shanmugam, T. Sathya, G. Viruthagiri, C. Kalyanasundaram, R. Gobi 2016 *Applied Surface Science* **360** 283-290
[7] Tiejun Xin, Mingliang Ma, Hepeng Zhang, Junwei Gu, Shuangjie Wang, Mengjiao Liu, Qiuyu Zhang 2014 *Applied Surface Science* **288** 51-59
[8] Zaihua Wang, Yongling Du, Fengyuan Zhang, Zhixiang Zheng, Xiaolong Zhang, Qingliang Feng, Chunming Wang 2013 *Materials Chemistry and Physics* **140** 373-381
[9] Abdollah Fallah Shojaei, Ali Shams-Nateri, Maryam Ghomashpasand 2015 *Superlattices and Microstructures* **88** 211-224
[10] Wenjuan Wang, Haipeng Zhang, Ling Zhang, Haiqin Wan, Shourong Zheng, Zhaoyi Xu 2015 *Colloids and Surfaces A: Physicochem. Eng. Aspects* **469** 100-106
[11] Hui Liu, Tingting Liu, Xiaonan Dong, Yuwu Lv, Zhenfeng Zhu 2014 *Matterials Letters* **126** 36-38
[12] A. Taufik, I. K. Susanto, R. Saleh 2015 *Materials Science Forum* **827** 37-42
[13] Yue Li, Yanqun Guo, Ruiqin Tan, Ping Cui, Yong Li, Weijie Song 2009 *Materials Letters* **63** 2085-2088
[14] Qingzhi Luo, Lin Wang, Desong Wang, Rong Yin, Xueyan Li, Jing An, Xiaolian Yang 2015 *Journal of Environmental Chemical Engineering* **3** 622-629
[15] Lling-Lling Tan, Wee-Jun Ong, Siang-Piao Chai, Abdul Rahman Mohamed 2013 *Nanoscale Research Letters* **8** 465
[16] Xiaojun Peng, Yang Li, Guoliang Zhang, Fengbao Zhang, Xiaobin Fan, *Jaournal of Nanomaterials* http://dx.doi.org/10.1155/2013/841789
[17] Lihua Hu, Zhongping Yang, Limei Cui, Yan Li, Huu Hao Ngo, Yaoguang Wang, Qin Wei, Hongmin Ma, Liangguo Yan, Bin Du 2016 *Chemical Engineering Journal* **287** 545-556
[18] V. Loryuenyong, K. Totepvimarn, P. Eimburanapravat, W. Boonchompoo, A. Buasri 2013 *Advances in Material Science and Engineering* **2013** 923403
[19] Yu-Ju Chiang, Chia-Chang Lin 2013 *Powder Technology* **246** 137-143

Combinatorial Algorithms for Portfolio Optimization Problems – Case of Risk Moderate Investor

A Juarna

Gunadarma University, Jakarta - Indonesia

Email: ajuarna@staff.gunadarma.ac.id

Abstract. Portfolio optimization problem is a problem of finding optimal combination of n stocks from $N \geq n$ available stocks that gives maximal aggregate return and minimal aggregate risk. In this paper given $N = 43$ from the *IDX* (*Indonesia Stock Exchange*) group of the 45 most-traded stocks, known as the *LQ45,* with $p = 24$ data of monthly returns for each stock, spanned over interval 2013-2014. This problem actually is a combinatorial one where its algorithm is constructed based on two considerations: risk moderate type of investor and maximum allowed correlation coefficient between every two eligible stocks. The main outputs resulted from implementation of the algorithms is a multiple curve of three portfolio's attributes, e.g. the size, the ratio of return to risk, and the percentage of negative correlation coefficient for every two chosen stocks, as function of maximum allowed correlation coefficient between each two stocks. The output curve shows that the portfolio contains three stocks with ratio of return to risk at 14.57 if the maximum allowed correlation coefficient between every two eligible stocks is negative and contains 19 stocks with maximum allowed correlation coefficient 0.17 to get maximum ratio of return to risk at 25.48.

1. Introduction

Given data of N stocks, where every stock has time series data of p monthly returns; the portfolio optimization problem is the problem of finding a combination of $n < N$ stocks giving maximal aggregate return and minimal aggregate risk. The famous model of portfolio theory is Markowitz model; this model studies the effects of assets risk, return, correlation, and diversification on probable investment portfolio returns [1]. Today the Markowitz model is still an interesting topic in many implementations and applications, with some modernization and improvement, as can be found in some research by Gasser [2], Zhan [3], and Jayakumar [4]. Some original consideration in Markowitz model is still adopted which is usually consists of investor behavior and correlation coefficient between every two eligible stocks.

There are three types of investors: (a) risk taker investor, tends to like the high return stocks, no matter with the risk; (b) risk averter investor, tends to like the low risk stocks, no matter the return; (c) risk moderate investor, tends to like combination of high return stocks and low risk ones. The risk of a stock is represented by the standard deviation of its expected return. The negative correlations between two eligible stocks are preferred because they together represent an inverse relation, i.e. if the first stock return tends to rise then its pair tends to fall, and vice versa. This correlation implements the famous idiom: "*don't put all* your *eggs in one basket", meaning don't* concentrate *all* your prospects or wee resources in *one* thing or place, or you could lose everything.

There are some previous researches about portfolio optimization; the review until 2009 has been given by Bolshakova [5] which are all based on mean-variance models for returns and for risk-neutral density estimation; in two last year there Fang [6] who solve the optimization problem involving value-at-risk (VaR), Bacanin [7] who implemented the fireworks algorithm which is belongs to the group of swarm intelligence algorithm, and Shadkam [8] who implemented cuckoo optimization algorithm which is belongs to the group of evolutionary algorithm. In this paper, the portfolio optimization problem is considered as a combinatorial one where its algorithm is constructed based on two considerations: the third type of investor, i.e. the risk moderate one, and maximum allowed correlation coefficient between every two eligible stocks. The aim of the research is to find the optimal combination of stocks giving maximal aggregate return and minimal aggregate risk, especially is a multiple curve consisted of three portfolio's attributes, e.g. the size of portfolio, the ratio of return to risk, and the percentage of negative correlation coefficient among all chosen stocks, as function of the maximum allowed correlation coefficient. The source data is stored as an MS Excel file; MATLAB then reads and processes this file to get the main output, i.e. a multiple curve for the three portfolio's attributes.

2. Literature Review

The LQ45 index consists of 45 companies that fulfill certain criteria, which is:

- Included in the top 60 companies with the highest market capitalization in the last 12 months
- included in the top 60 companies with the highest transaction value in a regular market in the last 12 months
- Have been listed in the IDX for at least 3 months
- Have good financial conditions, prospect of growth and high transaction value and frequency

The portfolio is developed from the stock seed, i.e. the stock has smallest value of ratio of risk to return among N stocks, as the first element of the portfolio. Furthermore, the candidates of next portfolio elements must be stocks having *correlation coefficient* less than or equal to given maximum value, and the chosen new element is the one of them having smallest value of ratio of risk to return.

In statistics, the correlation coefficient ρ is a measure of the linear correlation or dependence between two variables, giving a value between +1 and −1 inclusive, where 1 is total positive correlation, 0 is no correlation, and −1 is total negative correlation, as is shown in Figure 1.

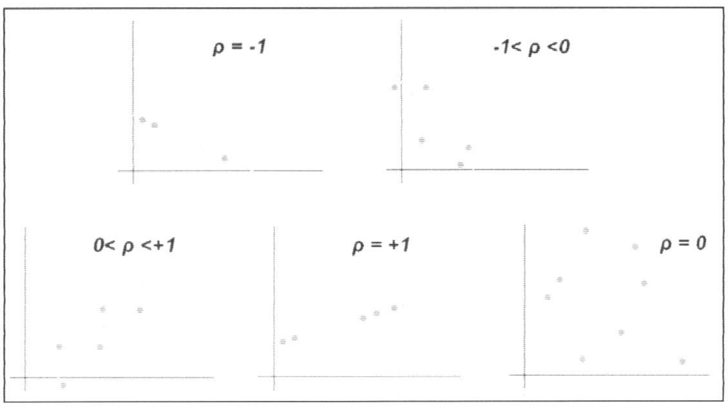

Figure 1. Correlation coefficients [9]

3. Research Methodology

In this paper given N = 43 from the *IDX* (*Indonesia Stock Exchange*) group of the 45 most-traded stocks, known as the *LQ45*, with p = 24 data of monthly returns spanned over interval 2013-2014. We take $N = 43$ instead of 45 because of incomplete of data. However this does not reduce the generality of model. The 24×43 data are stored in an MS Excel file named "source.xls", as is shown in Table 1. A MATLAB code file named "optimal_portfolio.m", stored in the same folder as the file "source.xls", accesses and processes the file this MS Excel file.

The algorithm of "optimal_portfolio.m" consisted of 7 following steps.

- Pre processing
  (i) Read the file "source.xls" and convert to the MATLAB environment.
  (ii) For each stock, compute the each following attributes: expected return, risk = standard deviation, ratio risk to expected return, covariance for each two stock, correlation coefficient for each two stock.
- Choosing elements of portfolio
  (i) Take the first stock of element "optimal portfolio", that is the stock with smallest value of ratio risk to expected return.
  (ii) Given the value of maximum allowed *correlation coefficient*; find candidates of new portfolio elements satisfying the maximum allowed correlation coefficient.
  (iii) Find the new portfolio element from the candidates that is the one has smallest ratio risk to expected return.
  (iv) Repeat the step (ii) and (iii) until there is no other new stock eligible found.
- Determining the weight of each element of portfolio
  (i) For each element of portfolio, determine its ratio of return to risk. Take the sum of all.
  (ii) The weight is the fraction of the ratio to the sum.
- Determining the expected return of portfolio
- Determining the risk value of portfolio
- Determining the ratio of expected return to risk of portfolio.
- Determine the percentage of negative correlation.

Table 1 "source.xls"

Periodicity	Stock 1	Stock 2	…	Stock 42	Stock 43
1	18,850	2,000	…	1,650	470
2	18,450	2,575	…	1,770	610
⋮	⋮	⋮	⋱	⋮	⋮
23	24,000	2,780	…	3,005	1,045
24	24,250	3,480	…	3,680	1,470

4. Result and Discussion

The main procedure in the code "optimal_portfolio.m" is the step 2 one, i.e. "choosing elements of portfolio". Sub step (i) of this procedure gives stock 21 as the first element of portfolio for all cases of maximum allowed correlation coefficient. Furthermore the second, third, and so on elements is depended on the maximum allowed correlation coefficient. The size ($|\cdot|$) and the elements of portfolio as function maximum allowed correlation coefficient is showed in Table 2.

The first three companies of portfolio elements for all cases of minimal correlation coefficient (CC) are: 21 (INDF - PT Indofood Sukses Makmur Tbk), 41(UNVR - PT Unilever Indonesia Tbk), 5 (ANTM - PT Aneka Tambang (Persero) Tbk), 40 (UNTR - PT United Tractors Tbk), 17 (EXCL - PT Excelcomindo Tbk), 31 (PGAS - PT Perusahaan Gas Negara Tbk), 6 (ASII - PT Astra International Tbk), and 22 (INTP - PT Indocement Tunggal Prakasa Tbk).

The percentage (%) of negative correlation coefficient and the ratio of return to risk (R2R) for all cases portfolio is presented in Table 3, while the weight of each stock in each portfolio is shown in Table 4. The main information about attributes for all portfolio composition is shown in multiple curve as in Figure 2.

Table 2. The size (|·|) and the stocks elements of portfolio as function of maximum allowed correlation coefficient (MaxCC).

MaxCC	|·|	Elements of portfolio
0	3	21, 41, 5
0.1	4	21, 41, 40, 5
0.2	4	21, 17, 41, 5
0.3	5	21, 31, 40, 5, 39
0.4	6	21, 31, 17, 5, 32, 1
0.5	10	21, 6, 31, 17, 29, 41, 13, 5, 32, 1
0.6	13	21, 6, 31, 17, 4, 29, 41, 40, 14, 13, 5, 32, 10
0.7	19	21, 6, 22, 31, 17, 4, 29, 41, 8, 40, 19, 14, 13, 5, 32, 35, 3, 23, 7
0.8	28	21, 6, 22, 31, 17, 4, 29, 41, 8, 40, 19, 14, 13, 15, 5, 25, 32, 18, 1, 28, 26, 3, 11, 23, 20, 27, 43, 39
0.9	36	21, 6, 22, 31, 17, 4, 29, 36, 41, 24, 12, 8, 40, 19, 14, 13, 15, 5, 25, 32, 18, 35, 1, 28, 26, 3, 34, 11, 38, 23, 2, 20, 27, 7, 37, 39

Table 3. The percentage of negative correlation coefficient, ratio return to risk, and the actual return of portfolio

MaxCC	%	R2R	AcR
0	100	14.6	8.3
0.1	83	17.5	9.9
0.2	83	16.2	4.8
0.3	50	18,0	10.4
0.4	47	17.7	4.2
0.5	38	21.5	9,8
0.6	28	23.1	12.8
0.7	30	25.5	11.3
0.8	31	20.0	10.1
0.9	26	19.4	10.7

Table 4. Weight of each element and the size of portfolio as function of maximum allowed correlation coefficient.

MaxCC	|·|	Weight of each elements of portfolio
0	3	0.5370, 0.2613, 0.2018
0.1	4	0.4309, 0.2096, 0.1975, 0.1619
0.2	4	0.4179, 0.2217, 0.2033, 0.1570
0.3	5	0.4119, 0.2190, 0.1888, 0.1548, 0.0255
0.4	6	0.3272, 0.1740, 0.1736, 0.1230, 0.1054, 0.0969
0.5	10	0.1947, 0.1361, 0.1035, 0.1033, 0.0966, 0.0947, 0.0777, 0.0732, 0.0627, 0.0576
0.6	13	0.1534, 0.1072, 0.0816, 0.0814, 0.0769, 0.0761, 0.0746, 0.0703, 0.0629, 0.0612, 0.0576, 0.0494, 0.0473
0.7	19	0.1159, 0.0810, 0.0628, 0.0616, 0.0615, 0.0581, 0.0575, 0.0564, 0.0534, 0.0531, 0.0521, 0.0476, 0.0462, 0.0436, 0.0373, 0.0357, 0.0304, 0.0273, 0.0183
0.8	28	0.0914, 0.0639, 0.0495, 0.0486, 0.0485, 0.0458, 0.0454, 0.0445, 0.0421, 0.0419, 0.0411, 0.0375, 0.0365, 0.0353, 0.0344, 0.0330, 0.0294, 0.0294,

MaxCC	|·|	Weight of each elements of portfolio
		0.0271, 0.0265, 0.0241, 0.0240, 0.0227, 0.0215, 0.0185, 0.0171, 0.0148, 0.0057
0.9	36	0.0740, 0.0517, 0.0401, 0.0394, 0.0393, 0.0371, 0.0367, 0.0367, 0.0360, 0.0353, 0.0343, 0.0341, 0.0339, 0.0333, 0.0304, 0.0295, 0.0286, 0.0278, 0.0267, 0.0238, 0.0238, 0.0228, 0.0219, 0.0215, 0.0195, 0.0194, 0.0189, 0.0183, 0.0182, 0.0174, 0.0157, 0.0150, 0.0138, 0.0117, 0.0088, 0.0046

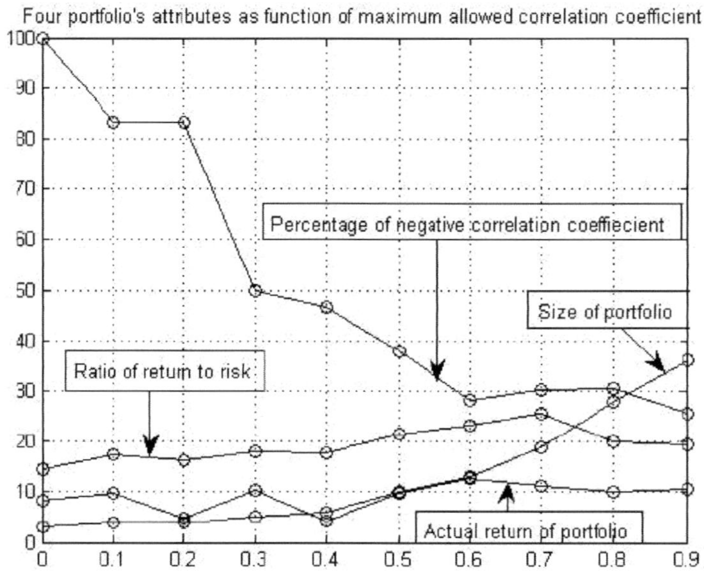

Figure 2. Three main outputs and the actual return

The final choices depend on two last consideration: If negative correlation composition is the primary consideration then the first composition consisting of three stocks is the choices: Spend 0.5370 portion of the money on the stock number 21 (PT Indofood Sukses Makmur Tbk), 0.2613 portion of the money on the stock number 41 (PT Unilever Indonesia Tbk), and 0.2018 portion of the money on the stock number 5 (PT Aneka Tambang (Persero) Tbk), to get ratio of return to risk at 14.6. If one wants the maximum ratio of return to risk then the eighth composition consisting of 19 stocks is the choices, to get the ratio of return to risk that is at 25.5.

5. Conclusion

Combinatorial algorithm for development of portfolio is constructed by two considerations: risk moderate type of investor and maximum allowed correlation coefficient between every two eligible stocks. The main result is multiple curves of three primary portfolio's attributes. If negative correlation composition is the primary consideration then the first composition consisting of three stocks is the choices, to get ratio of return to risk at 14.6. If ratio of return to risk is the primary consideration then the eighth composition consisting of 19 stocks is the choices, to get the maximum ratio of return to risk that is at 25.5.

Acknowledgment

This work is funded by Directorate of Higher Education, Department of Education and Culture, Republic of Indonesia, under Grant No. 790/K3/KM/SPK.LT/2016, June 14-th, 2016.

References

[1] Markowitz, H.M. (March 1952). "Portfolio Selection." *The Journal of Finance* **7** (1): 77-91.

[2] Gasser, Stephan M, et all (December, 2014). Markowitz Revisited: Social Portfolio Engineering". *Proceeding of 27th Australian Finance and Banking Conference*.

[3] Zhan, Haichen. (May 2015). "An Emphirical study on Markowitz Modern Portfolio." *http://polymer.bu.edu/hes/py538zhan15.pdf*.

[4] Jayakumar, G.S. David Sam and W. Samuel. (January 2015). "Modeling The Portfolio Selection Problem with Special Reference to Selected Industrial BSE". *International Journal of Management* (*IJM*) **6**(1): 1-15.

[5] Bolshakova, Irina and Mikhail Kovalev. (2009). "Portofolio Optimization Problems, - A Survey." *http://www.math.uni-magdeburg.de/~girlich/preprints/preprint0906.pdf*.

[6] Feng, Mingbin, Andreas Wachter, and Jeremy Staum. (2014). "Practical Algorithms for Value-at-Risk Portfolio Optimization Problems." *Quantitative Finance Letters* **3**:1-9.

[7] Bacanin, Nebojsa and Milan Tuba. (2015). "Fireworks Algorithm Applied to Constrained Portofolio Optimization Problem." *http://www.cil.pku.edu.cn/research/fwa/publication/Portfolio.pdf*.

[8] Shadkam, Elham, Reza Delavari, Farzad Memariani, and Morteza Poursaleh. (2015). "Portfolio Selection by the Means of Cuckoo Optimization Algorithm." *International Journal on Computational Science & Applications* (*IJCSA*) **5**(3):37-46.

[9] *https://en.wikipedia.org/wiki/Pearson_correlation_coefficient*.

The ranks of Indonesian and Japanese industrial sectors: A further study

Ubaidillah Zuhdi

Faculty of Management and Economics, Gdansk University of Technology, Gdansk 80-233, Poland

zuhdi@zie.pg.gda.pl

Abstract. The purpose of the current study is to extend the previous study which analyzes the industrial sectors ranks of specific countries. More specifically, the study aims to conduct a further analysis regarding the ranks by using the new data as well as new analysis periods. The study focuses on the cases of Indonesia and Japan. In this study, the analysis period of the Indonesian case is 2010 while for the case of Japan is 2011. Using the previous study as a reference, the study employs the Input-Output (IO) approach as an analysis device. As with the previous study, the results of this study show that the orders of the ranks depend on the methods used. Nevertheless, from the results, one can argue that the electricity and gas industry was a leading sector in the Indonesian economy in 2010. On the other hand, for the case of Japan, the competitive sector in the Japanese economy in 2011 was the manufacturing industry. In terms of the leading sectors, the current study is different from the previous one.

1. Introduction

Industrial sectors are important parts in the economic activities of one country. The importance of sectors can be seen on both macro and micro levels. Therefore, analyzing the industries of a particular country can be an alternative way to know the outlook of its economic situations. Further, the recommendations for improving the economic conditions of the country might be generated from the analysis.

The examples of previous studies which focus on the analysis are [1], [2], [3], [4], [5], and [6]. To the best of my knowledge, after learning the studies, the research which the topic is to determine the ranks of the industrial sectors of countries is still needed. The research is required in order to get the insight about the competitive sectors in the economic activities of focused countries. The current study attempts to fulfill the gap.

The purpose of the study is to extend the previous study which analyzes the industrial sectors ranks of specific countries. More specifically, the study aims to conduct a further analysis regarding the ranks by using the new data as well as new analysis periods. The study focuses on the cases of Indonesia and Japan. In this study, the analysis period of the Indonesian case is 2010 while for the case of Japan is 2011. Using the previous study as a reference, the study employs the Input-Output (IO) approach as an analysis device.

2. Methodology

The methodology of this study refers to the previous study which was conducted by [7]. The first step of the methodology is to expose the data sources. The data sources of this study are the 2010 Indonesian IO table, and Japanese IO table for 2011. The former table is obtained from [8] while the latter one is from [9]. The second step is to explore the industrial sectors of Indonesia and Japan used in this study. Tables 1 and 2 show the industries for the cases of Indonesia and Japan, respectively. The former table consists of seventeen industrial sectors while the latter one is thirteen industries.

The third step is to conduct the calculations in order to determine the ranks of the industrial sectors of discussed countries on the analyzed periods. The methods of backward and forward linkages, the analysis tools in the IO analysis, are employed in the calculations. Both linkages describe the relationship between a particular industrial sector and other industries. More specifically, the backward linkage focuses on the demands of the specific industry on other industrial sectors. These demands appear as a consequence of the industry as a purchaser in the economy. On the other hand, the forward linkage explains the consequence of the industrial sector as a seller.

The methods are suitable for specifying the ranks of industrial sectors on the specific time period. [10] affirm this argument through the following statement:

"*Measures have been proposed to quantify such backward and forward linkages, or economic "connectedness." Comparisons of the strengths of backward and forward linkages for the sectors in a single economy provide one mechanism for identifying "key" or "leading" sectors in that economy (those sectors that are most connected and therefore, in some sense, most "important") and for grouping sectors into spatial clusters.*"

One can argue that if the backward linkage value of sector *a* is higher than that of sector *b*, then sector *a* gives more beneficial effect because of the higher capability of the sector in attracting the economy through its demands. The similar argument which uses the supplier's point of view can be utilized in analyzing the forward linkage values of two or more industries.

This study applies two forms for each linkage, namely "direct" and "total". These forms can be seen on both demand-side and supply-side IO models. [11] propose the concepts of the former form on both models. To summarize, this form only indicates the straightforward impacts of the demands and supplies of one particular industry. Further, [10] explain the concept of the form on the backward linkage by using the following equation:

$$BL(d)_j = \sum_{i=1}^{n} a_{ij} \quad (1)$$

where $BL(d)_j$, a_{ij}, and n are the direct backward linkage of sector j, the direct input coefficient, and the number of industrial sectors, respectively. One can calculate a_{ij} by dividing the value of interindustry sale from sector i to sector j by the total outlay of sector j. On the other hand, [12] describes the concept of the form on the forward linkage through the following equation:

$$F(d)_i = \sum_{j=1}^{n} a_{ij} \quad (2)$$

where $F(d)_i$ is the direct forward linkage of sector i. Both equations use the demand-driven IO model as a base. The using of row sums in describing the direct forward linkage of the specific sector in the model is also mentioned by [10].

The insight of the "total" form on the backward linkage in the demand-driven IO model is proposed by Rasmussen (1957) in [10] while [10] explain this on the forward linkage. To summarize, the form describes the straightforward and indirect effects of the demands and supplies of one specific industry. [10] explore the former insight by using the following equation:

$$BL(t)_j = \sum_{i=1}^{n} l_{ij} \quad (3)$$

where $BL(t)_j$ and l_{ij} are the total backward linkage of sector j, and Leontief inverse or the total requirements matrix, respectively. The latter insight, on the other hand, is explained by [12] by using the following equation:

$$F(d+i)_i = \sum_{j=1}^{n} \alpha_{ij} \quad (4)$$

where $F(d+i)_i$ and α_{ij} are the total forward linkage of sector *i*, and Leontief inverse matrix, respectively. As with equations (1) and (2), the equations (3) and (4) also use the demand-driven IO model as a base.

[10] mention that the Ghosh inverse is suggested as a better tool in explaining the total forward linkage. As a consequence, the supply-driven IO model is a suitable device for analyzing the forward linkage. [10] describe the following equations in representing the linkage which uses the model:

$$FL(d)_i = \sum_{j=1}^{n} b_{ij} \quad (5)$$

$$FL(t)_i = \sum_{j=1}^{n} g_{ij} \quad (6)$$

where $FL(d)_i$, $FL(t)_i$, b_{ij}, and g_{ij} are the direct forward linkage of sector *i*, the total forward linkage of sector *i*, the coefficients matrix in a supply-driven IO model, and the Ghosh inverse, respectively. In this study, all above equations are implemented in the calculations step. The next step is to analyze the ranks of Indonesian and Japanese industrial sectors on the analysis periods. Conclusions of this study, and suggestions for further researches are explained on the final step.

Table 1. Indonesian industrial sectors used in this study.

Sector Number	Sector Name
1	Agriculture, forestry, and fishing
2	Mining and quarrying
3	Manufacturing
4	Electricity and gas
5	Water supply, sewerage, waste management, and remediation activities
6	Construction
7	Wholesale and retail trade; repair of motor vehicles and motorcycles
8	Transportation and storage
9	Accommodation and food service activities
10	Information and communication
11	Financial and insurance activities
12	Real estate activities
13	Business activities
14	Public administration and defence; compulsory social security
15	Education
16	Human health and social work activities
17	Other services activities

Table 2. Japanese industrial sectors used in this study.

Sector Number	Sector Name
1	Agriculture, forestry, and fishery
2	Mining
3	Manufacturing
4	Construction
5	Electricity, gas, and water supply
6	Commerce
7	Finance and insurance
8	Real estate
9	Transport and postal services
10	Information and communications
11	Public administration
12	Services
13	Activities not elsewhere classified

3. Results and analysis

Tables 3 and 4 describe the ranks of Indonesian industrial sectors on the analysis period which are viewed from the calculations results using direct and total backwards linkages, respectively. On the other hand, tables 5 and 6 expose the ranks which are viewed from the results using direct and total forwards linkages, respectively. Meanwhile, the results using the supply-driven IO model for the Indonesian case are explored in tables 7 and 8. The analysis then focuses on the top five sectors of each result. More specifically, one can argue that the sector number 4, electricity and gas, was a leading sector in the Indonesian economy in 2010 because it appears as one of the top five sectors in tables 3 – 8. In other words, on the analysis period, the sector had beneficial effects in the Indonesian economy.

Tables 9 and 10 expose the ranks of Japanese industrial sectors on the analysis period which are viewed from the results of calculations using direct and total backwards linkages, respectively. Meanwhile, tables 11 and 12 describe the ranks which are viewed from the results using direct and total forwards linkages, respectively. On the other hand, the results using the supply-driven IO model for the Japanese case are explored in tables 13 and 14. As with the previous case, the top five sectors on each result are also viewed in the analysis of the Japanese issue. Based on the analysis, one can say that the sector number 3, manufacturing, was a leading sector in the Japanese economy in 2011 because it appears as one of the top five sectors in almost all tables which explain the ranks. In other words, on the analysis period, the sector had beneficial effects in the Japanese economy.

Table 3. The ranks of Indonesian industrial sectors on the analysis period, based on the results of calculations using equation (1).

Sector Number	Sector Name	Calculated Value
4	Electricity and gas	0.765
6	Construction	0.649
3	Manufacturing	0.635
8	Transportation and storage	0.575
16	Human health and social work activities	0.545
9	Accommodation and food service activities	0.542

17	Other services activities	0.497
13	Business activities	0.405
14	Public administration and defence; compulsory social security	0.375
10	Information and communication	0.375
15	Education	0.348
7	Wholesale and retail trade; repair of motor vehicles and motorcycles	0.330
11	Financial and insurance activities	0.273
2	Mining and quarrying	0.264
1	Agriculture, forestry, and fishing	0.192
12	Real estate activities	0.172
5	Water supply, sewerage, waste management, and remediation activities	0.171

Table 4. The ranks of Indonesian industrial sectors on the analysis period, based on the results of calculations using equation (2).

Sector Number	Sector Name	Calculated Value
4	Electricity and gas	2.889
6	Construction	2.300
8	Transportation and storage	2.184
3	Manufacturing	2.150
16	Human health and social work activities	2.071
17	Other services activities	1.996
9	Accommodation and food service activities	1.994
13	Business activities	1.774
14	Public administration and defence; compulsory social security	1.752
10	Information and communication	1.695
15	Education	1.688
7	Wholesale and retail trade; repair of motor vehicles and motorcycles	1.651
11	Financial and insurance activities	1.486
2	Mining and quarrying	1.472
1	Agriculture, forestry, and fishing	1.357
12	Real estate activities	1.349
5	Water supply, sewerage, waste management, and remediation activities	1.348

Table 5. The ranks of Indonesian industrial sectors on the analysis period, based on the results of calculations using equation (3).

Sector Number	Sector Name	Calculated Value
3	Manufacturing	2.740
4	Electricity and gas	0.565
7	Wholesale and retail trade; repair of motor vehicles and motorcycles	0.542
10	Information and communication	0.462
1	Agriculture, forestry, and fishing	0.426
2	Mining and quarrying	0.374
8	Transportation and storage	0.362
11	Financial and insurance activities	0.359
17	Other services activities	0.297
6	Construction	0.268

13	Business activities	0.260
9	Accommodation and food service activities	0.168
16	Human health and social work activities	0.088
14	Public administration and defence; compulsory social security	0.071
5	Water supply, sewerage, waste management, and remediation activities	0.049
15	Education	0.043
12	Real estate activities	0.038

Table 6. The ranks of Indonesian industrial sectors on the analysis period, based on the results of calculations using equation (4).

Sector Number	Sector Name	Calculated Value
3	Manufacturing	6.436
1	Agriculture, forestry, and fishing	2.309
4	Electricity and gas	2.217
2	Mining and quarrying	2.203
7	Wholesale and retail trade; repair of motor vehicles and motorcycles	2.072
10	Information and communication	1.767
8	Transportation and storage	1.614
11	Financial and insurance activities	1.606
13	Business activities	1.450
17	Other services activities	1.427
6	Construction	1.410
9	Accommodation and food service activities	1.236
14	Public administration and defence; compulsory social security	1.114
16	Human health and social work activities	1.107
12	Real estate activities	1.068
5	Water supply, sewerage, waste management, and remediation activities	1.068
15	Education	1.052

Table 7. The ranks of Indonesian industrial sectors on the analysis period, based on the results of calculations using equation (5).

Sector Number	Sector Name	Calculated Value
13	Business activities	0.820
4	Electricity and gas	0.720
2	Mining and quarrying	0.677
11	Financial and insurance activities	0.673
1	Agriculture, forestry, and fishing	0.656
3	Manufacturing	0.654
5	Water supply, sewerage, waste management, and remediation activities	0.638
10	Information and communication	0.574
8	Transportation and storage	0.503
7	Wholesale and retail trade; repair of motor vehicles and motorcycles	0.417
17	Other services activities	0.355
9	Accommodation and food service activities	0.244
16	Human health and social work activities	0.142
12	Real estate activities	0.129

6	Construction	0.088
14	Public administration and defence; compulsory social security	0.084
15	Education	0.051

Table 8. The ranks of Indonesian industrial sectors on the analysis period, based on the results of calculations using equation (6).

Sector Number	Sector Name	Calculated Value
4	Electricity and gas	2.801
13	Business activities	2.598
2	Mining and quarrying	2.458
1	Agriculture, forestry, and fishing	2.367
5	Water supply, sewerage, waste management, and remediation activities	2.300
11	Financial and insurance activities	2.298
3	Manufacturing	2.186
10	Information and communication	2.108
8	Transportation and storage	1.890
7	Wholesale and retail trade; repair of motor vehicles and motorcycles	1.758
17	Other services activities	1.586
9	Accommodation and food service activities	1.369
12	Real estate activities	1.233
16	Human health and social work activities	1.218
6	Construction	1.165
14	Public administration and defence; compulsory social security	1.163
15	Education	1.071

Table 9. The ranks of Japanese industrial sectors on the analysis period, based on the results of calculations using equation (1).

Sector Number	Sector Name	Calculated Value
5	Electricity, gas, and water supply	0.730
3	Manufacturing	0.715
13	Activities not elsewhere classified	0.601
2	Mining	0.553
4	Construction	0.548
1	Agriculture, forestry, and fishery	0.515
9	Transport and postal services	0.497
10	Information and communications	0.474
12	Services	0.384
7	Finance and insurance	0.342
11	Public administration	0.317
6	Commerce	0.315
8	Real estate	0.194

Table 10. The ranks of Japanese industrial sectors on the analysis period, based on the results of calculations using equation (2).

Sector Number	Sector Name	Calculated Value
3	Manufacturing	2.769
5	Electricity, gas, and water supply	2.625
4	Construction	2.268
1	Agriculture, forestry, and fishery	2.210
2	Mining	2.181
13	Activities not elsewhere classified	2.160
9	Transport and postal services	2.077
10	Information and communications	1.935
12	Services	1.840
11	Public administration	1.664
7	Finance and insurance	1.658
6	Commerce	1.627
8	Real estate	1.361

IOP Conf. Series: Journal of Physics: Conf. Series **820** (2017) 012029 doi:10.1088/1742-6596/820/1/012029

Table 11. The ranks of Japanese industrial sectors on the analysis period, based on the results of calculations using equation (3).

Sector Number	Sector Name	Calculated Value
3	Manufacturing	1.664
12	Services	1.197
9	Transport and postal services	0.772
10	Information and communications	0.417
6	Commerce	0.364
2	Mining	0.335
7	Finance and insurance	0.312
5	Electricity, gas, and water supply	0.289
11	Public administration	0.227
8	Real estate	0.203
4	Construction	0.170
1	Agriculture, forestry, and fishery	0.155
13	Activities not elsewhere classified	0.080

Table 12. The ranks of Japanese industrial sectors on the analysis period, based on the results of calculations using equation (4).

Sector Number	Sector Name	Calculated Value
3	Manufacturing	5.629
12	Services	3.379
9	Transport and postal services	2.455
6	Commerce	1.871
10	Information and communications	1.767
2	Mining	1.763
5	Electricity, gas, and water supply	1.582
7	Finance and insurance	1.521
8	Real estate	1.363
1	Agriculture, forestry, and fishery	1.336
4	Construction	1.297
11	Public administration	1.261
13	Activities not elsewhere classified	1.151

Table 13. The ranks of Japanese industrial sectors on the analysis period, based on the results of calculations using equation (5).

Sector Number	Sector Name	Calculated Value
2	Mining	31.701
13	Activities not elsewhere classified	1.003
1	Agriculture, forestry, and fishery	0.887
5	Electricity, gas, and water supply	0.690
3	Manufacturing	0.668
9	Transport and postal services	0.645
10	Information and communications	0.552
7	Finance and insurance	0.517
6	Commerce	0.378
12	Services	0.360
4	Construction	0.186
8	Real estate	0.167
11	Public administration	0.029

Table 14. The ranks of Japanese industrial sectors on the analysis period, based on the results of calculations using equation (6).

Sector Number	Sector Name	Calculated Value
2	Mining	82.632
1	Agriculture, forestry, and fishery	3.276
13	Activities not elsewhere classified	3.072
9	Transport and postal services	2.722
5	Electricity, gas, and water supply	2.597
3	Manufacturing	2.583
10	Information and communications	2.087
7	Finance and insurance	1.970
6	Commerce	1.837
12	Services	1.755
4	Construction	1.358
8	Real estate	1.329
11	Public administration	1.089

4. Conclusions and further researches

This study, as a continuation study of the previous one, determines the ranks of Indonesian and Japanese industries by using the analysis methods in the IO analysis, backward and forward linkages, and new data as well as new analysis periods for the focused countries. As with the previous study, the results of the current study show that the orders of the ranks depend on the methods used. Nevertheless, from the results, one can argue that the electricity and gas industry was a leading sector in the Indonesian economy in 2010. On the other hand, for the Japanese case, the sector which had the beneficial effects in the Japanese economy in 2011 was the manufacturing industry. In terms of the leading sectors, the current study is different from the previous one.

The suggested further research from the current study is to expand the analyzed industrial sectors for each analyzed country. This suggestion is mentioned in order to get the deeper understanding regarding the economic conditions of analyzed countries on the specific period of analysis, especially about the ranks of their industries. This deeper understanding might be useful in determining the prioritized sectors on their economic activities in the future. The other suggested further research is to expand the area of analysis using the same methodology so the industrial sectors ranks of, for example, the specific region can also be analyzed. One of the examples is to expand the analysis for the other ASEAN countries.

References

[1] Zuhdi U 2014 The impacts of final demand changes on total output of Indonesian ICT sectors: an analysis using input-output approach *IOP Conference Series: Materials Science and Engineering* **58** 012011

[2] Zuhdi U, Prasetyo A D and Sianipar C P M 2013 Analyzing the dynamics of total output of Japanese creative industry sectors: an input-output approach *Procedia Economics and Finance* **5** 827–35

[3] Zuhdi U 2012 Analyzing the influence of creative industry sector to the national economic structural changes by decomposition analysis: the case of Indonesia *Procedia-Social and Behavioral Sciences* **65** 980–5

[4] Zuhdi U, Utomo D S and Alamanda D T 2011 Analyzing the role of ICT sector to the national economic structural changes: the case of Indonesia *Jurnal Manajemen Teknologi* **10** 299–307

[5] Zuhdi U 2015 An application of input-output analysis in analyzing the impacts of final demands changes on the total outputs of Japanese energy sectors: a further study *Journal of Physics: Conference Series* **622** 012041

[6] Zuhdi U, Mori S and Kamegai K 2014 Analysis of influences of GDP and ICT on Indonesian industrial structural changes using statistical analysis: 1990-2005 *Journal of Finance and Accountancy* **17** 1–19

[7] Zuhdi U 2016 The ranks of Indonesian and Japanese indsutrial sectors *IOP Conference Series: Earth and Environmental Science* **38** 012008

[8] BPS-Statistics Indonesia 2016 *Statistical Yearbook of Indonesia 2016* [online] https://www.bps.go.id/website/pdf_publikasi/Statistik-Indonesia-2016--_rev.pdf (accessed December 30, 2016)

[9] Japanese Ministry of Internal Affairs and Communications 2016 *2011 Input-Output Tables for Japan* [online] http://www.soumu.go.jp/main_content/000443188.pdf (accessed December 30, 2016)

[10] Miller R E and Blair P D 2009 *Input-Output Analysis: Foundations and Extensions* (Cambridge: University Press)

[11] Chenery H B and Watanabe T 1958 International comparisons of the structure of productions *Econometrica* **4** 487–521

[12] Nazara S 2005 *Input-Output Analysis* [in Indonesian] (Jakarta: The Faculty of Economics of University of Indonesia)

[13] Zuhdi U 2014 Analyzing the impacts of final demand changes on total output using input-output approach: the case of Japanese ICT sectors *IOP Conference Series: Earth and Environmental Science* **19** 012016

[14] Zuhdi U 2014 Analyzing the role of creative industries in national economy of Japan: 1995-2005 *Open Journal of Applied Sciences* **4** 197–211

[15] Zuhdi U 2014 An input-output approach to analyze the ways to increase total output of energy sectors: the case of Japan *IOP Conference Series: Earth and Environmental Science* **19** 012015

[16] Zuhdi U 2014 The dynamics of total output of Indonesian information and communication technology sector when final demand changes occur: an analysis using input-output approach *Advanced Science Letters* **20** 2254–57
[17] Zuhdi U 2014 The other perspective related to the role of information and communication technologies sectors in national economy: the case of Japan *Advanced Science Letters* **20** 483–6
[18] Zuhdi U 2014 The role of information and communication technology sectors in Indonesian national economy from 1990 through 2008: an analysis using input-output approach *Advanced Science Letters* **20** 1932–5
[19] Zuhdi U 2014 Using multipliers analysis in order to get another perspective related to the role of ICT sectors in national economy of Indonesia: 1990-2005 *Journal of Physics: Conference Series* **495** 012051
[20] Zuhdi U 2015 An analysis of the role of information and communication technology sectors on Japanese national economy from 1995 through 2005: an application of multiplier analysis *IOP Conference Series: Earth and Environmental Science* **23** 012014
[21] Zuhdi U 2015 An application of multiplier analysis in analyzing the role of information and communication technology sectors on Indonesian national economy: 1990-2005 *IOP Conference Series: Earth and Environmental Science* **23** 012015
[22] Zuhdi U 2015 The dynamics of Indonesian creative industry sectors: an analysis using input-output approach *Journal of the Knowledge Economy* **6** 1177–90
[23] Zuhdi U 2016 The dynamics of the total output of the fishery sector: the case of Indonesia *Journal of Physics: Conference Series* **710** 012039
[24] Zuhdi U 2016 The dynamics of the total output of the Japanese fisheries sector: an analysis using input-output approach *Journal of Physics: Conference Series* **710** 012040
[25] Zuhdi U 2016 The dynamics of the total outputs of Japanese information and communication technology sectors: a further study *Journal of Physics: Conference Series* **710** 012041
[26] Zuhdi U 2016 The Indonesian economy in 2005: an analysis using the input-output approach *Proceedings of 20th EBES Conference-Vienna* **3** 1825–42
[27] Zuhdi U, Mori S and Kamegai K 2012 Analyzing the role of ICT sector to the national economic structural changes by decomposition analysis: the case of Indonesia and Japan *Procedia-Social and Behavioral Sciences* **65** 749–54
[28] Zuhdi U, Mori S and Kamegai K 2013 Analysis of influences of ICT on structural changes in Japanese commerce, business services and office supplies, and personal services sectors using multivariate analysis: 1985-2005 *The Asian Journal of Technology Management* **6** 102–11
[29] Zuhdi U, Mori S and Kamegai K 2014 Statistical analysis of influences of ICT on industrial structure changes from 1985 through 2005: the case of Japan *Journal of Computers* **9** 1291–9
[30] Zuhdi U, Mori S and Kamegai K 2015 Forecasting the influences of information and communication technology on the structural changes of Japanese industrial sectors: a study using statistical analysis *International Journal of Social, Behavioral, Educational, Economic, Business and Industrial Engineering* **9** 531–7
[31] Zuhdi U and Prasetyo A D 2014 Examining the total output changes of ICT sectors of Japan: an approach of input-output *Procedia-Social and Behavioral Sciences* **109** 659–63
[32] Zuhdi U, Prasetyo A D and Putranto N A R 2014 Analyzing the changes of total output of Japanese livestock sector: an input-output approach *Procedia-Social and Behavioral Sciences* **109** 649–53
[33] Zuhdi U, Putranto N A R and Prasetyo A D 2014 An input-output approach to know the dynamics of total output of livestock sectors: the case of Indonesia *Procedia-Social and Behavioral Sciences* **109** 634–8
[34] Zuhdi U, Putranto N A R and Prasetyo A D 2014 Encouraging information and communication technology sectors using input-output approach: the case of Indonesia *Advanced Science Letters* **20** 199–202

IOP Conf. Series: Journal of Physics: Conf. Series **820** (2017) 012030 doi:10.1088/1742-6596/820/1/012030

The ranks of Japanese industrial sectors: 2005-2011

Ubaidillah Zuhdi

Faculty of Management and Economics, Gdansk University of Technology, Gdansk 80-233, Poland

zuhdi@zie.pg.gda.pl

Abstract. The purpose of the current study is to expand the previous study which analyzes the industrial sectors ranks of specific countries. More specifically, the study aims to conduct a comparison in terms of the ranks of Japanese industrial sectors on the specific analysis period. The analysis period of the study is 2005-2011. Using the previous study as a reference, the study employs the Input-Output (IO) approach as an analysis instrument. From the results of the study, one can argue that the competitive industries in the Japanese economies in 2005 and 2011 were transport and manufacturing sectors, respectively. Besides, the results also show that, in terms of the movements of calculated values of Japanese industrial sectors, the increasing and decreasing patterns appeared on the analysis period.

1. Introduction

One of the important parts in the economic activities is industrial sectors. One can argue that the importance of industries can be seen from their trade activities in the economy. The importance can also be seen on both macro and micro levels. Therefore, an analysis of industries of one country is an important thing in describing the outlook of its economic situations. Further, the recommendations for enhancing the economic aspects of the country might be generated from the analysis.

The examples of previous studies which focus on the analysis are [1], [2], [3], [4], [5], and [6]. To the best of my knowledge, after learning the studies, the research which the subject is to identify the ranks of the industrial sectors of countries is still needed. The research is required in order to get the understanding about the competitive industries in the economic activities of focused countries. The current study tries to fulfill the gap.

The purpose of the study is to expand the previous study which analyzes the industrial sectors ranks of specific countries. More specifically, the study aims to conduct a comparison in terms of the ranks of Japanese industrial sectors on the specific analysis period. The analysis period of the study is 2005-2011. Using the previous study as a reference, the study employs the Input-Output (IO) approach as an analysis instrument.

2. Methodology

The methodology of this study refers to the previous study which was conducted by [7]. The first step of the methodology is to explain the data sources. The data sources of this study are Japanese IO tables for 2005 and 2011. The former table is obtained from [8] while the latter one is from [9]. The second step is to expose the industrial sectors of Japan used in this study. Table 1 shows the industries. The main difference between 2005 and 2011 Japanese industrial sectors is located on the sector number 9. More specifically, in the former year, the name of the industry is "Transport" while in the latter one is "Transport and postal services".

The third step is to conduct the calculations in order to identify the ranks of the industrial sectors of Japan on the analyzed years. The methods of backward and forward linkages, the analysis instruments in the IO analysis, are employed in the calculations. Both linkages describe the relationship between a specific industrial sector and other industries. More specifically, the backward linkage focuses on the demands of the particular industry on other industrial sectors. These demands appear as a consequence of the industry as a purchaser in the economic activities. On the other hand, the forward linkage describes the consequence of the industrial sector as a seller.

The methods are suitable for identifying the ranks of industrial sectors on the specific time period. [10] affirm this thought through the following statement:

"Measures have been proposed to quantify such backward and forward linkages, or economic "connectedness." Comparisons of the strengths of backward and forward linkages for the sectors in a single economy provide one mechanism for identifying "key" or "leading" sectors in that economy (those sectors that are most connected and therefore, in some sense, most "important") and for grouping sectors into spatial clusters."

One can say that if the backward linkage value of sector a is higher than that of sector b, then sector a gives more beneficial effect because of the higher capability of the sector in attracting the economy through its demands. The similar argument which uses the supplier's point of view can be imposed in analyzing the forward linkage values of two or more industries.

This study implements two forms for each linkage, namely "direct" and "total". These forms can be seen on both demand-side and supply-side IO models. [11] propose the concepts of the former form on both models. To summarize, this form only describes the straightforward impacts of the demands and supplies of one specific industry. Further, [10] expose the concept of the form on the backward linkage by using the following equation:

$$BL(d)_j = \sum_{i=1}^{n} a_{ij} \tag{1}$$

where $BL(d)_j$, a_{ij}, and n are the direct backward linkage of sector j, the direct input coefficient, and the number of industrial sectors, respectively. On the other hand, [12] explains the concept of the form on the forward linkage through the following equation:

$$F(d)_i = \sum_{j=1}^{n} a_{ij} \tag{2}$$

where $F(d)_i$ is the direct forward linkage of sector i. Both equations use the demand-driven IO model as a base. The using of row sums in describing the direct forward linkage of the particular sector in the model is also described by [10].

The insight of the "total" form on the backward linkage in the demand-driven IO model is proposed by Rasmussen (1957) in [10] while [10] expose this on the forward linkage. To summarize, the form describes the straightforward and indirect effects of the demands and supplies of one particular industry. [10] explain the former insight by using the following equation:

$$BL(t)_j = \sum_{i=1}^{n} l_{ij} \tag{3}$$

where $BL(t)_j$ and l_{ij} are the total backward linkage of sector j, and Leontief inverse or the total requirements matrix, respectively. The latter insight, on the other hand, is exposed by [12] by using the following equation:

$$F(d+i)_i = \sum_{j=1}^{n} \alpha_{ij} \tag{4}$$

where $F(d+i)_i$ and α_{ij} are the total forward linkage of sector *i*, and Leontief inverse matrix, respectively. As with equations (1) and (2), the equations (3) and (4) also use the demand-driven IO model as a base.

[10] mention that the Ghosh inverse is suggested as a better device in explaining the total forward linkage. As a consequence, the supply-driven IO model is a suitable tool for analyzing the forward linkage. [10] expose the following equations in representing the linkage which uses the model:

$$FL(d)_i = \sum_{j=1}^{n} b_{ij} \tag{5}$$

$$FL(t)_i = \sum_{j=1}^{n} g_{ij} \tag{6}$$

where $FL(d)_i$, $FL(t)_i$, b_{ij}, and g_{ij} are the direct forward linkage of sector *i*, the total forward linkage of sector *i*, the coefficients matrix in a supply-driven IO model, and the Ghosh inverse, respectively. In this study, all above equations are applied in the calculations step. The next step is to analyze the ranks and movements of Japanese industrial sectors on the analysis period. Conclusions of this study, and suggestions for further researches are exposed on the final step.

Table 1. Japanese industrial sectors used in this study.

Sector Number	Sector Name
1	Agriculture, forestry, and fishery
2	Mining
3	Manufacturing
4	Construction
5	Electricity, gas, and water supply
6	Commerce
7	Finance and insurance
8	Real estate
9	Transport and postal services
10	Information and communications
11	Public administration
12	Services
13	Activities not elsewhere classified

3. Results and analysis

Tables 2 and 3 describe the ranks of Japanese industrial sectors on the analyzed years which the results in the tables are based on the calculations using equation (1). From the tables, the sector which had the highest different on the calculated values during the analysis period was sector 13. Figure 1 shows the calculated values movement of the sector during the period of analysis. Based on the figure, one can argue that, from 2005 through 2011, the sector had a decreasing pattern.

Table 2. The ranks of Japanese industrial sectors which are based on the results of calculations using equation (1), 2005.

Sector Number	Sector Name	Calculated Value
13	Activities not elsewhere classified	1.161
3	Manufacturing	0.698
2	Mining	0.570
4	Construction	0.538
9	Transport	0.522
5	Electricity, gas, and water supply	0.510
1	Agriculture, forestry, and fishery	0.472
10	Information and communications	0.409
12	Services	0.380
7	Finance and insurance	0.363
6	Commerce	0.315
11	Public administration	0.263
8	Real estate	0.146

Table 3. The ranks of Japanese industrial sectors which are based on the results of calculations using equation (1), 2011.

Sector Number	Sector Name	Calculated Value
5	Electricity, gas, and water supply	0.730
3	Manufacturing	0.715
13	Activities not elsewhere classified	0.601
2	Mining	0.553
4	Construction	0.548
1	Agriculture, forestry, and fishery	0.515
9	Transport and postal services	0.497
10	Information and communications	0.474
12	Services	0.384
7	Finance and insurance	0.342
11	Public administration	0.317
6	Commerce	0.315
8	Real estate	0.194

(Source: [7])

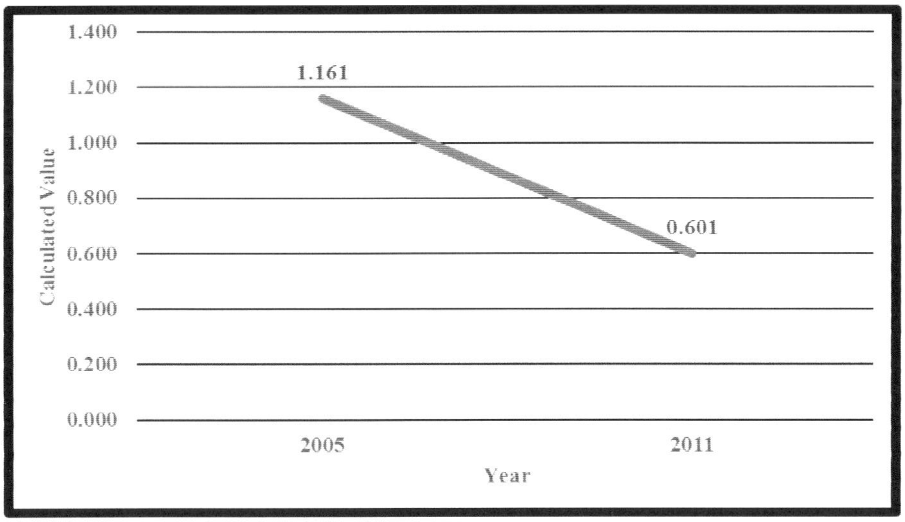

Figure 1. The calculated values movement of sector 13 based on the calculations using equation (1), 2005-2011.

On the other hand, tables 4 and 5 explain the ranks of Japanese industrial sectors on the analyzed years which the results in the tables are based on the calculations using equation (2). From the tables, the sector which had the highest different on the calculated values during the analysis period was sector 13. Figure 2 shows the calculated values movement of the sector during the period of analysis. Based on the figure, one can say that, from 2005 through 2011, the sector had a decreasing pattern.

Table 4. The ranks of Japanese industrial sectors which are based on the results of calculations using equation (2), 2005.

Sector Number	Sector Name	Calculated Value
13	Activities not elsewhere classified	3.044
3	Manufacturing	2.649
4	Construction	2.224
2	Mining	2.168
9	Transport	2.081
1	Agriculture, forestry, and fishery	2.071
5	Electricity, gas, and water supply	2.063
12	Services	1.799
10	Information and communications	1.797
7	Finance and insurance	1.669
6	Commerce	1.600
11	Public administration	1.558
8	Real estate	1.279

Table 5. The ranks of Japanese industrial sectors which are based on the results of calculations using equation (2), 2011.

Sector Number	Sector Name	Calculated Value
3	Manufacturing	2.769
5	Electricity, gas, and water supply	2.625
4	Construction	2.268
1	Agriculture, forestry, and fishery	2.210
2	Mining	2.181
13	Activities not elsewhere classified	2.160
9	Transport and postal services	2.077
10	Information and communications	1.935
12	Services	1.840
11	Public administration	1.664
7	Finance and insurance	1.658
6	Commerce	1.627
8	Real estate	1.361

(Source: [7])

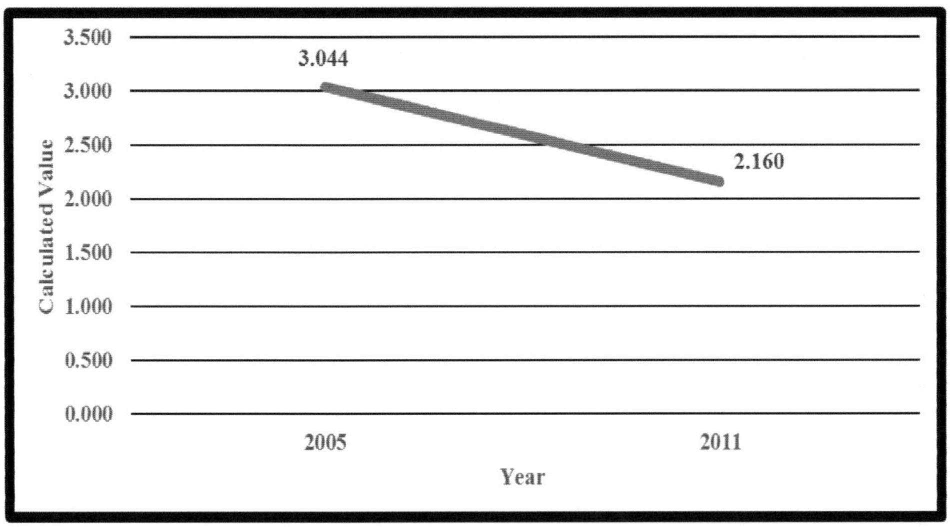

Figure 2. The calculated values movement of sector 13 based on the calculations using equation (2), 2005-2011.

Tables 6 and 7 explain the ranks of Japanese industrial sectors on the analyzed years which the results in the tables are based on the calculations using equation (3). From the tables, the sector which had the highest different on the calculated values during the analysis period was sector 7. Figure 3 shows the calculated values movement of the sector during the period of analysis. Based on the figure, one can say that, from 2005 through 2011, the sector had a decreasing pattern.

Table 6. The ranks of Japanese industrial sectors which are based on the results of calculations using equation (3), 2005.

Sector Number	Sector Name	Calculated Value
3	Manufacturing	1.614
7	Finance and insurance	1.028
12	Services	1.009
9	Transport	0.742
10	Information and communications	0.373
6	Commerce	0.358
11	Public administration	0.280
5	Electricity, gas, and water supply	0.269
2	Mining	0.175
1	Agriculture, forestry, and fishery	0.159
4	Construction	0.157
8	Real estate	0.113
13	Activities not elsewhere classified	0.071

Table 7. The ranks of Japanese industrial sectors which are based on the results of calculations using equation (3), 2011.

Sector Number	Sector Name	Calculated Value
3	Manufacturing	1.664
12	Services	1.197
9	Transport and postal services	0.772
10	Information and communications	0.417
6	Commerce	0.364
2	Mining	0.335
7	Finance and insurance	0.312
5	Electricity, gas, and water supply	0.289
11	Public administration	0.227
8	Real estate	0.203
4	Construction	0.170
1	Agriculture, forestry, and fishery	0.155
13	Activities not elsewhere classified	0.080

(Source: [7])

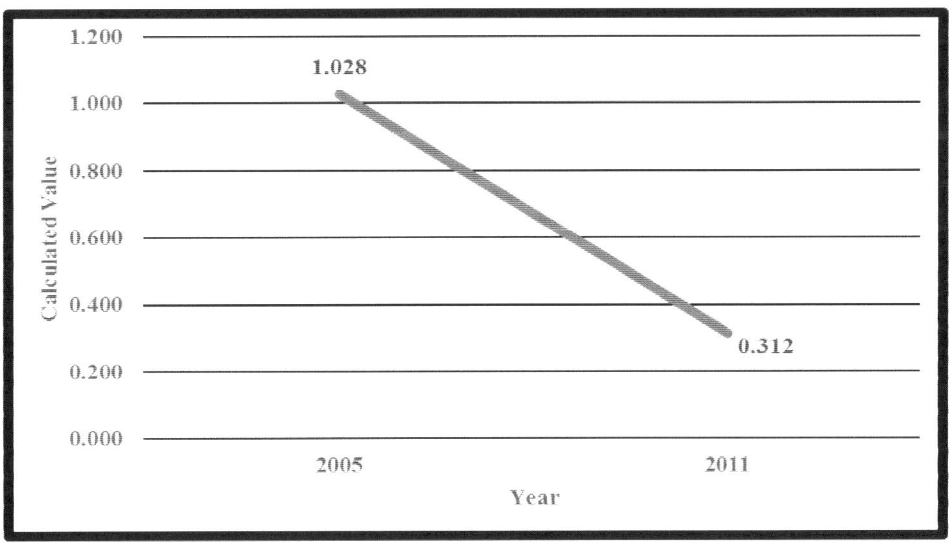

Figure 3. The calculated values movement of sector 7 based on the calculations using equation (3), 2005-2011.

Tables 8 and 9 expose the ranks of Japanese industrial sectors on the analyzed years which the results in the tables are based on the calculations using equation (4). From the tables, the sector which had the highest different on the calculated values during the analysis period was sector 7. Figure 4 shows the calculated values movement of the sector during the period of analysis. Based on the figure, one can argue that, from 2005 through 2011, the sector had a decreasing pattern.

Table 8. The ranks of Japanese industrial sectors which are based on the results of calculations using equation (4), 2005.

Sector Number	Sector Name	Calculated Value
3	Manufacturing	5.249
12	Services	3.107
7	Finance and insurance	2.549
9	Transport	2.322
6	Commerce	1.828
10	Information and communications	1.751
5	Electricity, gas, and water supply	1.523
2	Mining	1.417
1	Agriculture, forestry, and fishery	1.321
11	Public administration	1.317
4	Construction	1.260
8	Real estate	1.227
13	Activities not elsewhere classified	1.133

Table 9. The ranks of Japanese industrial sectors which are based on the results of calculations using equation (4), 2011.

Sector Number	Sector Name	Calculated Value
3	Manufacturing	5.629
12	Services	3.379
9	Transport and postal services	2.455
6	Commerce	1.871
10	Information and communications	1.767
2	Mining	1.763
5	Electricity, gas, and water supply	1.582
7	Finance and insurance	1.521
8	Real estate	1.363
1	Agriculture, forestry, and fishery	1.336
4	Construction	1.297
11	Public administration	1.261
13	Activities not elsewhere classified	1.151

(Source: [7])

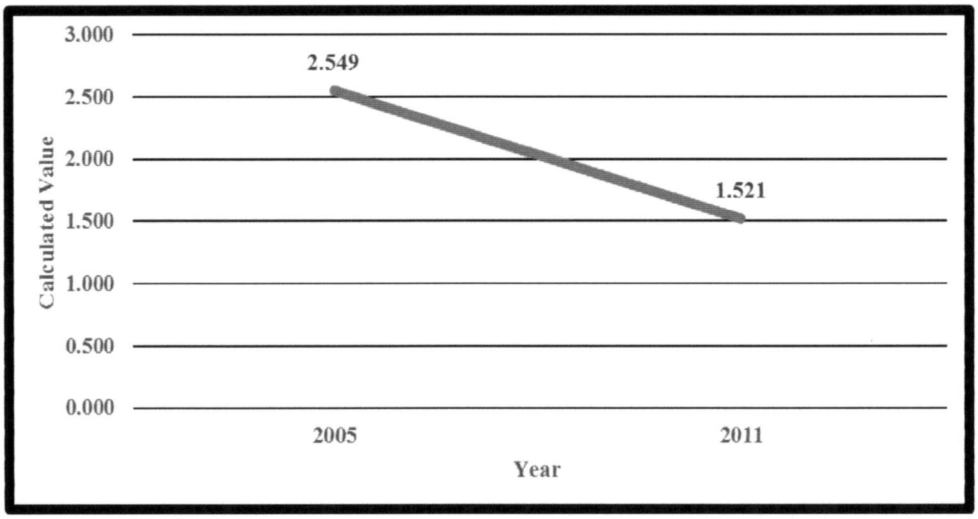

Figure 4. The calculated values movement of sector 7 based on the calculations using equation (4), 2005-2011.

Tables 10 and 11 describe the ranks of Japanese industrial sectors on the analyzed years which the results in the tables are based on the calculations using equation (5). From the tables, the sector which had the highest different on the calculated values during the analysis period was sector 2. Figure 5 exposes the calculated values movement of the sector during the period of analysis. Based on the figure, one can say that, from 2005 through 2011, the sector had an increasing pattern.

Table 10. The ranks of Japanese industrial sectors which are based on the results of calculations using equation (5), 2005.

Sector Number	Sector Name	Calculated Value
2	Mining	16.321
13	Activities not elsewhere classified	1.167
1	Agriculture, forestry, and fishery	0.825
7	Finance and insurance	0.709
5	Electricity, gas, and water supply	0.677
3	Manufacturing	0.647
9	Transport	0.641
10	Information and communications	0.581
12	Services	0.353
6	Commerce	0.346
4	Construction	0.144
8	Real estate	0.125
11	Public administration	0.029

Table 11. The ranks of Japanese industrial sectors which are based on the results of calculations using equation (5), 2011.

Sector Number	Sector Name	Calculated Value
2	Mining	31.701
13	Activities not elsewhere classified	1.003
1	Agriculture, forestry, and fishery	0.887
5	Electricity, gas, and water supply	0.690
3	Manufacturing	0.668
9	Transport and postal services	0.645
10	Information and communications	0.552
7	Finance and insurance	0.517
6	Commerce	0.378
12	Services	0.360
4	Construction	0.186
8	Real estate	0.167
11	Public administration	0.029

(Source: [7])

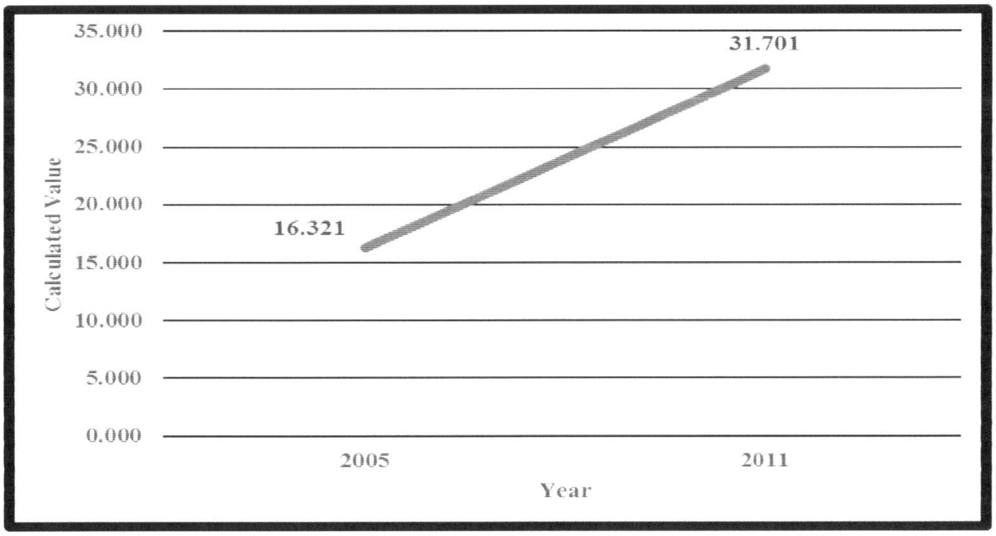

Figure 5. The calculated values movement of sector 2 based on the calculations using equation (5), 2005-2011.

Meanwhile, tables 12 and 13 expose the ranks of Japanese industrial sectors on the analyzed years which the results in the tables are based on the calculations using equation (6). From the tables, the sector which had the highest different on the calculated values during the analysis period was sector 2. Figure 6 explains the calculated values movement of the sector during the period of analysis. Based on the figure, one can argue that, from 2005 through 2011, the sector had an increasing pattern.

Table 12. The ranks of Japanese industrial sectors which are based on the results of calculations using equation (6), 2005.

Sector Number	Sector Name	Calculated Value
2	Mining	40.454
13	Activities not elsewhere classified	3.405
1	Agriculture, forestry, and fishery	3.014
7	Finance and insurance	2.556
9	Transport	2.544
3	Manufacturing	2.449
5	Electricity, gas, and water supply	2.437
10	Information and communications	2.139
12	Services	1.746
6	Commerce	1.735
4	Construction	1.263
8	Real estate	1.247
11	Public administration	1.098

(Source: [7])

Table 13. The ranks of Japanese industrial sectors which are based on the results of calculations using equation (6), 2011.

Sector Number	Sector Name	Calculated Value
2	Mining	82.632
1	Agriculture, forestry, and fishery	3.276
13	Activities not elsewhere classified	3.072
9	Transport and postal services	2.722
5	Electricity, gas, and water supply	2.597
3	Manufacturing	2.583
10	Information and communications	2.087
7	Finance and insurance	1.970
6	Commerce	1.837
12	Services	1.755
4	Construction	1.358
8	Real estate	1.329
11	Public administration	1.089

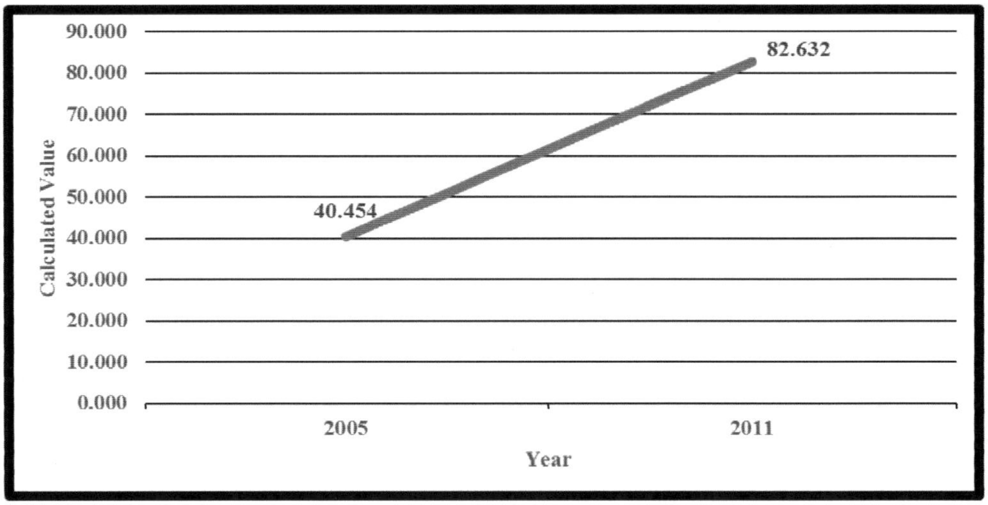

Figure 6. The calculated values movement of sector 2 based on the calculations using equation (6), 2005-2011.

4. Conclusions and further researches

The current study, as a continuation study of the previous one, compares the ranks of Japanese industries by using the analysis methods in the IO analysis, backward and forward linkages, on the analyzed years. From the results of the study, one can argue that the industries which had the beneficial effects in the Japanese economies in 2005 and 2011 were transport and manufacturing sectors, respectively. Besides, the results also show that, in terms of the movements of calculated values of Japanese industrial sectors, the increasing and decreasing patterns appeared on the analysis period.

The suggested further research from the study is to enrich the analyzed industrial sectors for the focused country using the same analysis period. This suggestion is exposed in order to get the deeper insight regarding the economic conditions of Japan on the period, especially about the ranks of its industries. This deeper insight might be useful in determining the focused sectors on its economic activities in the future. The other suggested further research is to expand the area of analysis using the same methodology so the industrial sectors ranks of other countries can also be analyzed. The examples are to conduct the analyses for the ASEAN and EU countries.

References

[1] Kramer W J, Jenkins B and Katz R S 2007 *The Role of the Information and Communications Technology Sector in Expanding Economic Opportunity* [online] http://www.hks.harvard.edu/m-rcbg/CSRI/publications/report_22_EO%20ICT%20Final.pdf (accessed 5 January 2016)
[2] Zuhdi U, Prasetyo A D and Sianipar C P M 2013 Analyzing the dynamics of total output of Japanese creative industry sectors: an input-output approach *Procedia Economics and Finance* **5** 827–35
[3] Zuhdi U 2012 Analyzing the influence of creative industry sector to the national economic structural changes by decomposition analysis: the case of Indonesia *Procedia-Social and Behavioral Sciences* **65** 980–5
[4] Zuhdi U, Utomo D S and Alamanda D T 2011 Analyzing the role of ICT sector to the national economic structural changes: the case of Indonesia *Jurnal Manajemen Teknologi* **10** 299–307
[5] Zuhdi U 2015 An application of input-output analysis in analyzing the impacts of final demands changes on the total outputs of Japanese energy sectors: a further study *Journal of Physics: Conference Series* **622** 012041
[6] Zuhdi U, Mori S and Kamegai K 2014 Analysis of influences of GDP and ICT on Indonesian industrial structural changes using statistical analysis: 1990-2005 *Journal of Finance and Accountancy* **17** 1–19
[7] Zuhdi U 2016 The ranks of Indonesian and Japanese indsutrial sectors *IOP Conference Series: Earth and Environmental Science* **38** 012008
[8] Japanese Ministry of Internal Affairs and Communications 2009 *Introduction: The Japanese Economy and the 2005 Input-Output Tables* [online] http://www.soumu.go.jp/main_content/000327476.pdf (accessed 26 April 2016)
[9] Japanese Ministry of Internal Affairs and Communications 2016 *2011 Input-Output Tables for Japan* [online] http://www.soumu.go.jp/main_content/000443188.pdf (accessed December 30, 2016)
[10] Miller R E and Blair P D 2009 *Input-Output Analysis: Foundations and Extensions* (Cambridge: University Press)
[11] Chenery H B and Watanabe T 1958 International comparisons of the structure of productions *Econometrica* **4** 487–521
[12] Nazara S 2005 *Input-Output Analysis* [in Indonesian] (Jakarta: The Faculty of Economics of University of Indonesia)
[13] Zuhdi U 2014 Analyzing the impacts of final demand changes on total output using input-output approach: the case of Japanese ICT sectors *IOP Conference Series: Earth and Environmental Science* **19** 012016
[14] Zuhdi U 2014 Analyzing the role of creative industries in national economy of Japan: 1995-2005 *Open Journal of Applied Sciences* **4** 197–211
[15] Zuhdi U 2014 An input-output approach to analyze the ways to increase total output of energy sectors: the case of Japan *IOP Conference Series: Earth and Environmental Science* **19** 012015
[16] Zuhdi U 2014 The dynamics of total output of Indonesian information and communication technology sector when final demand changes occur: an analysis using input-output approach *Advanced Science Letters* **20** 2254–57
[17] Zuhdi U 2014 The role of information and communication technology sectors in Indonesian national economy from 1990 through 2008: an analysis using input-output approach *Advanced Science Letters* **20** 1932–5
[18] Zuhdi U 2015 An analysis of the role of information and communication technology sectors on Japanese national economy from 1995 through 2005: an application of multiplier analysis *IOP Conference Series: Earth and Environmental Science* **23** 012014
[19] Zuhdi U 2015 An application of multiplier analysis in analyzing the role of information and communication technology sectors on Indonesian national economy: 1990-2005 *IOP Conference Series: Earth and Environmental Science* **23** 012015
[20] Zuhdi U 2015 The dynamics of Indonesian creative industry sectors: an analysis using input-output approach *Journal of the Knowledge Economy* **6** 1177–90

IOP Conf. Series: Journal of Physics: Conf. Series **820** (2017) 012031 doi:10.1088/1742-6596/820/1/012031

The dynamics of total outputs of Indonesian industrial sectors: A further study

Ubaidillah Zuhdi

Faculty of Management and Economics, Gdansk University of Technology, Gdansk 80-233, Poland

zuhdi@zie.pg.gda.pl

Abstract. The purpose of the current study is to extend the previous studies which analyze the impacts of final demands modifications on the total outputs of industrial sectors of a particular country. More specifically, the study conducts the analysis regarding the impacts on the total outputs of Indonesian industrial sectors. The study employs a demand-pull Input-Output (IO) quantity model, one of the calculation instruments in the IO analysis. The study focuses on seventeen industries. There are two scenarios used in this study, namely other final demands and import modifications. The "whole sector change" condition is implemented in the calculations. An initial period in this study is 2010. The results show that the positive impacts on the total outputs of focused sectors are distributed by scenario 1, the change of other final demands. On the contrary, the negative impacts are delivered by scenario 2, the modification of imports. The suggestions for improving the total outputs of discussed industries are based on the results.

1. Introduction

One of the important aspects in the economic activities of one country is industrial sectors. The importance can be seen not only on the micro level but also macro. Therefore, an analysis of industries cannot be separated from the investigation of economic activities of one or several countries. Further, the suggestions for improving the economic conditions of focused countries might be generated from the analysis.

The examples of previous studies which conduct the analysis are [1], [2], [3], [4], [5], and [6]. These previous studies focus on the impacts of final demands changes on the total outputs of industrial sectors of analyzed countries. The previous studies, however, only focus on the specific industries. In other words, the study focuses on identifying the impacts on the whole sector of particular countries is still needed. This identification is required in order to know the characteristics of industries when the final demands changes appear so the suggestions for improving the sectors can be formalized properly. The current study is conducted in order to fulfill the gap.

The purpose of the study is to extend the previous studies which analyze the impacts of final demands modifications on the total outputs of industrial sectors of a particular country. More specifically, the study aims to conduct a further analysis regarding the impacts by using the new data as well as new approach, namely to analyze all industrial sectors. The study focuses on the case of Indonesia, and employs the Input-Output (IO) analysis as an analysis tool.

2. Methodology

The methodology of the current study refers to the previous studies which were conducted, for example, by [1] and [2]. The first step of the methodology is to describe the data source. The data source of this study is the 2010 Indonesian IO table. The table is obtained from [7]. The second step is to explore the industrial sectors of Indonesia used in this study. Table 1 shows the industries. The table consists of seventeen industrial sectors.

Table 1. Indonesian industrial sectors used in this study.

Sector Number	Sector Name
1	Agriculture, forestry, and fishing
2	Mining and quarrying
3	Manufacturing
4	Electricity and gas
5	Water supply, sewerage, waste management, and remediation activities
6	Construction
7	Wholesale and retail trade; repair of motor vehicles and motorcycles
8	Transportation and storage
9	Accommodation and food service activities
10	Information and communication
11	Financial and insurance activities
12	Real estate activities
13	Business activities
14	Public administration and defence; compulsory social security
15	Education
16	Human health and social work activities
17	Other services activities

The third step is to conduct the calculations in order to determine the impacts of final demands changes on the total outputs of analyzed industries. A demand-pull IO quantity model, one of the calculation instruments in the IO analysis, is employed in the calculations. [8] explain that the following equation is a representation of the model:

$$\mathbf{x}^1 = \mathbf{L}^0\mathbf{f}^1 \quad (1)$$

where **x**, **L**, and **f** are the matrices of the total outputs of sectors, the Leontief inverse, and the final demands of sectors, respectively. 0 and 1 describe initial and future periods, respectively. An initial period in this study is 2010. The scenarios of final demand modification are exposed in table 2. There are two scenarios used in this study, namely other final demands and import modifications.

The condition of "whole sector change" is considered in the calculations. In this condition, the change on each scenario is addressed to all Indonesian industries. The analysis regarding the impacts is focused on the next step. Conclusions of this study, and suggestions for further researches are explained on the final step.

Table 2. The scenarios of final demand modification used in this study.

The component of the final demand	Scenario	
	1	2
	Other final demands modification	Import modification
Other final demands	Increase 30%	Constant
Import	Constant	Increase 30%

IOP Conf. Series: Journal of Physics: Conf. Series **820** (2017) 012031 doi:10.1088/1742-6596/820/1/012031

3. Results and analysis

Table 3 describes the total outputs of analyzed sectors for each scenario. Figures 1-17 explain in more details the dynamics of the total outputs of discussed sectors for each scenario. Based on the information in the tables and figures, the positive impacts on the total outputs of analyzed sectors are distributed by scenario 1, the modification of other final demands. On the contrary, the negative impacts are given by scenario 2, the change of import.

The suggestions for improving the total outputs of focused sectors are based on above results. One can argue that the Indonesian government should encourage the national economic activities so Indonesian industrial sectors have a trigger to produce more. One of the actions that the Indonesian government can do is to increase the exports goals. Conversely, they should set carefully the rate of imports.

Table 3. The total outputs of analyzed sectors for each scenario (billion Rupiah).

Sector Number	Sector Name	(X_t), t = 2010	(X_{t+1}), Scenario 1	(X_{t+1}), Scenario 2
1	Agriculture, forestry, and fishing	1,187,980.10	1,648,768.93	1,083,584.83
2	Mining and quarrying	941,316.80	1,317,181.73	847,846.61
3	Manufacturing	4,370,817.00	6,245,824.25	3,807,054.70
4	Electricity and gas	309,493.10	414,711.10	297,123.00
5	Water supply, sewerage, waste management, and remediation activities	23,812.60	34,274.55	20,493.95
6	Construction	1,724,302.60	2,251,344.60	1,714,550.91
7	Wholesale and retail trade; repair of motor vehicles and motorcycles	1,369,971.40	1,823,341.44	1,327,592.74
8	Transportation and storage	550,887.80	740,514.52	526,527.12
9	Accommodation and food service activities	440,129.50	589,359.29	422,939.01
10	Information and communication	409,067.60	554,793.66	386,061.77
11	Financial and insurance activities	333,056.60	449,446.55	316,583.11
12	Real estate activities	246,308.00	322,624.98	243,883.64
13	Business activities	202,971.50	285,019.26	181,815.15
14	Public administration and defence; compulsory social security	418,489.30	547,896.01	414,629.37
15	Education	311,211.20	405,825.55	309,960.20
16	Human health and social work activities	149,315.20	196,603.07	146,821.64
17	Other services activities	119,990.10	158,969.25	117,008.30

IOP Conf. Series: Journal of Physics: Conf. Series **820** (2017) 012031 doi:10.1088/1742-6596/820/1/012031

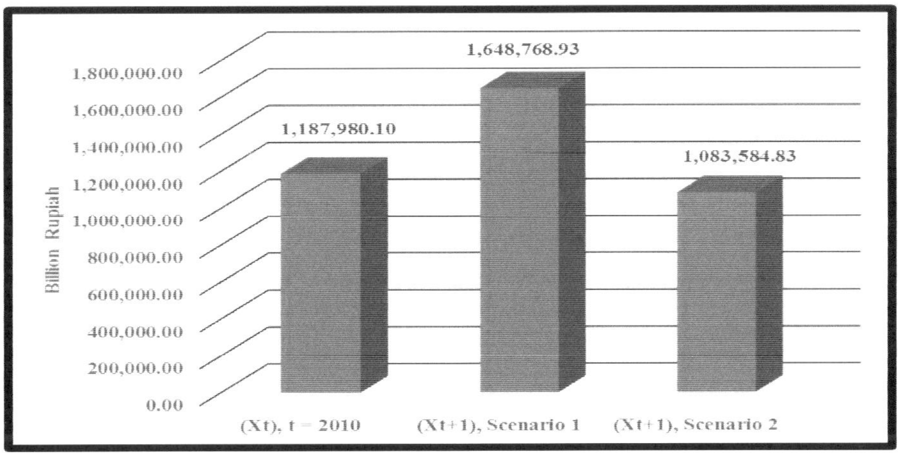

Figure 1. The dynamics of the total output of the agriculture, forestry, and fishing sector.

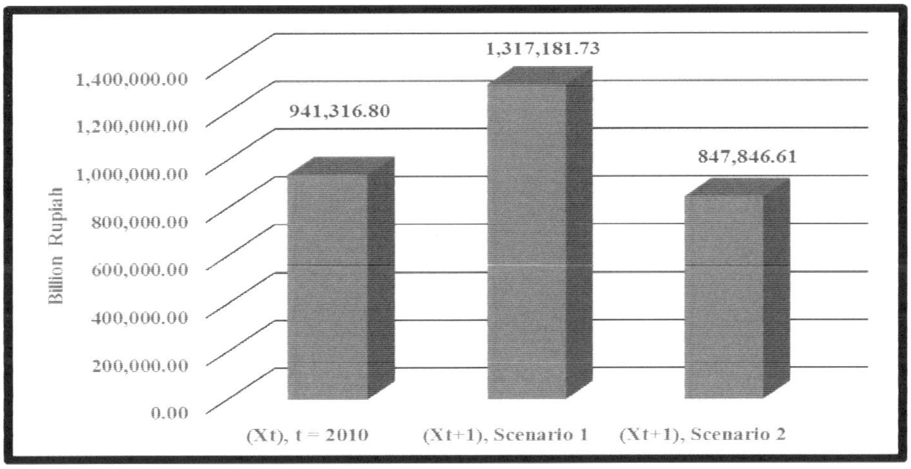

Figure 2. The dynamics of the total output of the mining and quarrying sector.

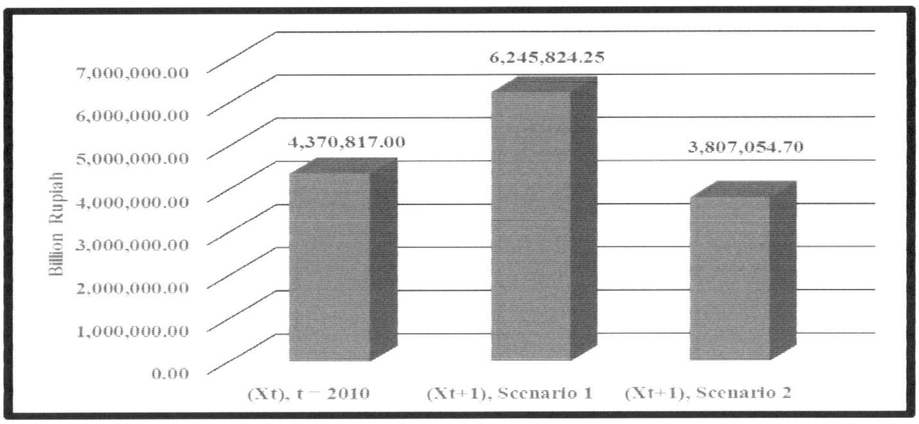

Figure 3. The dynamics of the total output of the manufacturing sector.

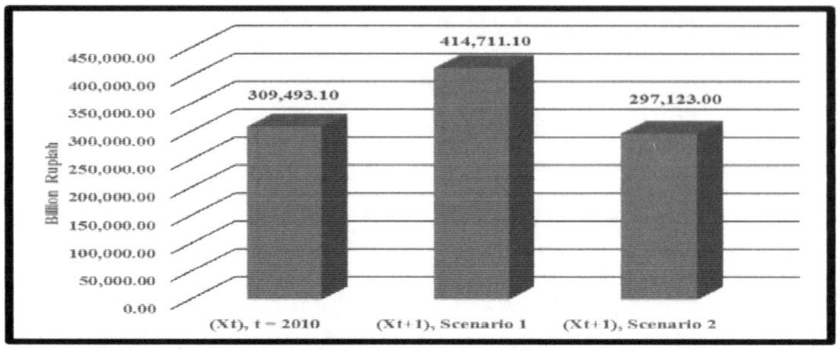

Figure 4. The dynamics of the total output of the electricity and gas sector.

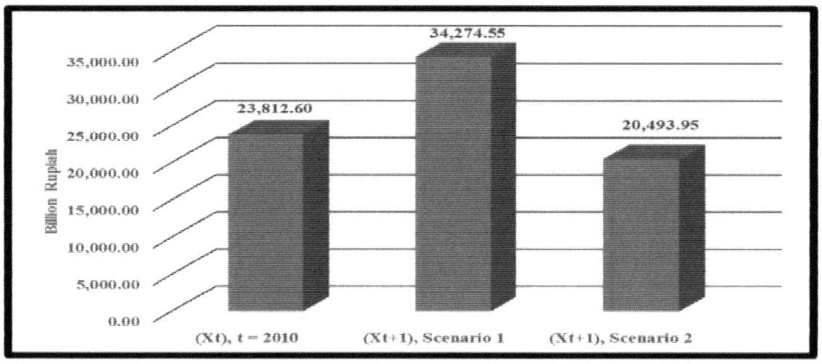

Figure 5. The dynamics of the total output of the water supply, sewerage, waste management, and remediation activities sector.

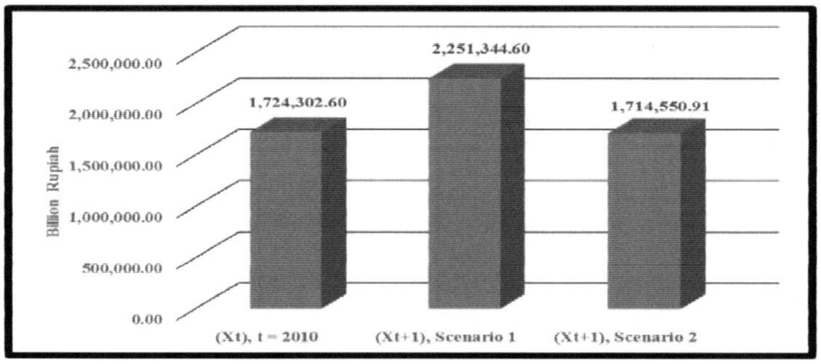

Figure 6. The dynamics of the total output of the construction sector.

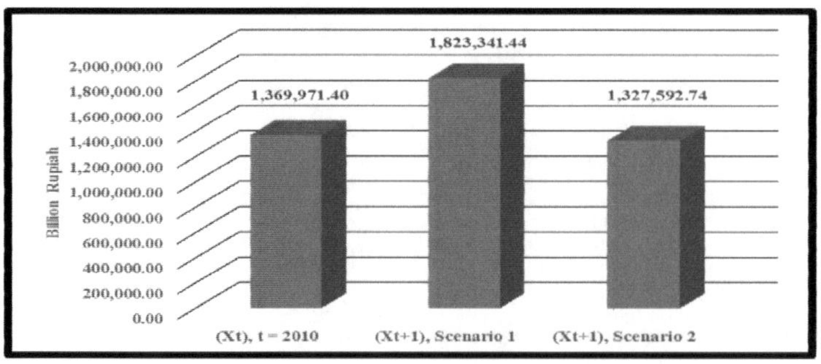

Figure 7. The dynamics of the total output of the wholesale and retail trade; repair of motor vehicles and motorcycles sector.

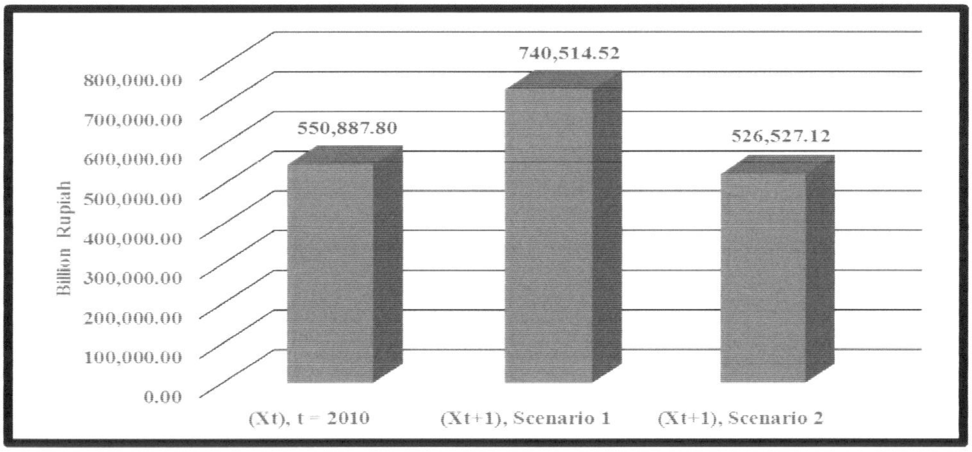

Figure 8. The dynamics of the total output of the transportation and storage sector.

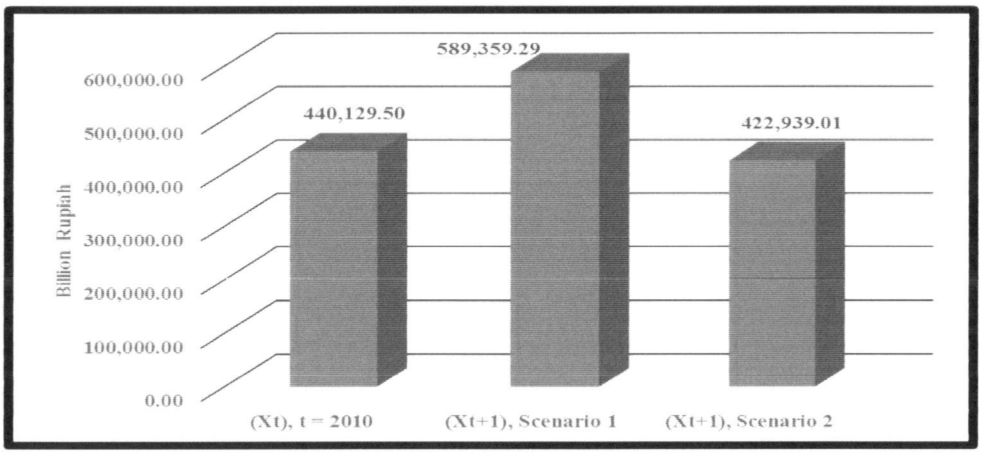

Figure 9. The dynamics of the total output of the accommodation and food service activities sector.

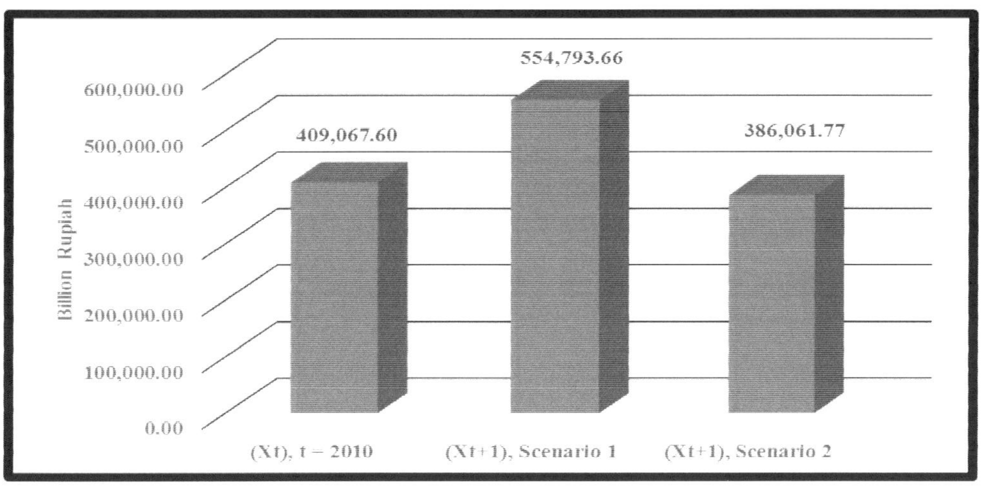

Figure 10. The dynamics of the total output of the information and communication sector.

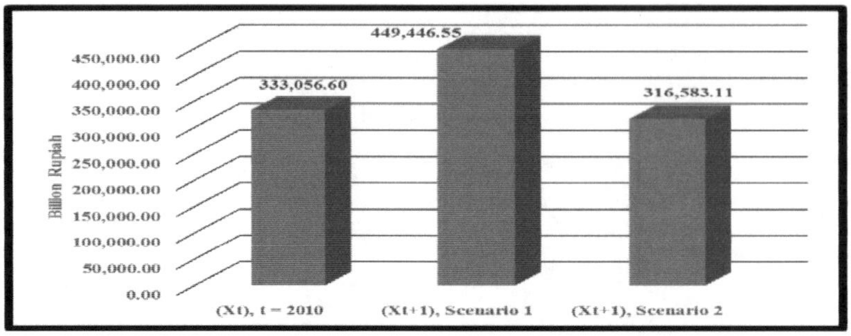

Figure 11. The dynamics of the total output of the financial and insurance activities sector.

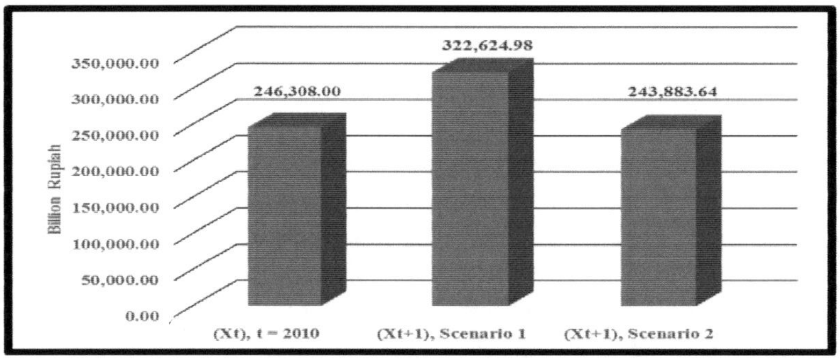

Figure 12. The dynamics of the total output of the real estate activities sector.

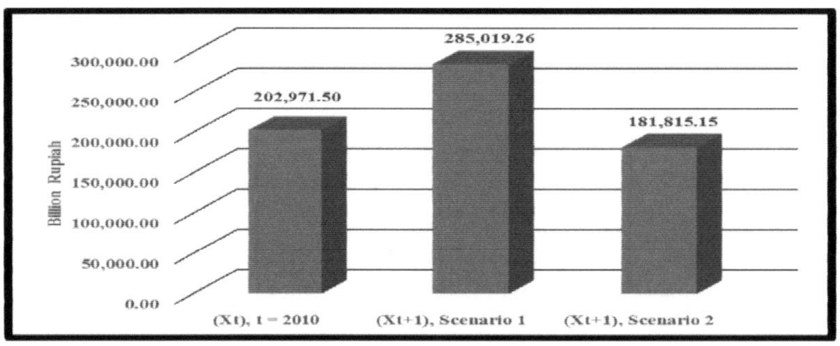

Figure 13. The dynamics of the total output of the business activities sector.

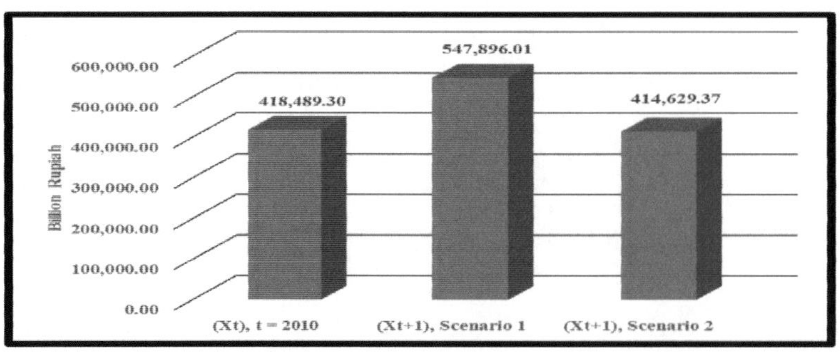

Figure 14. The dynamics of the total output of the public administration and defence; compulsory social security sector.

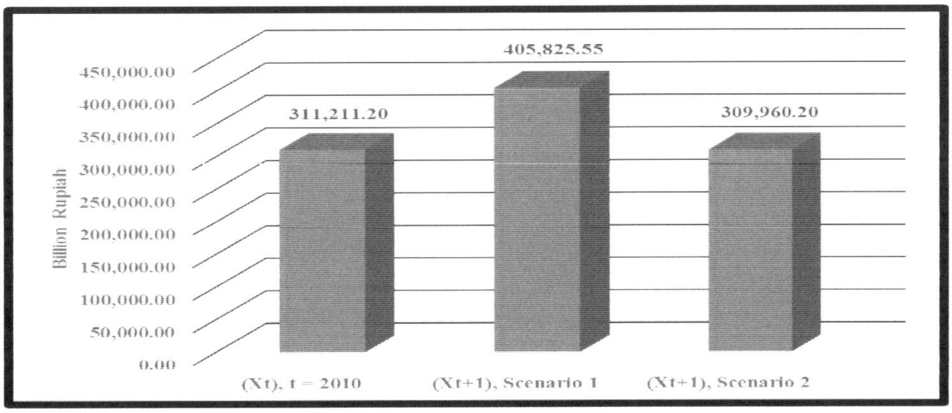

Figure 15. The dynamics of the total output of the education sector.

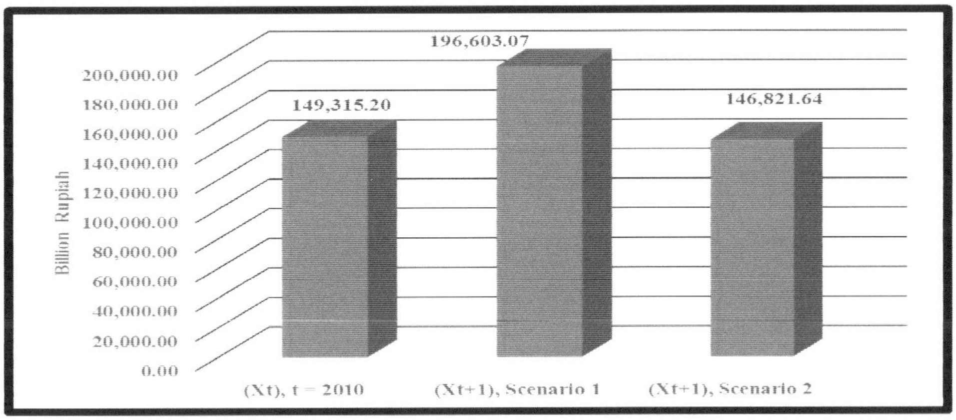

Figure 16. The dynamics of the total output of the human health and social work activities sector.

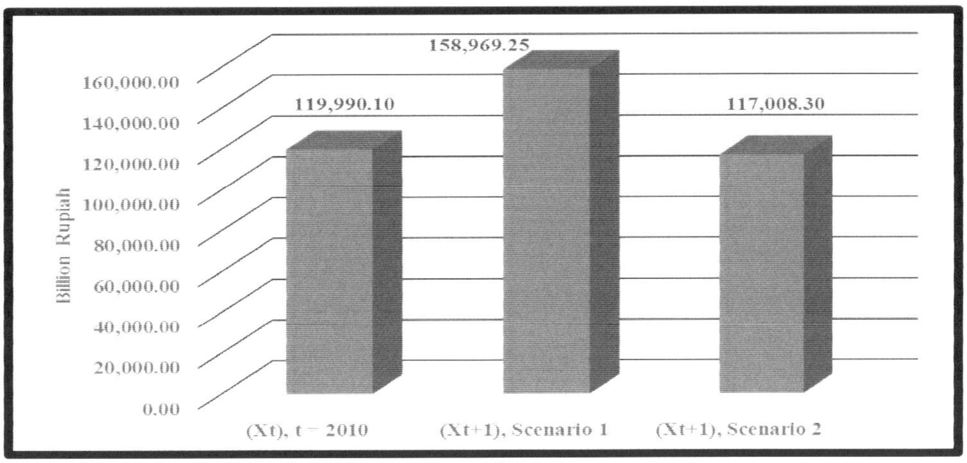

Figure 17. The dynamics of the total output of the other services activities sector.

4. Conclusions and further researches

This study, as a continuation study of previous ones, conducts the analysis regarding the impacts of the modifications of final demands on the total outputs of Indonesian industrial sectors. This study employs a demand-pull IO quantity model, one of the calculation instruments in the IO analysis. This study focuses on seventeen industries. The "whole sector change" condition is implemented in the calculations. In this condition, the modification on each scenario is addressed to all analysed industries. An initial period in this study is 2010.

The results show that the positive impacts on the total outputs of focused sectors are distributed by scenario 1, the change of other final demands. On the contrary, the negative impacts are delivered by scenario 2, the modification of imports. The suggestions for improving the total outputs of discussed industries are based on the results. One can say that the Indonesian government should push the activities of national economy so Indonesian industrial sectors have a trigger in producing more products. One of the steps that the Indonesian government can do is to accelerate the exports targets. On the contrary, they should assign carefully the rate of imports.

The deeper insight regarding the impacts of the changes of final demands on the total outputs of Indonesian industries is obtained from this study. However, this study focuses on the aggregated industrial sectors. In other words, this study is still far in explaining the whole view about the impacts on the Indonesian national economy. This view is needed in order to know better the circumstances of Indonesian national economy so the comprehensive regulations for increasing those in the future can be formalized. Therefore, as a further research, this study proposes the same analysis for the disaggregated Indonesian industries. The other suggested further research from this study is to implement the same analysis on the specific regional area. This analysis might be useful in exposing the characteristics of countries in the area in responding the impacts. The examples are to implement the analyses for the ASEAN and EU countries.

References

[1] Zuhdi U 2016 The dynamics of the total output of the fishery sector: the case of Indonesia *Journal of Physics: Conference Series* **710** 012039

[2] Zuhdi U 2016 The dynamics of the total outputs of Japanese information and communication technology sectors: a further study *Journal of Physics: Conference Series* **710** 012041

[3] Zuhdi U 2015 An application of input-output analysis in analyzing the impacts of final demands changes on the total outputs of Japanese energy sectors: a further study *Journal of Physics: Conference Series* **622** 012041

[4] Kramer W J, Jenkins B and Katz R S 2007 *The Role of the Information and Communications Technology Sector in Expanding Economic Opportunity* [online] http://www.hks.harvard.edu/m-rcbg/CSRI/publications/report_22_EO%20ICT%20Final.pdf (accessed 5 January 2016)

[5] Zuhdi U 2015 The dynamics of Indonesian creative industry sectors: an analysis using input-output approach *Journal of the Knowledge Economy* **6** 1177–90

[6] Zuhdi U 2014 The impacts of final demand changes on total output of Indonesian ICT sectors: an analysis using input-output approach *IOP Conference Series: Materials Science and Engineering* **58** 012011

[7] BPS-Statistics Indonesia 2016 *Statistical Yearbook of Indonesia 2016* [online] https://www.bps.go.id/website/pdf_publikasi/Statistik-Indonesia-2016--_rev.pdf (accessed December 30, 2016)

[8] Miller R E and Blair P D 2009 *Input-Output Analysis: Foundations and Extensions* (Cambridge: University Press)

[9] Zuhdi U, Prasetyo A D and Sianipar C P M 2013 Analyzing the dynamics of total output of Japanese creative industry sectors: an input-output approach *Procedia Economics and Finance* **5** 827–35

[10] Zuhdi U 2012 Analyzing the influence of creative industry sector to the national economic structural changes by decomposition analysis: the case of Indonesia *Procedia-Social and Behavioral Sciences* **65** 980–5

[11] Zuhdi U, Utomo D S and Alamanda D T 2011 Analyzing the role of ICT sector to the national economic structural changes: the case of Indonesia *Jurnal Manajemen Teknologi* **10** 299–307

[12] Zuhdi U, Mori S and Kamegai K 2014 Analysis of influences of GDP and ICT on Indonesian industrial structural changes using statistical analysis: 1990-2005 *Journal of Finance and Accountancy* **17** 1–19

[13] Zuhdi U 2016 The ranks of Indonesian and Japanese indsutrial sectors *IOP Conference Series: Earth and Environmental Science* **38** 012008
[14] Chenery H B and Watanabe T 1958 International comparisons of the structure of productions *Econometrica* **4** 487–521
[15] Nazara S 2005 *Input-Output Analysis* [in Indonesian] (Jakarta: The Faculty of Economics of University of Indonesia)
[16] Zuhdi U 2014 Analyzing the impacts of final demand changes on total output using input-output approach: the case of Japanese ICT sectors *IOP Conference Series: Earth and Environmental Science* **19** 012016
[17] Zuhdi U 2014 Analyzing the role of creative industries in national economy of Japan: 1995-2005 *Open Journal of Applied Sciences* **4** 197–211
[18] Zuhdi U 2014 An input-output approach to analyze the ways to increase total output of energy sectors: the case of Japan *IOP Conference Series: Earth and Environmental Science* **19** 012015
[19] Zuhdi U 2014 The dynamics of total output of Indonesian information and communication technology sector when final demand changes occur: an analysis using input-output approach *Advanced Science Letters* **20** 2254–57
[20] Zuhdi U 2014 The other perspective related to the role of information and communication technologies sectors in national economy: the case of Japan *Advanced Science Letters* **20** 483–6
[21] Zuhdi U 2014 The role of information and communication technology sectors in Indonesian national economy from 1990 through 2008: an analysis using input-output approach *Advanced Science Letters* **20** 1932–5
[22] Zuhdi U 2014 Using multipliers analysis in order to get another perspective related to the role of ICT sectors in national economy of Indonesia: 1990-2005 *Journal of Physics: Conference Series* **495** 012051
[23] Zuhdi U 2015 An analysis of the role of information and communication technology sectors on Japanese national economy from 1995 through 2005: an application of multiplier analysis *IOP Conference Series: Earth and Environmental Science* **23** 012014
[24] Zuhdi U 2016 The dynamics of the total output of the Japanese fisheries sector: an analysis using input-output approach *Journal of Physics: Conference Series* **710** 012040
[25] Zuhdi U 2016 The Indonesian economy in 2005: an analysis using the input-output approach *Proceedings of 20th EBES Conference-Vienna* **3** 1825–42
[26] Zuhdi U, Mori S and Kamegai K 2012 Analyzing the role of ICT sector to the national economic structural changes by decomposition analysis: the case of Indonesia and Japan *Procedia-Social and Behavioral Sciences* **65** 749–54
[27] Zuhdi U, Mori S and Kamegai K 2013 Analysis of influences of ICT on structural changes in Japanese commerce, business services and office supplies, and personal services sectors using multivariate analysis: 1985-2005 *The Asian Journal of Technology Management* **6** 102–11
[28] Zuhdi U, Mori S and Kamegai K 2014 Statistical analysis of influences of ICT on industrial structure changes from 1985 through 2005: the case of Japan *Journal of Computers* **9** 1291–9
[29] Zuhdi U, Mori S and Kamegai K 2015 Forecasting the influences of information and communication technology on the structural changes of Japanese industrial sectors: a study using statistical analysis *International Journal of Social, Behavioral, Educational, Economic, Business and Industrial Engineering* **9** 531–7
[30] Zuhdi U and Prasetyo A D 2014 Examining the total output changes of ICT sectors of Japan: an approach of input-output *Procedia-Social and Behavioral Sciences* **109** 659–63
[31] Zuhdi U, Prasetyo A D and Putranto N A R 2014 Analyzing the changes of total output of Japanese livestock sector: an input-output approach *Procedia-Social and Behavioral Sciences* **109** 649–53
[32] Zuhdi U, Putranto N A R and Prasetyo A D 2014 An input-output approach to know the dynamics of total output of livestock sectors: the case of Indonesia *Procedia-Social and Behavioral Sciences* **109** 634–8
[33] Zuhdi U, Putranto N A R and Prasetyo A D 2014 Encouraging information and communication technology sectors using input-output approach: the case of Indonesia *Advanced Science Letters* **20** 199–202

The impacts of final demands changes on the total outputs of Japanese industrial sectors: A further study

Ubaidillah Zuhdi

Faculty of Management and Economics, Gdansk University of Technology, Gdansk 80-233, Poland

zuhdi@zie.pg.gda.pl

Abstract. The purpose of the current study is to expand the previous studies which analyze the impacts of final demands changes on the total outputs of industrial sectors of a specific country. More specifically, the study conducts the analysis regarding the impacts on the total outputs of Japanese industries. The study employs a demand-pull Input-Output (IO) quantity model, one of the calculation tools in the IO analysis. The study focuses on thirteen industries. There are two scenarios used in this study, namely exports and imports modifications. The "whole sector change" condition is considered in the calculations. An initial period in this study is 2011. The results show that the positive impacts on the total outputs of focused sectors are delivered by scenario 1, the change of exports. On the contrary, the negative impacts are distributed by scenario 2, the modification of imports. The suggestions for improving the total outputs of discussed sectors are based on the results.

1. Introduction

Industrial sectors have an important role in the economic activities of one country. The importance can be seen on the macro and micro scopes. Therefore, an analysis of industries cannot be separated from the exploration of economic situations of one or several countries. Further, the proposals for improving the economic conditions of focused countries might be generated from the analysis.

The examples of previous studies which focus on the analysis are [1], [2], [3], [4], [5], and [6]. These previous studies investigate the impacts of final demands changes on the total outputs of industrial sectors of analyzed countries. The previous studies, however, only focus on the specific industries. In other words, the study focuses on examining the impacts on the whole sector of particular countries is still needed. This examination is required in order to know the responds of industries when the final demands changes appear so the suggestions for improving the sectors can be made properly. The current study is performed in order to fulfill the gap.

The purpose of the study is to expand the previous studies which analyze the impacts of final demands modifications on the total outputs of industries of a specific country. More specifically, the study aims to conduct a further investigation regarding the impacts by using the new data as well as new approach, namely to analyze all industrial sectors. The study focuses on the case of Japan, and employs the Input-Output (IO) analysis as an analysis instrument.

2. Methodology

The methodology of the current study refers to the previous studies which were conducted, for example, by [1] and [2]. The first step of the methodology is to expose the data source. The data source of this study is the 2011 Japanese IO table. The table is obtained from [7]. The second step is to describe the industrial sectors of Japan used in this study. Table 1 shows the industries. The table consists of thirteen industrial sectors.

Table 1. Japanese industrial sectors used in this study.

Sector Number	Sector Name
1	Agriculture, forestry, and fishery
2	Mining
3	Manufacturing
4	Construction
5	Electricity, gas, and water supply
6	Commerce
7	Finance and insurance
8	Real estate
9	Transport and postal services
10	Information and communications
11	Public administration
12	Services
13	Activities not elsewhere classified

The third step is to conduct the calculations in order to identify the impacts of final demands changes on the total outputs of discussed industries. A demand-pull IO quantity model, one of the calculation tools in the IO analysis, is employed in the calculations. [8] expose that the following equation is a representation of the model:

$$\mathbf{x}^1 = \mathbf{L}^0\mathbf{f}^1 \tag{1}$$

where **x**, **L**, and **f** are the matrices of the total outputs of industries, the Leontief inverse, and the final demands of industries, respectively. 0 and 1 explain initial and future periods, respectively. An initial period in this study is 2011. The scenarios of final demand modification are described in table 2. There are two scenarios used in this study, namely exports and imports modifications.

The condition of "whole sector change" is considered in the calculations. In this condition, the change on each scenario is addressed to all Japanese industries. The analysis regarding the impacts is done on the next step. Conclusions of this study, and suggestions for further researches are described on the final step.

Table 2. The scenarios of final demand modification used in this study.

The component of the final demand	Scenario	
	1	2
	Exports modification	Imports modification
Exports	Increase 30%	Constant
Import	Constant	Increase 30%

3. Results and analysis

Table 3 describes the total outputs of analyzed sectors for each scenario. Figures 1-13 explain in more details the dynamics of the total outputs of focused sectors for each scenario. Based on the information in the tables and figures, the positive impacts on the total outputs of analyzed sectors are given by scenario 1, the modification of exports. On the contrary, the negative impacts are distributed by scenario 2, the change of imports.

The suggestions for improving the total outputs of discussed sectors are based on above results. One can say that the Japanese government should reinforce the activities of national economy so Japanese industries have a stimulus in producing more outputs. One of the actions that the Japanese government can conduct is to increase the exports targets. On the contrary, they should set carefully the rate of imports.

Table 3. The total outputs of analyzed sectors for each scenario (100 million Yen).

Sector Number	Sector Name	(X_t), t = 2011	(X_{t+1}), Scenario 1	(X_{t+1}), Scenario 2
1	Agriculture, forestry, and fishery	120,360.00	131,012.89	101,435.11
2	Mining	7,600.00	30,140.90	(84,918.00)
3	Manufacturing	2,899,045.00	3,230,899.09	2,579,848.77
4	Construction	525,145.00	528,990.91	520,455.86
5	Electricity, gas, and water supply	257,547.00	268,981.06	243,309.95
6	Commerce	936,558.00	983,963.55	906,919.68
7	Finance and insurance	320,939.00	329,502.58	309,384.65
8	Real estate	711,875.00	716,714.09	706,463.88
9	Transport and postal services	482,340.00	524,157.06	428,526.11
10	Information and communications	461,603.00	471,273.74	449,612.73
11	Public administration	394,052.00	394,605.77	393,319.93
12	Services	2,229,582.00	2,276,303.92	2,173,435.88
13	Activities not elsewhere classified	50,103.00	52,544.12	46,875.94

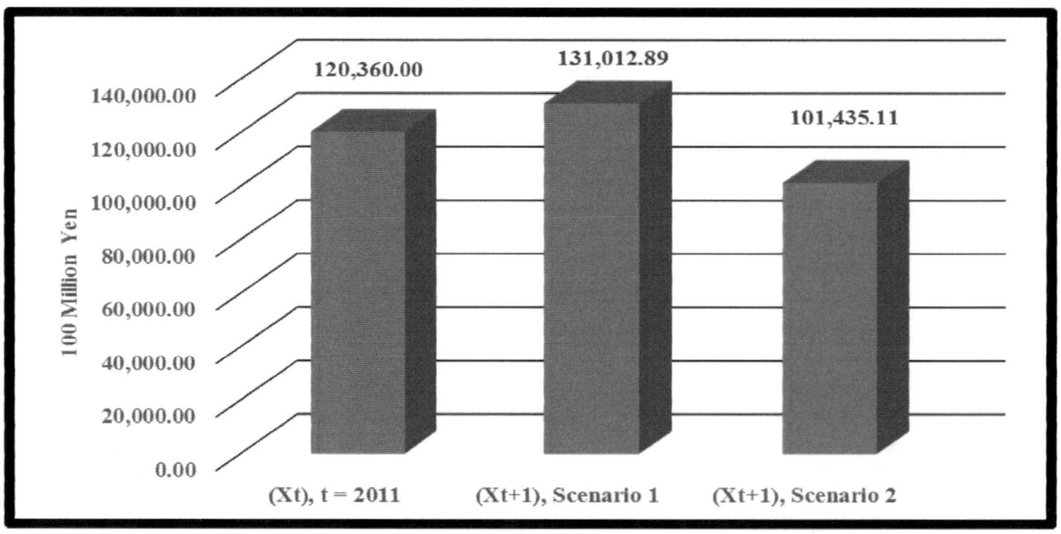

Figure 1. The dynamics of the total output of the agriculture, forestry, and fishery sector.

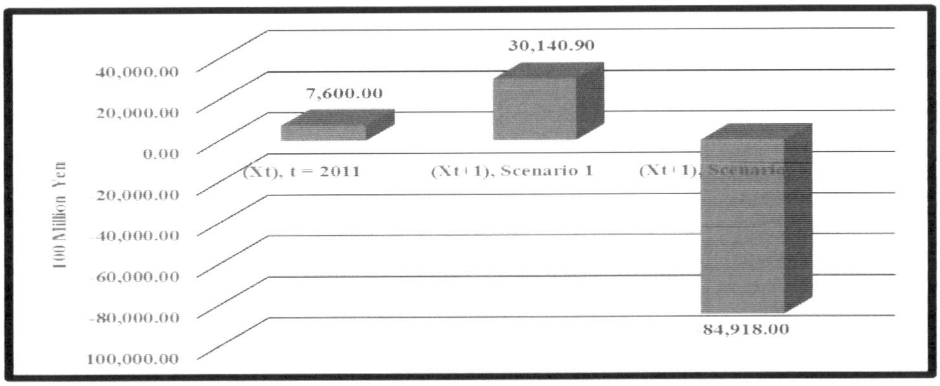

Figure 2. The dynamics of the total output of the mining sector.

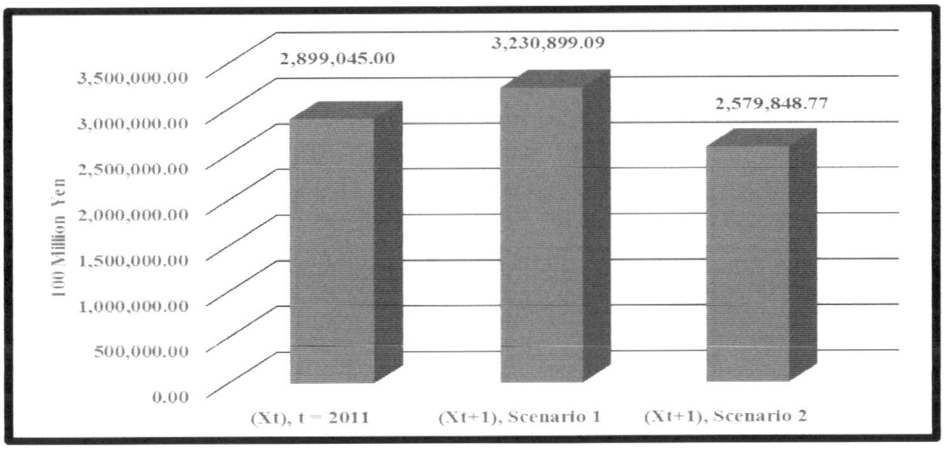

Figure 3. The dynamics of the total output of the manufacturing sector.

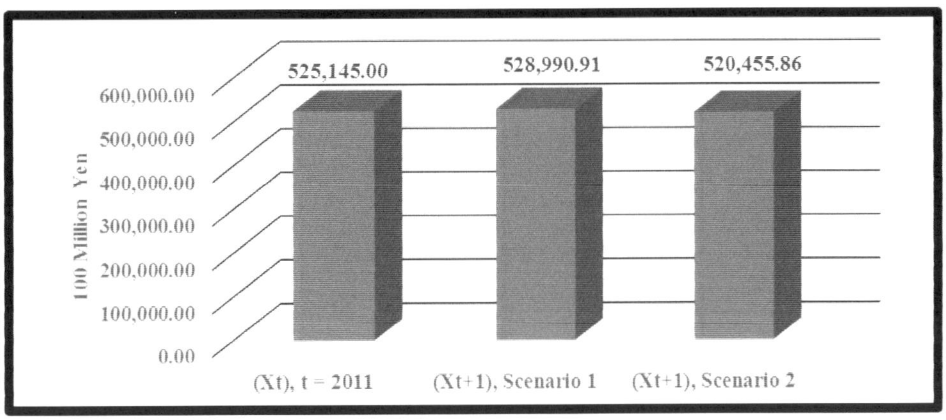

Figure 4. The dynamics of the total output of the construction sector.

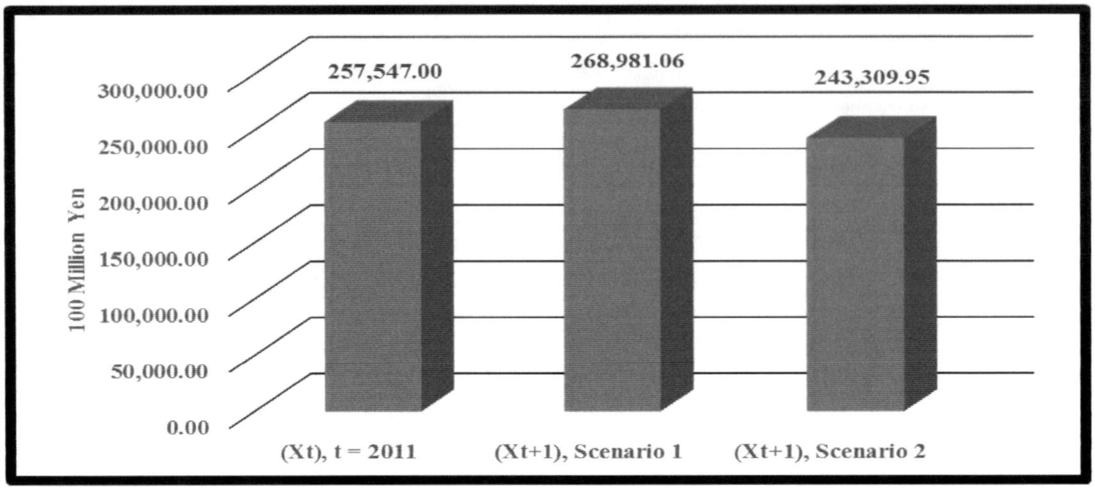

Figure 5. The dynamics of the total output of the electricity, gas, and water supply sector.

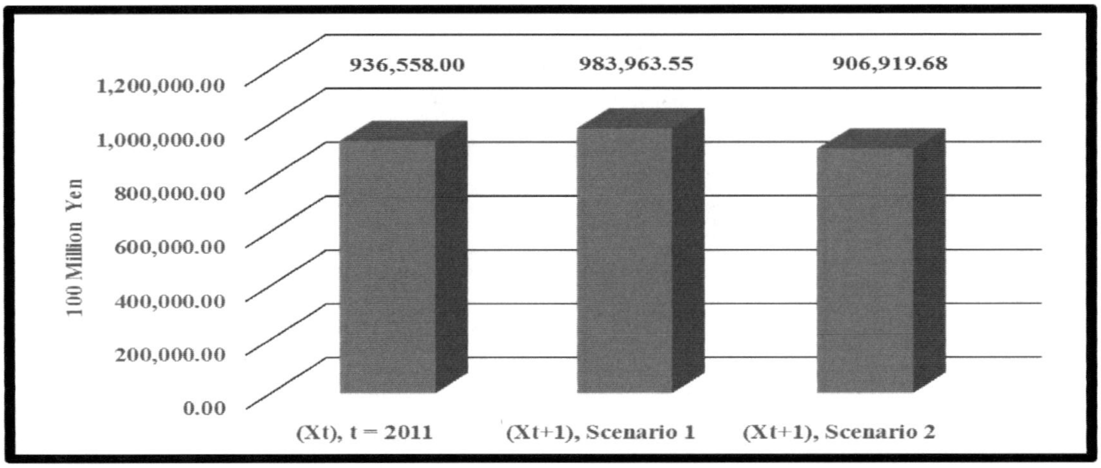

Figure 6. The dynamics of the total output of the commerce sector.

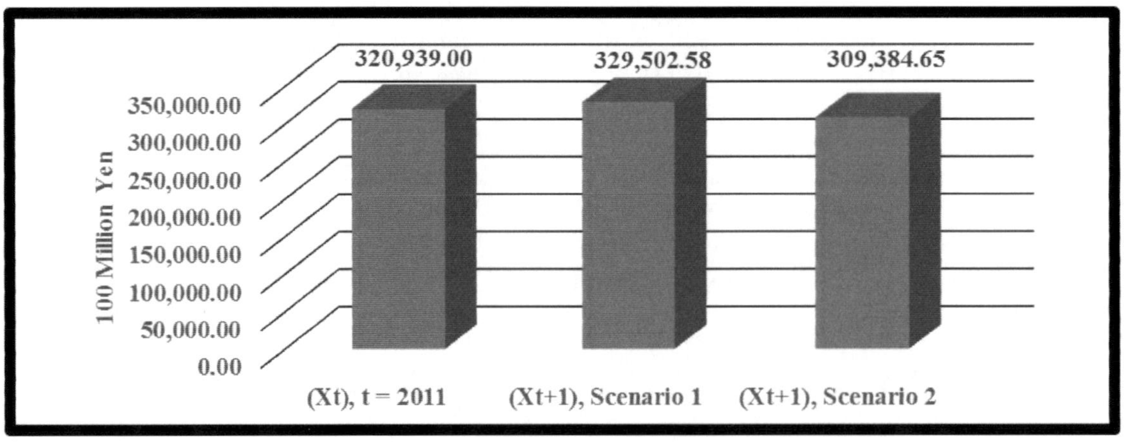

Figure 7. The dynamics of the total output of the finance and insurance sector.

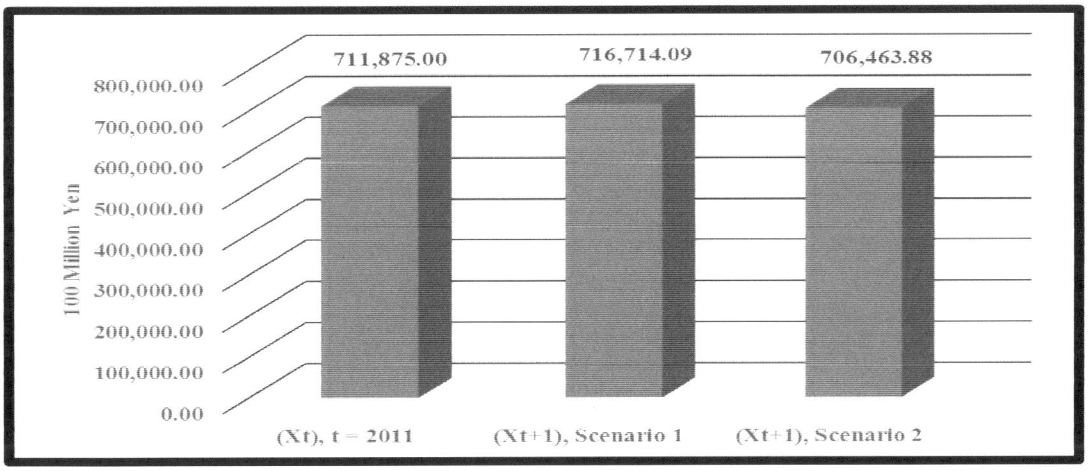

Figure 8. The dynamics of the total output of the real estate sector.

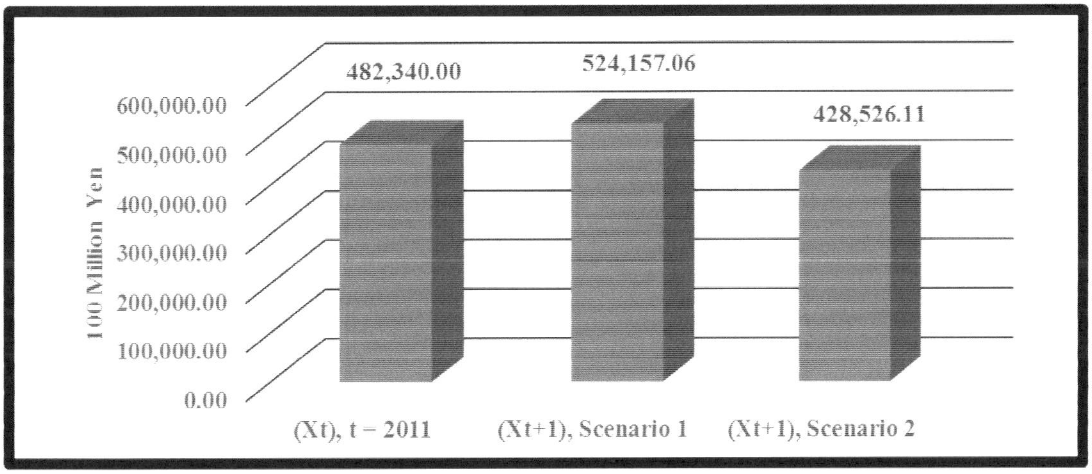

Figure 9. The dynamics of the total output of the transport and postal services sector.

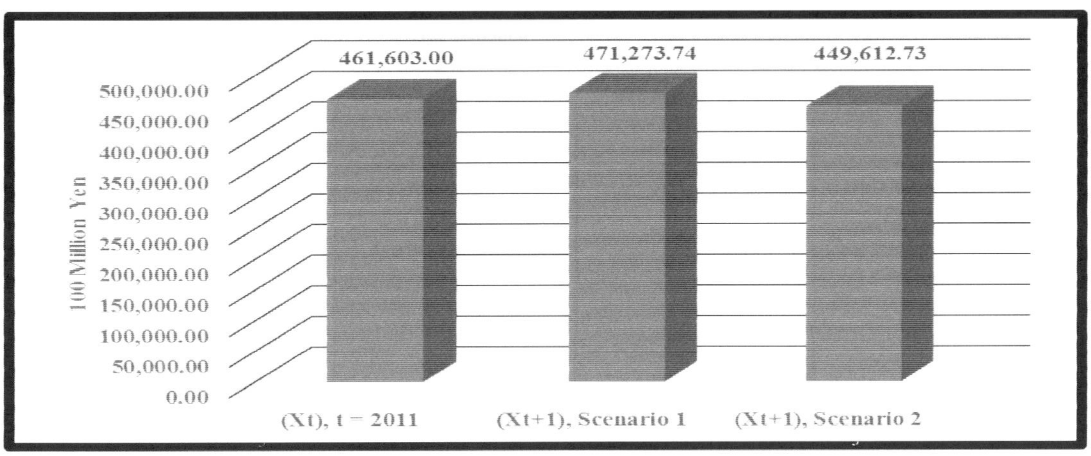

Figure 10. The dynamics of the total output of the information and communications sector.

IOP Conf. Series: Journal of Physics: Conf. Series **820** (2017) 012032 doi:10.1088/1742-6596/820/1/012032

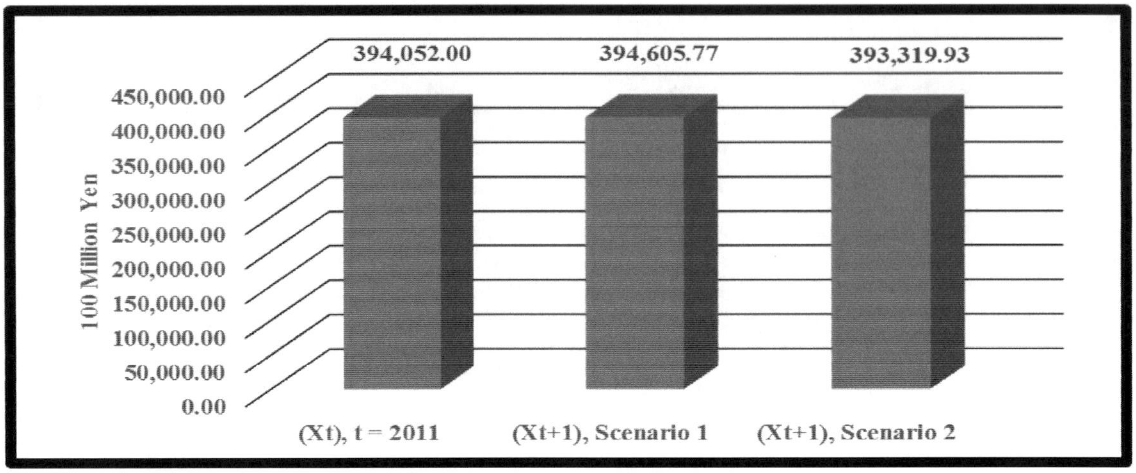

Figure 11. The dynamics of the total output of the public administration sector.

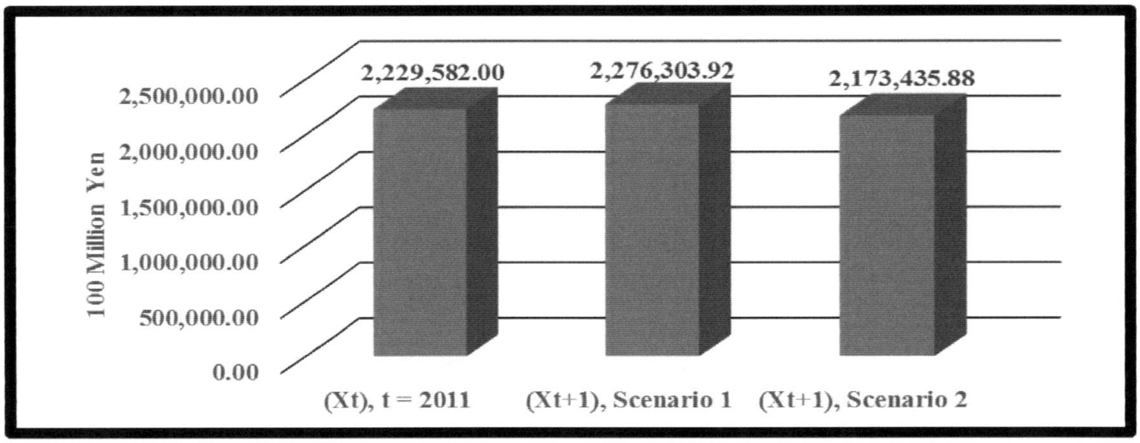

Figure 12. The dynamics of the total output of the services sector.

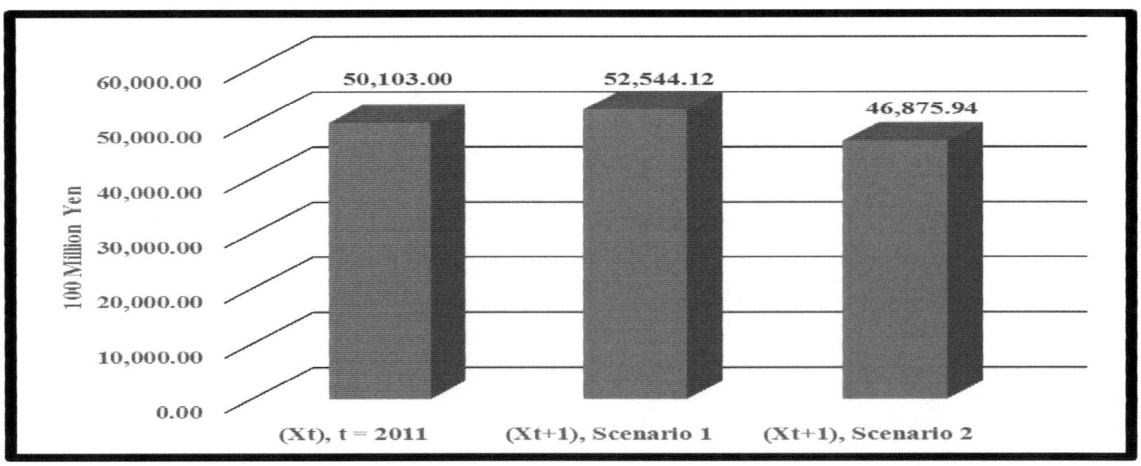

Figure 13. The dynamics of the total output of the activities not elsewhere classified sector.

4. Conclusions and further researches

This study, as a continuation study of previous ones, conducts the analysis regarding the impacts of the modifications of final demands on the total outputs of Japanese industrial sectors. This study employs a demand-pull IO quantity model, one of the calculation instruments in the IO analysis, as an analysis device. This study focuses on thirteen industries. The "whole sector change" condition is considered in the calculations. In this condition, the modification on each scenario is addressed to all focused industries. An initial period in this study is 2011.

The results show that the positive impacts on the total outputs of focused sectors are delivered by scenario 1, the change of exports. On the contrary, the negative impacts are distributed by scenario 2, the modification of imports. The suggestions for improving the total outputs of discussed sectors are based on the results. One can argue that the Japanese government should push the activities of national economy so Japanese industries have a stimulus in producing more products. One of the push actions that the Japanese government can do is to increase the exports targets. On the contrary, they should stipulate carefully the rate of imports.

The deeper understanding regarding the impacts of the changes of final demands on the total outputs of Japanese industrial sectors is obtained from this study. However, this study focuses on the aggregated industries. In other words, this study is still far in describing the whole view about the impacts on the Japanese national economy. This view is needed in order to know better the conditions of Japanese national economy so the comprehensive policies for increasing those in the future can be made. Therefore, as a further research, this study proposes the same analysis for the disaggregated Japanese industries.

The other suggested further research from this study is to conduct the same analysis on the specific regional area. This analysis might be useful in describing the characteristics of countries in the area in responding the impacts. An interesting example is to conduct the analysis on the ASEAN countries. Besides, the analysis for the EU countries might be another interesting discussion.

References

[1] Zuhdi U 2016 The dynamics of the total output of the fishery sector: the case of Indonesia *Journal of Physics: Conference Series* **710** 012039

[2] Zuhdi U 2016 The dynamics of the total outputs of Japanese information and communication technology sectors: a further study *Journal of Physics: Conference Series* **710** 012041

[3] Zuhdi U 2015 An application of input-output analysis in analyzing the impacts of final demands changes on the total outputs of Japanese energy sectors: a further study *Journal of Physics: Conference Series* **622** 012041

[4] Zuhdi U 2015 An application of multiplier analysis in analyzing the role of information and communication technology sectors on Indonesian national economy: 1990-2005 *IOP Conference Series: Earth and Environmental Science* **23** 012015

[5] Zuhdi U 2015 The dynamics of Indonesian creative industry sectors: an analysis using input-output approach *Journal of the Knowledge Economy* **6** 1177–90

[6] Kramer W J, Jenkins B and Katz R S 2007 *The Role of the Information and Communications Technology Sector in Expanding Economic Opportunity* [online] http://www.hks.harvard.edu/m-rcbg/CSRI/publications/report_22_EO%20ICT%20Final.pdf (accessed 5 January 2016)

[7] Japanese Ministry of Internal Affairs and Communications 2016 *2011 Input-Output Tables for Japan* [online] http://www.soumu.go.jp/main_content/000443188.pdf (accessed December 30, 2016)

[8] Miller R E and Blair P D 2009 *Input-Output Analysis: Foundations and Extensions* (Cambridge: University Press)

[9] Zuhdi U, Prasetyo A D and Sianipar C P M 2013 Analyzing the dynamics of total output of Japanese creative industry sectors: an input-output approach *Procedia Economics and Finance* **5** 827–35

[10] Zuhdi U 2012 Analyzing the influence of creative industry sector to the national economic structural changes by decomposition analysis: the case of Indonesia *Procedia-Social and Behavioral Sciences* **65** 980–5

[11] Zuhdi U, Utomo D S and Alamanda D T 2011 Analyzing the role of ICT sector to the national economic structural changes: the case of Indonesia *Jurnal Manajemen Teknologi* **10** 299–307

[12] Zuhdi U, Mori S and Kamegai K 2014 Analysis of influences of GDP and ICT on Indonesian industrial structural changes using statistical analysis: 1990-2005 *Journal of Finance and Accountancy* **17** 1–19

[13] Zuhdi U 2016 The ranks of Indonesian and Japanese indsutrial sectors *IOP Conference Series: Earth and Environmental Science* **38** 012008

[14] Chenery H B and Watanabe T 1958 International comparisons of the structure of productions *Econometrica* **4** 487–521

[15] Nazara S 2005 *Input-Output Analysis* [in Indonesian] (Jakarta: The Faculty of Economics of University of Indonesia)

[16] Zuhdi U 2014 Analyzing the impacts of final demand changes on total output using input-output approach: the case of Japanese ICT sectors *IOP Conference Series: Earth and Environmental Science* **19** 012016

[17] Zuhdi U 2014 Analyzing the role of creative industries in national economy of Japan: 1995-2005 *Open Journal of Applied Sciences* **4** 197–211

[18] Zuhdi U 2014 An input-output approach to analyze the ways to increase total output of energy sectors: the case of Japan *IOP Conference Series: Earth and Environmental Science* **19** 012015

[19] Zuhdi U 2014 The dynamics of total output of Indonesian information and communication technology sector when final demand changes occur: an analysis using input-output approach *Advanced Science Letters* **20** 2254–57

[20] Zuhdi U 2014 The other perspective related to the role of information and communication technologies sectors in national economy: the case of Japan *Advanced Science Letters* **20** 483–6

[21] Zuhdi U 2014 The role of information and communication technology sectors in Indonesian national economy from 1990 through 2008: an analysis using input-output approach *Advanced Science Letters* **20** 1932–5

[22] Zuhdi U 2014 Using multipliers analysis in order to get another perspective related to the role of ICT sectors in national economy of Indonesia: 1990-2005 *Journal of Physics: Conference Series* **495** 012051

[23] Zuhdi U 2015 An analysis of the role of information and communication technology sectors on Japanese national economy from 1995 through 2005: an application of multiplier analysis *IOP Conference Series: Earth and Environmental Science* **23** 012014

[24] Zuhdi U 2016 The dynamics of the total output of the Japanese fisheries sector: an analysis using input-output approach *Journal of Physics: Conference Series* **710** 012040

[25] Zuhdi U 2016 The Indonesian economy in 2005: an analysis using the input-output approach *Proceedings of 20th EBES Conference-Vienna* **3** 1825–42

[26] Zuhdi U, Mori S and Kamegai K 2012 Analyzing the role of ICT sector to the national economic structural changes by decomposition analysis: the case of Indonesia and Japan *Procedia-Social and Behavioral Sciences* **65** 749–54

[27] Zuhdi U, Mori S and Kamegai K 2013 Analysis of influences of ICT on structural changes in Japanese commerce, business services and office supplies, and personal services sectors using multivariate analysis: 1985-2005 *The Asian Journal of Technology Management* **6** 102–11

[28] Zuhdi U, Mori S and Kamegai K 2014 Statistical analysis of influences of ICT on industrial structure changes from 1985 through 2005: the case of Japan *Journal of Computers* **9** 1291–9

[29] Zuhdi U, Mori S and Kamegai K 2015 Forecasting the influences of information and communication technology on the structural changes of Japanese industrial sectors: a study using statistical analysis *International Journal of Social, Behavioral, Educational, Economic, Business and Industrial Engineering* **9** 531–7

[30] Zuhdi U and Prasetyo A D 2014 Examining the total output changes of ICT sectors of Japan: an approach of input-output *Procedia-Social and Behavioral Sciences* **109** 659–63

[31] Zuhdi U, Prasetyo A D and Putranto N A R 2014 Analyzing the changes of total output of Japanese livestock sector: an input-output approach *Procedia-Social and Behavioral Sciences* **109** 649–53

[32] Zuhdi U, Putranto N A R and Prasetyo A D 2014 An input-output approach to know the dynamics of total output of livestock sectors: the case of Indonesia *Procedia-Social and Behavioral Sciences* **109** 634–8

[33] Zuhdi U, Putranto N A R and Prasetyo A D 2014 Encouraging information and communication technology sectors using input-output approach: the case of Indonesia *Advanced Science Letters* **20** 199–202

[34] Zuhdi U 2014 The impacts of final demand changes on total output of Indonesian ICT sectors: an analysis using input-output approach *IOP Conference Series: Materials Science and Engineering* **58** 012011

IOP Conf. Series: Journal of Physics: Conf. Series **820** (2017) 012033 doi:10.1088/1742-6596/820/1/012033

Generation of Source Data for Experiments with Network Attack Detection Software

Igor Kotenko [1,a], **Andrey Chechulin** [2,b] and **Alexander Branitskiy** [3,c]
[1]Head of the Laboratory, St. Petersburg Institute for Informatics and Automation of the Russian Academy of Sciences (SPIIRAS). St.Petersburg, Russia.
[2]Senior researcher, St. Petersburg Institute for Informatics and Automation of the Russian Academy of Sciences (SPIIRAS). St.Petersburg, Russia.
[3]Junior Researcher, St. Petersburg Institute for Informatics and Automation of the Russian Academy of Sciences (SPIIRAS). St.Petersburg, Russia.

E-mail: [a] ivkote@comsec.spb.ru, [b] chechulin@comsec.spb.ru,
[c] branitskiy@comsec.spb.ru

Abstract. The paper suggests a new approach for traffic generation and the architecture of the software tool for evaluation of attack detection and response mechanisms. To assess the proposed approach the automatic network attack detection and response mechanism Threshold Random Walk (TRW) was chosen and implemented. The results of evaluation of this mechanism by the proposed software tool are presented.

1. Introduction

Currently generation of network traffic is an important issue when testing the distributed systems and conducting the research related to the detection of network attacks, including forecasting of anomalous activity in enterprise networks and explanation of self-similarity properties of network traffic.

The complexity of such problems is due to a clear lack of open data sets that are suitable for the tasks of traffic studies and real application. Moreover, existing data sets usually contain a limited sample of network attack classes of interest. In these circumstances, it becomes necessary to generate such network traffic, that is close to the real and contains a number of attacks of current interest.

There are several ways to generate the network traffic for experiments. The first (W1) is to use the real networks for traffic collection. This option gives the closest to real world examples, but most likely this traffic will not contain attacks or will contain very small amount and variety of attacks. The second (W2) is to use a honeynet for traffic collection. It is not always possible to create such network and to attract malefactors. The third (W3) is to use a network simulator. This approach can be very effective, but it is hard in installation and tuning. The fourth (W4) is to use existing traffic dumps. This is the most common way, but the traffic will be static, so there will be no possibility to change the environment. The fifth (W5) consists in usage of network traffic generators. Such generators usually can provide traffic with a lot of preferences, but the result can be very far from real world examples. Finally (W6) it is possible to use combination of above mentioned techniques. This option can combine the advantages of several ways and will be described in this paper.

The ideal source for experiments related to the analysis of network traffic and detecting the attacks is a live traffic, which is obtained as a result of the recording of network packets circulating in a large corporate system. However, since such a possibility is not always available, one possible solution to this problem is the use of network traffic generators.

The designed generator of network traffic must satisfy a number of requirements: (R1) it should take into account all the features of any necessary network application and typical attacks for it; (R2) it should be minimized to the presence of any network traffic characteristics resulting from its artificial generation; (R3) an automatic generation mechanism of network traffic for a given configuration of network equipment and the whole environment in general should be provided; (R4) a small amount of resources should be consumed when generating traffic; (R5) the labels of anomalous packets or connections should be provided.

The ways to generate the traffic and the considered requirements are summarized in Table 1. The symbol '+' indicated that the requirement Ri presented in the column *i* is satisfied by the way for generating the traffic Wj located in the row *j*. Otherwise, the requirement is not satisfied.

Table 1. Ways to generate traffic and considered requirements.

	R1	R2	R3	R4	R5
W1	+	+	–	+	–
W2	–	+	–	+	+
W3	+	+	–	–	–
W4	–	+	–	+	+
W5	–	–	+	+	+
W6	+	+/–	+	+	+

It should be noticed, that the traffic can be divided in two groups: the mixed set of packets and the cohesive packet sequences. The peculiarity of the former is the simplicity of operation and manual control over the contents of each packet, while the feature of the second is the ability to create the logical connections using a variety of underlying protocols. This sequences are characterized, for example, by the number of sent packets, a time between packets, the models of user behaviour, etc.

The main contribution of the paper lies in the fact that it offers a new approach for evaluation of network attack detection and response mechanisms that works with the traffic. This approach allows us to combine advantages of the real world traffic (traces) with possibilities of tuning of malware activities. The paper presents the approach for traffic generation and the architecture of the software tool to evaluate different attack detection and response mechanisms. To evaluate the proposed approach the automatic network attack detection and response mechanism Threshold Random Walk (TRW) was chosen and implemented. The results of evaluation by the proposed software tool are presented. The organization of the paper is as follows. Second section provides the analysis of existing approaches for traffic generation or collection. In third section we describe the developed testbed for experiments based on the proposed approach for traffic generation. Fourth section considers the mechanism for attack detection and response based on traffic analysis. In fifth section the evaluation of the attack detection and response mechanism by the developed testbed is presented. In sixth section the conclusions and future research directions are discussed.

2. Related works

There are several network traffic data sets (dumps) with attacks traces. We consider five examples (MAWI, DARPA, CAIDA, DRDC and CDX) the most popular for research in cyber security.

Data Set MAWI represents files with images of real connections transmitted through the backbone Pacific ocean backbone between Japan and the United States [1]. To ensure the confidentiality of transmitted data, the network packets are captured using tcpdump analyzer and truncated so that they contain only headers. A few different types of attacks were presented, the overwhelming majority of which includes port scanning and DoS attacks. For marking the connections four detectors are used

[2], and the severity level of the classified connection depends on the number of detectors, which recognize the connection as anomalous.

The data set DARPA of Linkoln Laboratory [3] contains network traces obtained by simulating the network interaction between hosts within one of the DARPA projects. In 1998 such data were collected for 7 weeks with a total volume including log files greater than 7 GB. In 1999 data were collected for 5 weeks (total volume of about 10 GB). This set is one of the most popular for scientific research, but it has a significant drawback – it is outdated.

The data set CAIDA [4] contains one hour traffic which includes DDoS attacks with total volume 21 GB. This set represents only anomalous traffic from the attacker and the corresponding responses from the victim. Traffic that does not contain attacks is removed as much as possible.

The data set DRDC (Defence Research and Development Canada) [5] is composed of more than 7000 documented attack scenarios and provides network traffic in pcap format, parameters in XML format and the output of commands in text form. A total of 94 exploitability programs are presented, there are scripts and program sources for the implementation of the attacks. The data set comprises 49 different vulnerabilities using 17 network services, 108 different target operating systems. A significant drawback of this data set that it is not publicly available.

The data set CDX (Cyber Defense Exercise) [6] includes 23 pcap-files generated in April 2009 as the result of the competition for building the secure network environments and protecting them from the network attacks. Its special feature is the presence of a set of log files obtained as a result of functioning the server applications (Web and DNS) and IDS Snort.

Network traffic generators can be implemented as software or hardware components. Using these generators one can set various network traffic parameters, including traffic direction (addresses of a source and a receiver), packet length, protocol type, packet rate, etc.

As it was mentioned before the network traffic generators can be divided into two categories: generators on the packet level and generators based on cohesive packet sequences.

One of the most famous examples of the first class utilities is *hping3*. Some of the popular tools of the second class include *ostinato*, which supports many different protocols from the channel to application layer of the OSI model. Since the generators of the second class are of the greatest interest, we will consider briefly the corresponding relevant work concerning them.

Harpoon [7] is a research prototype generating the network traffic on TCP and UDP level. The features of this software tool is the ability to generate such traffic that is largely statistically similar to the real traffic. The downside is the lack of opportunity to create network flows that would be able to simulate the data transmission between the network applications (for example, between the Internet browser and the Web-server). This system has a three-layer architecture, represented by the file level, session level and user level.

Swing [8] is without drawbacks of Harpoon and can take into account the properties of packets, depending on the highest level protocol. Swing's architecture considers different features of the network traffic, typical users, sessions, connections and network characteristics.

In [9] the fast algorithm of fractional Gaussian noise was developed for improving the quality characteristics of the network traffic and approximating the parameters of self-similar stochastic processes of the network traffic.

After analyzing the existing data sets and traffic generators, it can be concluded that the existing solutions do not satisfy all the requirements we considered and are focused on the generation of certain classes of data according to the protocol type, application, etc. Since the simultaneous and complete fulfilment of these conditions is often unfeasible in practice with the help of such generators, we propose to develop a combined approach, which allows to take into account, if not all, then at least the maximum number of these requirements.

This approach is based on combining different sources of traffic, including existing data sets (real traffic traces) as well as output of normal and (or) attack traffic generators. Such approach is intended to combine advantages of the real world traffic (traces) with possibilities of tuning of traffic of normal applications and malware activities.

3. Proposed approach

The combined approach consists in *application of various scenarios of traffic traces with the insertion of a needed traffic*. The needed traffic in this case is the attack traffic and (or) the traffic of some applications, for example, "fast" applications, such as NetBIOS/NS or P2P. |It is proposed to combine both the real traffic traces and the generated traffic (plugging various modules of traffic models).

The proposed common architecture of the tool for generation and analysis of the legitimate and network attack traffic is in Figure 1.

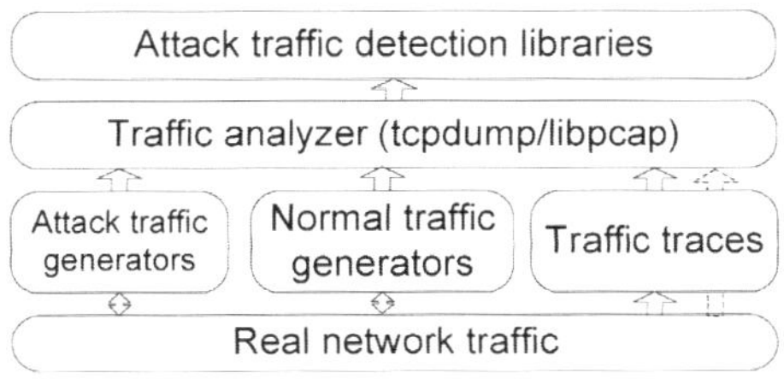

Figure 1. Traffic generation and analysis tool architecture.

The tool is based on using the real network traffic. This traffic can be used to create traffic traces, extract parameters for legitimate and attack traffic generators and their tuning. Traffic traces can be both downloaded from the corresponding repositories and recorded due to tcpdump/libpcap. Libpcap is the application program interface for packet capturing from the network interface. Tcpdump is intended for traffic recording due to libpcap. Traffic analyzer is based upon tcpdump/libpcap and allows receiving and synchronizing the traffic from various sources.

The main idea of evaluation methodology is to run a series of experiments using the software tool for various values of input parameters measuring the defense parameters. It is supposed that measuring the effectiveness and efficiency parameters of automatic detection and response techniques and their combination as well as fulfilling their analysis enables to select the rational (or optimum) parameters of functioning and (or) developing the dynamic strategies of operation.

To simulate the impact of the defense methods the traffic filter (if defense mechanism blocks some traffic than it will not reach the detection mechanisms) is added to the traffic generator (Figure 2) [10].

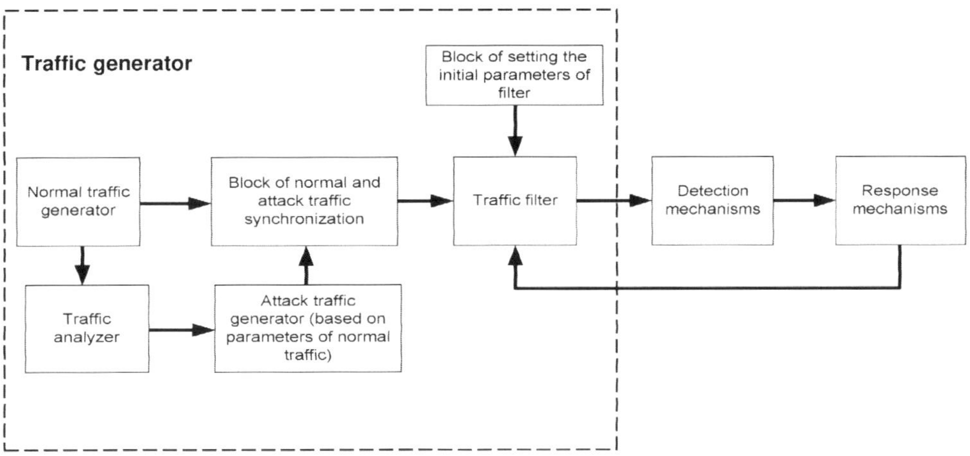

Figure 2. General scheme of traffic generator and its interrelation with defense mechanisms [10].

The normal traffic generator implements reading traffic traces from files. *The block of normal and attack traffic synchronization* serves for transferring the network packets to defense mechanisms in the time ordered sequence under simultaneous use of several traffic sources. *The traffic analyzer* calculates traffic parameters. These are parameters are used for attack traffic generation, and also to investigate the correspondence of traffic parameters and the defense mechanisms which work in the best way at the given traffic. *The attack traffic generator* allows generating the traffic for both already known attacks, and unknown attacks which can appear in the future. *The traffic filter* – as traffic traces have answers to the requests blocked by response methods, the traffic filter is used to simulate dynamics in traffic records. *The block of setting the initial parameters of the filter* allows using access control list (ACL) to ignore connections to certain addresses, ports and protocols. *Mechanisms of detection and response* are the defense mechanisms under investigation. They make decisions on presence of attacks on hosts and protecting against them.

Traffic is supposed to be preprocessed to raise the defense effectiveness. The preprocessing is in ACL filtering, "dark addresses" determination and processing the honeypot messages, etc. ACL is assigned with the traffic filtering rules, and traffic is filtered before it comes to defense mechanism. The rules are defined with addresses and ports of such hosts that have legitimate applications installed creating a lot of connections.

If traffic source is represented by the real traffic, then the real time interval of the experiment is set. In case of the recorded traffic, the time interval inside the recorded traffic is set.

Let us consider the one of the attack types – network scanning. *TCP scanning* is performed by initiating a TCP connection. Scanning tool creates a thread and sends TCP-SYN packet to the victim host. If it will not reply, the packet will be resent after 3 and 6 seconds accordingly. The scanning stops after 3 unsuccessful tries. Scanning speed depends on the amount of allowed threads and simultaneous connections on this host. It is defined by the operating system. *UDP scanners* usually send only one packet without retries. Scanning speed is limited by the channel bandwidth. Various scanning strategies are defined to choose the scanned hosts.

Table 2 represents parameters of different scanners.

Table 2. Parameters for scanners simulation

	Scanner 1	Scanner 2	Scanner 3
Protocol	TCP	UDP	UDP
Packet generation frequency (per second)	5.65	4000	357
Length (bytes)	40	404	796-307
Ports (source/destination)	any/80	any/1434	4000/random
Scanning strategy	1/8 of time the random IPs, 1/2 of time IPs from the same /8 subnet (X.*.*.*), 3/8 of time IPs from the same /16 subnet (X.Y.*.*)	the source formula (to receive the next pseudo-random address from the previous) is as follows: x' = (x * 214013 + 2531011) mod 2^32	Random distribution

There are more parameters that can be specified for scanner: connection type - connections can be either TCP-based or UDP-based; frequency of packet generation (number of packets (connections) per second); scan speed variation - the speed at which the scanner scans can be constant or varied; scan type (destination address and port choice technique) - random-scanning, sequential-scanning, permutation-scanning, partition-scanning, local-preference-scanning, topological-scanning, hitlist-scanning, combination; probability of successful TCP connection; packet size; probability of RST packet receiving as response to TCP connection request; complete or incomplete closing of TCP connection; probability of UDP packet receiving as response to sent UDP packet; address and port spoofing technique; number of addresses used; etc.

4. Scanning detection technique for experiments

The analysis of different techniques for attack traffic detection was presented in previous authors' papers [10-12]. In our research we investigated several techniques: virus throttling (VT) mechanisms, mechanisms based on failed connections (FC) analysis, mechanisms based on Threshold Random Walk (TRW), credit based (CB) mechanisms, DNS-based mechanisms; and combined detection and response mechanisms. In this paper we describe only basic TRW technique for attack detection and response just to show how it can be evaluated by the proposed approach.

Threshold Random Walk (TRW) technique [13] uses a random walk to decide whether a new connection initiated by a host is benign or malicious. It keeps a ratio for each host and in the case of a successful connection started by that host, increases its ratio by some value, making the ratio farther from a fixed threshold (and vice versa in the case of failed connection attempt).

The following supposition is offered to discover hosts which took part in the attacks (here it is the scanning attack). This method uses TCP protocol. TCP SYN, TCP RST, TCP SYN ASK packets and fields "source IP" and "destination IP" are analyzed.

The method of sequential hypothesis testing is used to analyze the host that shows high network activity and might be source of scanning:

- Let H_1 be a hypothesis that the host r shows high network activity (it is the source of scanning), H_0 - a hypothesis that the host does not show the high network activity (it isn't the source of scanning). Let Y_i be a variable describing the connection attempt to the *i*-th host. This variable can have the following values:

$$Y_i = \begin{cases} 0, \text{ if connection is established} \\ 1, \text{ if connection is not established} \end{cases}.$$

- Let us now assume that conditional on the hypothesis H_j the random variables $Y_i \mid h_j = 1,2,...$ are independent and identically distributed. Then we can express the distribution of the Bernoulli random variable Y_i as:

$$\begin{aligned} P_r[Y_i = 0 \mid H_0] = \theta_0, \quad P_r[Y_i = 1 \mid H_0] = 1 - \theta_0 \\ P_r[Y_i = 0 \mid H_1] = \theta_1, \quad P_r[Y_i = 1 \mid H_1] = 1 - \theta_1 \end{aligned}. \quad (1)$$

- The observation that a connection attempt is more likely to be a success from a benign source than a malicious one implies the condition: $\theta_0 > \theta_1$.
- According to these two hypotheses and the real state four possible decisions can be selected (Table 3):

Table 3. Possible decisions

	H_1 is selected	H_0 is selected
There is an attack in the traffic	Scanning detection	False negative
There is no attacks in the traffic	False positive	No attack

- Let us evaluate the performance of detection by two variables: detection probability P_D and false positive probability P_F. So the algorithm has appropriate quality if the following conditions are correct:

$$P_D \geq \beta \text{ and } P_F \leq \alpha, \ \alpha\text{=0.01 } \beta\text{=0.99}. \quad (2)$$

- As each event is observed we calculate the likelihood ratio:

$$\Lambda(Y) \equiv \frac{\Pr[Y \mid H_1]}{\Pr[Y \mid H_0]} = \prod_{i=1}^{n} \frac{\Pr[Y_i \mid H_1]}{\Pr[Y_i \mid H_0]}, \tag{3}$$

where Y is the vector of events that was analyzed and $\Pr[Y_i \mid H_i]$ is the conditional probability mass function of the event stream Y given that model H_i is true; and where the second equality in (3) follows from the independent and identically-distributed random variables assumption.

- Then Λ is compared to an *upper* threshold η_1 and a *lower* threshold η_0. The results of comparison show the presence or absence of an attack actions on the host:

$$\Lambda(Y) \le \eta_0 \Rightarrow H_0, \quad \Lambda(Y) \ge \eta_1 \Rightarrow H_1.$$

If $\eta_0 < \Lambda(Y) < \eta_1$ then we should wait till the next observation and update $\Lambda(Y)$.

In [13] it is shown that

$$\eta_1 = \frac{\beta}{\alpha}, \quad \eta_0 = \frac{1-\beta}{1-\alpha}.$$

The proposed method of sequential testing hypotheses can be represented graphically as follows (Figure 3).

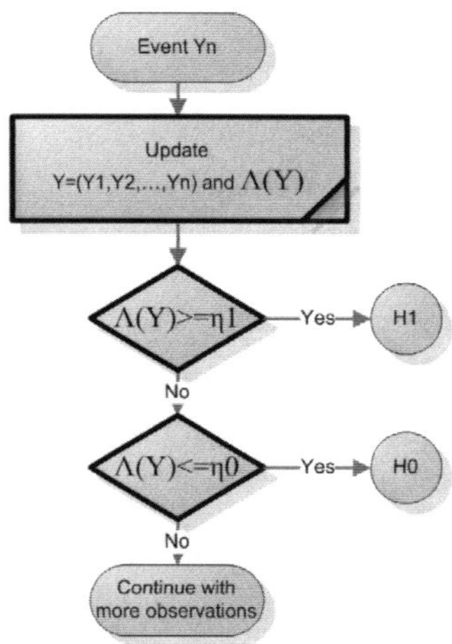

Figure 3. Sequential Hypothesis Testing scheme [13].

Input parameters of TRW algorithm: source IP-address; destination IP-address.

Control parameters of TRW algorithm: Previous Contacted Hosts (PCH) list for each host; First Contact Connection (FCC) queue.

For each host s TRW maintains three variables: D_s - is the set of distinct IP addresses to which s has previously made connections; S_s - reflects the decision state (PENDING, H_0, H_1); L_s - likelihood rate.

The following steps are executed for each record:

1. The sequence is not analyzed, if S_s is not PENDING.
2. Determine whether the connection is successful or not (a connection is successful if it elicited the SYN ASK response).
3. Check whether the address d is already in D_s. If so, skip the further processing and go to the next request.

IOP Conf. Series: Journal of Physics: Conf. Series **820** (2017) 012033 doi:10.1088/1742-6596/820/1/012033

4. Add d in D_s and update L_s using (3).

5. If $L_s \geq \eta_1$ set host status to H_1. If $L_s \leq \eta_0$ set host status to H_0.

TRW parameters dialog is presented in Figure 4. When the parameters are set, user presses the "OK" button and thus finishes defense method customization. In the main form, when traffic source is also selected, the "Start" button for simulation start-up.

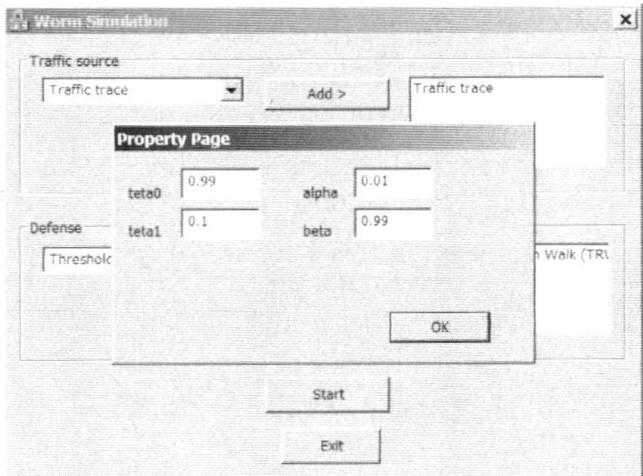

Figure 4. TRW parameters dialog.

5. Experiments results

For reasoned comparison of defense mechanisms it is necessary to select optimal parameters of launching of every one of them. Selection of parameters was performed on two types of traffics, recorded in files, with artificially generated attack traffic, admixed to them.

For selection of the best defense mechanisms parameters, two traffic types were used: (1) enterprise-level network traffic with big set of different applications (including vulnerability scanners) without prevalence of any traffic; (2) local area network traffic with prevalence of P2P applications.

The specified traffics were mixed with attack traffic.

Selection of such traffic types is reasoned by the fact that the mechanisms of detection and response against network attacks should precisely detect them both in traffic of legitimate applications that are used for networks purposes, and in traffic of applications that create great number of connections with different nodes. The latter is especially important, because traffic of applications that create great number of connections with different nodes resembles the behavior of network attacks.

The following traffic records were used for parameters selection:

- *Record of traffic lbl-internal.20041216-1618.port008.dump.anon* (181 Mb) with scanner traffic lbl-internal.20041216-1618.port008.dump.anon-scanners (870 Kb). It is one of the traffic records from the library [14] that includes more than 100 hours of records of enterprise level network traffic for the period from October 2004 till January 2005. The specified record is one of the biggest in the library with corresponding to it scanners traffic. By different applications traffic content this record is typical for LBNL library. Exception makes a choice of a file for adjustment of the DNS-based mechanism: the traffic *lbl-internal.20050107-1323.port026.dump.anon* (163 Mb) with the traffic of the scanner *lbl-internal.20050107-1323.port026.dump.anon-scanners* (2,81 Mb) (which contains sufficient amounts of DNS-traffic) was investigated.
- *Record of traffic p2p_emule_web_upload_download* (101 Mb). It is one of records, specially created on our testbed, that include a number of traffic records for a local area network, in which traffic of P2P file exchange networks of different types prevails. The specified record is a traffic formed by a client of P2P network E-donkey called Emule. At the same time WWW traffic was also created in the network. This traffic resembles a attack traffic by its structure,

because this P2P client attempts and creates multiple connections with different nodes. Selection of this record is also associated with the fact that Emule is one of the most widely spread p2P clients, as well as by that WWW traffic is present in the record.

Three types of scanners were admixed to the specified traffics:

- Scanner 1 has the following parameters: generation time—during the whole traffic time recorded in the file (connTime=-1), protocol—TCP (protocol=TCP), connection completion type—full (endOfConnection=FULL), addresses scanning—random (scanType=RANDOM), frequency—50 packets a second (scanSpeed=50), packet size—500 bytes (packetSize=500), successful connection probability— 0.3 (connProb=0.3), TCP-RST receive probability— 0.3 (rstProb=0.3).
- Scanner 2 is characterized by the following parameters: generation time—during the whole traffic time recorded in the file (connTime=-1), protocol—TCP (protocol=TCP), connection completion type—full (endOfConnection=FULL), addresses scanning—random (scanType=RANDOM), frequency—6 packets a second (scanSpeed=6), packet size—50 Kbytes (packetSize=50*1024), successful connection probability— 0.2 (connProb=0.2), TCP-RST receive probability— 0.6 (rstProb=0.6).
- Scanner 3 has the following parameters: generation time—400 sec (connTime=400), protocol—UDP (protocol=UDP), addresses scanning—random (scanType=RANDOM), frequency—50 packets a second (scanSpeed=50), packet size—404 bytes (packetSize=404), successful connection probability— 0.2 (connProb=0.2).

Selection of the given scanners types is reasoned by the following. It was necessary to estimate work of mechanisms both for known scanners, using different protocols, and for potentially possible or future ones.

For each method one-two main parameters were selected, changing of which to the greatest extent effects the number of errors: permits the attack traffic and block the normal traffic. In the case of one parameter the search of local minimum of errors was performed, that is of such parameter value, shifts to the greater or to the lesser side from which lead to increase of errors. In the case of two parameters the search was performed iteratively on each of them.

The aim of experiments with TRW was to find the optimal values for θ_0 and α. There were performed more than 500 experiments, and the experiments showed that the optimal parameters values for θ_0 and α in TRW mechanism are: Normal traffic, mixed with scanner 1, 2 or 3 traffic — 0.52 and 0.001 respectively; Peer-To-Peer traffic, mixed with scanner 1 traffic — 0.5005 and 0.001 respectively; Peer-To-Peer traffic, mixed with scanner 2 traffic — 0.502 and 0.0011 respectively; Normal traffic, mixed with scanner 3 traffic — 0.5005 and 0.01 respectively.

The suggested approach for traffic generation allowed us to provide controlled series of experiments and select the optimal parameters for usage of the TRW mechanism. We experimented also with other defense mechanisms (for example, VT, FC, CB, DNS-based and combined mechanisms) and also selected the optimal parameters. Thus , we validated the suggested approach and demonstrated that it can be implemented for design, adjustment and evaluation of different network attack detection and response mechanisms.

6. Conclusion

The main results of the research presented in the paper are as follows. The challenge was formulated, and the main requirements for network traffic generation for evaluation of detection and response techniques against network attacks were developed. The current status of network traffic generation was considered. The new approach to evaluate the mechanisms for network attack detection and response was proposed. It includes the technique for simulation of network attacks by mixed traffic generation. The proposed simulation tool architecture was presented. This tools allows to evaluate the main parameters of detection and response techniques. The series of experiments (using variety of scenarios) for different values of input parameters was conducted. The evaluation of automatic detection and response mechanism was performed by the software which implements the proposed approach. Suggestions on application and further improvement of the proposed approach to assess automatic detection and response mechanisms against network attacks were developed.

The future work will be devoted to the development of the software benchmarking system for evaluation of attack detection and response mechanisms on the basis of the proposed approach for traffic generation. It is suggested to create the set of various traffic sources (creation and use of already generated traffic traces, generators of malicious and legitimate traffics) that will cover all aspects of network functioning in normal mode and attack actions. Each benchmark could consist of testing the mechanism on various traffic combinations, in various modes and with various parameters.

Acknowledgments
The work is performed by the grant of RSF #15-11-30029 in SPIIRAS.

References

1. Fontugne R, Borgnat P, Abry P and Fukuda K 2010 MAWILab: Combining Diverse Anomaly Detectors for Automated Anomaly Labeling and Performance Benchmarking *Proc. Int. Conf. on emerging Networking EXperiments and Technologies* (Philadelphia, USA) pp 1–12 http://conferences.sigcomm.org/co-next/2010/CoNEXT_papers/08-Fontugne.pdf
2. MAWILab http://www.fukuda-lab.org/mawilab/documentation.html
3. DARPA Intrusion Detection Data Sets https://www.ll.mit.edu/ideval/data/
4. The CAIDA "DDoS Attack 2007" Dataset http://www.caida.org/data/passive/ddos-20070804_dataset.xml
5. McKenna J and Treurniet J Network Attack Reference Data Set 2004 *Technical Memorandum* (Defense R&D, Canada, Ottawa) http://cradpdf.drdc-rddc.gc.ca/PDFS/unc88/p523954.pdf
6. The United States Military Academy Data Sets http://www.usma.edu/crc/SitePages/DataSets.aspx
7. Sommers J, Kim H, and Barford P 2004 Harpoon: A Flow-Level Traffic Generator for Router and Network Tests *Proc. of the Joint Int. Conf. on Measurement and Modeling of Computer Systems* vol 32 (New York, NY, USA, ACM) pp 392–392.
8. Vishwanath K and Vahdat A 2016 Realistic and Responsive Network Traffic Generation *Proc. of the 2006 Conference on Applications, Technologies, Architectures, and Protocols for Computer Communications* (New York, NY, USA, ACM) pp 111–122.
9. Paxson V 1997 Fast, Approximate Synthesis of Fractional Gaussian Noise for Generating Self-Similar Network Traffic *Proc. of the 1997 Conference on Applications, Technologies, Architectures, and Protocols for Computer Communications* (New York, NY, USA, ACM) pp 5–18.
10. Kotenko I 2009 Framework for Integrated Proactive Network Worm Detection and Response *Proc. of the 17th Euromicro International Conference on Parallel, Distributed and network-based Processing* (Los Alamitos, CA, USA IEEE) pp 379-386.
11. Kotenko I V, Chechulin A A, Doynikova E V 2011 Combining of Scanning Protection Mechanisms in GIS and Corporate Information Systems *Proc. of the 5th International Workshop on Information Fusion and Geographical Information Systems* vol 5 (Berlin Heidelberg Springer) pp 45–58.
12. Branitskiy A A, Kotenko I V 2015 Network attack detection based on combination of neural, immune and neuro-fuzzy classifiers *Proc. of the 18th IEEE International Conference on Computational Science and Engineering* (Los Alamitos, CA, USA IEEE) pp 152–159.
13. Jung J, Paxson V, Berger A W and Balakrishnan H 2004 Fast portscan detection using sequential hypothesis testing *Proc. of the 2004 IEEE Symposium on Security and Privacy* (Los Alamitos, CA, USA IEEE) pp 211-225.
14. LBNL/ICSI Enterprise Tracing Project http://www.icir.org/enterprise-tracing/index.html

Metric Tensor Approach in Solar Wind for Rapidly Rotating Stellar

Wihayati[1], **F W Wibowo**[2]
[1]Astrophysics and Mathematics Group, Sekip Utara Bulaksumur, Yogyakarta 55281, Indonesia
[2] Department of Informatics Engineering, STMIK AMIKOM Yogyakarta, Jl. Ring Road Utara, Condong Catur, Depok, Sleman, Yogyakarta 55283, Indonesia
E-mail: wihayati@gmail.com[1], ferry.w@amikom.ac.id[2]

Abstract. Many researchers have observed the phenomena associated with the solar wind. One of which became a reference in this field is Parker's solar wind. The development of ideas initiated by Parker led to many cases related to Parker's solar wind. This paper aims at finding the metric tensor of the solar wind by using the idea of Parker's solar wind in rapidly rotating stellar. By the discovery of the metric tensor can be used to find the distance and characteristics of the coordinate system in approaching oblateness because a rapidly rotating stellar has an oblate shape at the poles.

1. Introduction

In the mid-ninth century, Richard Carrington, an amateur astronomer from British discovered sunspots during solar image mapping project is practiced. When observing, suddenly he saw two spots that appear and then disappear for 5 minutes, then it is called solar flares. Solar flares are a huge explosion of energy in the sun, which affects the Earth's magnetic field as well. Large geomagnetic disturbances were seen in the telegraph system, when the aurora spread out in the space. Carrington thought that it was related to something he found, when observing the solar [1]. While it was true, but Carrington beware of his statement because he also found the frequency of flame that isn't always followed by geomagnetic disturbances and auroras. Carrington wasn't the only person who knew it, previously also been observed by Aristotle, but less obvious exposure incident [2]. Relations solar and terrestrial magnetic interference becomes serious posed by physicists until the late 19th century George Fitzgerald suggested, that the material came from the solar that had adequate burst of speed and acceleration several times of the Sun's gravity, so as to reach the earth in a few days like electrode with a giant vacuum tubes [3].

The concept of solar emissions continues to support the facts with a phenomenon of comets [4]. Comets are classified into two, namely the comet has a curved tail and comets have tails almost straight, made of gas dust. Dust tail trajectory shape is produced by solar radiation pressure and gravity influenced the dust grains. At the gas tail appeared a problem, observations show that the tail is always away from the sun with the deviation angle and showed acceleration irregularities. Explanation of the pressure of the solar radiation that affects the comet gas has failed to explain, because of the comet sized provisions. In 1950, Ludwig Biermann, a German physicist explained the pressure of solar radiation that affected the comet gas, by stating that it isn't affecting the solar photons, but the solar corpuscular radiation. Biermann developed a model of comet particles interaction with a permanent magnet flux for a comet's gas tail [5]. Biermann argument was then deemed invalid, resulting in a fundamental view change because the orbits of comets passing all heliolatitudes. The conclusion that couldn't be explained is that the sun emits particles in all directions at a time. Half a century of findings Birkeland, the concept of continuous solar emissions reappears with the support of the stronger theory and observations. But at the same time, different conclusions found by the British physicist, Sydney Chapman, through a different approach on the basis that the sun's outer atmosphere, the corona, known to be very hot [6]. Chapman became a calculation pioneer of gas kinetic properties and found that heat of the ionized atmosphere able to conduct heat well until a far distance. As a result, the particles have great heat speeds despite far from the solar and must resist the pull of gravitational, so making a density is down very slowly. Because it is too slow, making the solar atmosphere must pass through the Earth's orbit, in other words, the Earth immersed in an atmosphere of solar static.

In 1958 the results a great approach obtained by Eugene Parker, by supporting the idea of Biermann and Chapman. Parker referred to the theory of continuous flux of solar particles by Biermann and expands the theory of the solar atmosphere that stretches across the sky as a supersonic flow by Chapman. Because the atmosphere is very hot, even in places that are very far away from the sun, and the gravitational pull of the solar isn't an interstellar medium pressure, finally came the modern concept of the solar wind [7]. The Parker theory wasn't only finding a solution that brought evidence but made a number of predictions that could be tested, one of which is the wind flow of several hundred kilometers per second radially from the solar. The Parker theory was followed by a long debate, whether the supersonic solar wind there. Observations were made to limit this debate, but to measure the solar wind was a study a few years when it began to develop space technology. This made the two superpowers to show ability in space technology. Russia on a mission four people, while the USA sent seven people, among them many who failed when it took off. Russian aircraft successfully is Lunik II in 1959, just two years after the first Russian Sputnik went and the US launched its first satellite, showed the result of positive ion flux detection. Despite this observations weren't conclusions, because the direction of the particle velocity was unknown. The enough stronger evidence presented in 1962 from Marine American 2 that explored Venus and able to withstand the failure. Marcia Neugebauer declared success retrieve the data as evidence of the existence and properties of the solar wind. Thus ended the first era story of the solar wind and the idea of Parker became known. After a few years, the development of a disagreement about the concept of the solar wind to get a position in the world of Physics, and until now the idea of Parker as a reference in this field.

Pulsar is magnetized neutron stellar that rotates very fast [8]. The neutron stellar is one of the stellars categorized fast spinning, and for this state, the stellar that spins very quickly have spheroidal shape [9]. So considering to this shape how to find the tensor for this, because it is important to provide a framework of mathematics in formulation and completion of physics problems. From this information and the idea of Parker's solar wind, this paper aims at finding a metric tensor of the Solar wind in the rapidly rotating stellar that have spheroid shape.

2. Related works

In 1967, Weber has developed the theory of Parker, explaining the model of the solar wind to flow equilibrium steady state in the equatorial plane. The equation of motion is experienced in terms involving the influence of gravity, viscosity, pressure and magnetic field gradient, while the state of the simplest systems to be reviewed with regard electrical conductivity of plasma is infinite. The model involves isotropic pressure, lack of viscosity and the power system is disclosed in polytrope equation. The review resulted in the answer to the radial motion which must pass three critical points, and significantly described in the speed of the propagation characteristics of small amplitude disturbances. Solving for azimuth motion should pass the radial Alfvénic critical point with radial Alfvénic Mach number in the same equation. Terms of this point, in relation to the answer to the radial motion determined by determining the value of the azimuthal velocity and magnetic field at various places on the surface of the sun. Numerical completion obtained for specific parameters and showed that the magnetic field only slightly affects the motion of the solar wind co-rotation. The Azimuth velocity of 1 kms^{-1} are in the area 1 A.U., but the magnetic strain using torque to the solar equal to the force required to produce an effective co-rotation at a critical point of radial Alfvénik. Velocity rate of the solar wind will occur in the area that is between 15 and 50 times of the radius of the solar, resulting in substantially less for the angular momentum of the solar. A polytropic model doesn't produce a real model for the supply of energy in the solar wind. The energy transfer in the solar revealed through modeling of the solar wind. The distribution of the density between the photosphere and $10R_{\odot}$ was determined by observation, while $10R_{\odot}$ and 1 A.U. heat derived from the thermal conduction. With this model, a heat number due to the waves in the solar corona could be determined. Thus, the motion doesn't affect the azimuthal direction of radial motion, while the stability of the azimuth motion investigated at the radial completion which is a function of the distance. This shows the disturbance travel conditions throughout the characteristics and for a stable model disturbances. The same assumption is obtained for the radial velocity, the effect of viscosity, and anisotropy in pressure on the azimuth motion. The results showed that the total torque on the solar quite a bit rather than the results obtained from the non-viscosity model, but the torque on the solar wind associated with the results of additional force in the azimuthal velocity at 1 A.U. or about 5 kms^{-1} [10].

The results of a study of the Alfvén waves is time-dependent, and the magnetohydrodynamic (MHD) equation of a single fluid for the solar wind with the code model of a modified ZEUS MHD. Wind models were tested had a radial symmetry and magnetized. Early exit flow is described by the standard reply of Parker's wind. This focuses on the impact of flow out Alfvén waves and is based on the completion of a nonlinear ideal MHD equation. In contrast to previous studies, there were assumptions that were ignored namely the wave linearity, damping wave, and interaction of wave flow, so that the wave modeling is calculating a wind backlash that occurs as non-linear interactions between different types of MHD waves. The result was able to explain the beginning of the rapid flow of the solar corona hole that preceded a momentum deposition by Alfvén waves in the solar wind, also discussed the distance of the wave amplitude needed to obtain answers to the rapid flow of the corona hole [11].

In 2004, Cranmer presents answers to some problems in the solar wind with the full analytical approach that had previously not completed. Answer completion of the solar wind velocity and the value of the missing mass reduction in plasma particles in the solar wind derived analytically with a particular model approach. Calculation of the solar wind velocity that depends on the radial direction implementing Parker's equation in circumstances of the isothermal and Bondi equation for spherical accretion. Solution to this problem involves a transcendental equation that generates the variables in the logarithmic form. Furthermore, the equation obtained solved by using the Lambert W function [12]. It has also been developed a model of the solar wind to within 1 A.U. The basic equation used is the equation of conservation of mass, momentum, and energy. The equation involved the mechanism of a corona heat wave and the wind acceleration. The model used calculated the Alfvén wave fluctuating power spectrum. The computational model was done by limiting the temperature state of the electron, proton, and solar wind velocity in area 1 A.U. This modeling is able to explain the warming phenomenon of protons in the corona and solar wind acceleration [13].

At the beginning, the space between the solar and the planets or known as interplanetary space is considered as a vacuum. But the reality is not so, an interplanetary space contains various particles that move and develop interplanetary medium. These particles move from one another and each brings energy to the ranges of a great difference. The solar wind particles with a density change or flux that brings out as an electron, proton, a helium nucleus, including heavy material that is constantly emitted by the solar at an average velocity of 1.5 million kmh^{-1}. The density of particles in the earth close to 7 $ionscm^{-3}$ so that the flow velocity in the radial direction to reach about 450 $kmcm^{-1}$. The source of the

IOP Conf. Series: Journal of Physics: Conf. Series **820** (2017) 012034 doi:10.1088/1742-6596/820/1/012034

solar wind is the solar corona heat. The corona's temperature is very high such that the solar's gravity isn't able to compensate for the release of the solar wind streams out. However, not known in detail about why and where the corona gas can be accelerated so that it has an extremely high velocity. The issue is related to the question that hasn't answered on heating the corona. Furthermore, the solar wind across the interplanetary space and meet planets. The solar wind isn't uniform in both time and space, though it has tendency directed away from the solar. In around corona's hole is very high wind velocity about 500-800 kms^{-1}, and in the low-velocity area is about 300-400 kms^{-1}[14]. The solar wind velocity depends on the geographical latitude. The fluctuations that were observed in the wind at the orbit of the earth have scales about 10^2 km, because the fluctuations with scales less than about 10^6 km may not survive from the solar [15].

3. Parker's Solar Wind Model

The solar as a stellar is the only object in the solar system that emits light. The solar has a center line of 1.4 million km, but among the billions of stellar, the solar isn't biggest even could be said it's small. However, the solar is a stellar shines steadily for a long time in astronomical standards with energy obtained from burning the hydrogen. Among the other stellar in the universe, the solar is stellar whose distance is closest to the earth which is about 150 million km (1 astronomical unit) [16]. That's way, the information about the other stellar aren't as complete research results than the solar. With some other properties of stellar that has similar to the solar properties, these stellar winds can be studied indirectly by observing the interaction between these winds and the interstellar medium (ISM) [17].

The solar wind occurs in the corona. The corona isn't in a static equilibrium, but wander and expanded continuously. It is assumed that the plasma wandering as the solar wind the state is an isothermal within the steady flow. Equations arranging the state can be derived from the magnetohydrodinamics (MHD) equation, with assumed that $d/dt = 0$ as shown in equation 1 [18].

$$\nabla(\rho u) = 0 \tag{1}$$

then it could be derived as in equation 2.

$$\rho(u\nabla)u = -\nabla p + \rho g \tag{2}$$

with $p = \rho RT$ and $T = T_0$. This completion restricted using spherical symmetries. The wind velocity u is in the radial direction, $u = u\hat{e}_r$ and the gravity acceleration $g = g\hat{e}_r$ inversely with the square of the distance (see equation 3), temperature and speed of sound (see equation 4) is assumed to be constant [19].

$$g = -\frac{GM_\bullet}{r^2} \tag{3}$$

$$C_s^2 = \frac{p}{\rho} \tag{4}$$

From [19] it is obtained the equation 5 with the critical radius shows the position where the solar wind velocity reaches the speed of sound $u = C_s$, $r_c = GM_*/(2C_s^2)$. The critical radius r_c for coronal sound speed that is about 10^5 ms^{-1} will get around 6×10^9 m or 9-10R_*.

$$\left(u - \frac{C_s^2}{u}\right)\frac{du}{dr} = 2\frac{C_s^2}{r^2}(r - r_c) \tag{5}$$

The equation 5 is ordinary differential equation (ODE), so if it is integrated will get the equation 6.

$$\int\left(u-\frac{C_s^2}{u}\right)du=\int 2\frac{C_s^2}{r^2}\left(r-r_c\right)dr \tag{6}$$

it will obtain equation 7.

$$\left(\frac{u}{C_s}\right)^2-\log\left(\frac{u}{C_s}\right)^2=4\log\left(\frac{r}{r_c}\right)+4\frac{r_c}{r}+C \tag{7}$$

In this case, there are four classes of solutions for existing of the solar wind that is shown in Figure 1. These classes are:
Class 1 – Initially stream is subsonic. Velocities approach zero close to the solar and have huge distances that reaches a maximum less than C_s at r_c.
Class 2 – Initially stream is subsonic. Low velocity close to the solar and high velocity at huge distances, reaches a sound speed at r_c becomes supersonic.
Class 3 – Initially stream is supersonic. High velocity close to the solar and low velocity at huge distances, decreases to reaches a sound speed at r_c.
Class 4 – Initially stream is supersonic. High velocity close to the solar and has huge distances, reaches a supersonic minimum at r_c then increases speed.

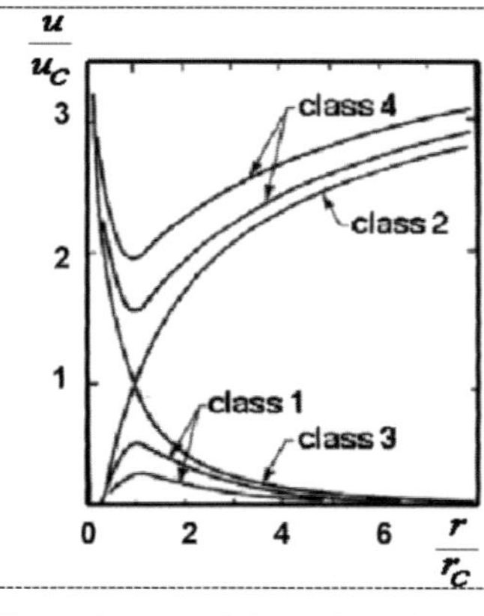

Figure 1. Four classes of the solar wind solution [19].

From these classes could be divided into three properties that $du/dr > 0$, $u < c_s$, and $r < r_c$ are properties of subsonic, $du/dr > 0$, $u > c_s$, and $r > r_c$ are properties of supersonic, and when $r = r_c$ and $u = c_s$ for a mathematically valid solution.

4. Result and Discussion

It has been found using program application of Mathematics that Parker's model in the subsonic area for the solar breeze, $u \leq c_s$, using values of $G = 2$, $M_* = 1$, $a = 1$, and $c_s = 1$. The figure from this program application is shown in Figure 2.

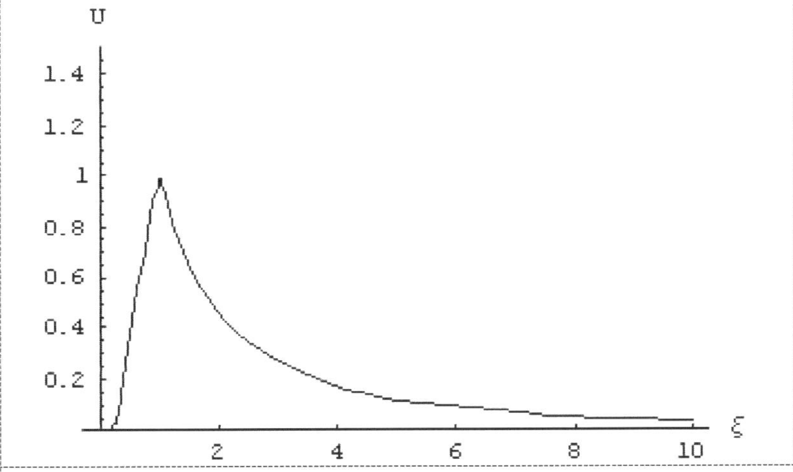

Figure 2. Parker's solar wind equation completion for subsonic velocity.

The solar wind velocity at the Earth's orbit could be obtained by substituting r=214R_* to equation 7 that will be obtained u=310 kms^{-1}. There are two important issues in the study of the solar wind i.e. a relatively low-velocity flow, u<350 kms^{-1}, or known as the slow solar wind, and a high-velocity flow, u>800 kms^{-1}, or known as the rapid solar wind. The rotation of the solar on a radial flow makes the solar magnetic field twisted upward becomes spiral as shown in Figure 3.

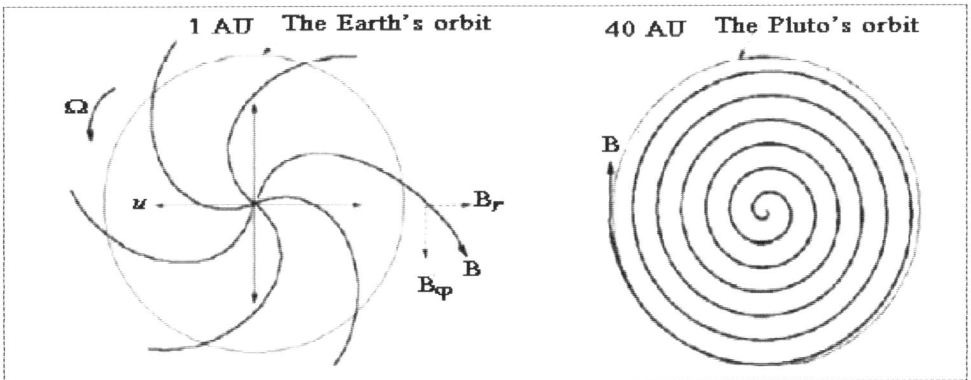

Figure 3. Magnetic field lines at a distance of 1 AU and more than 10 AU [20].

The component of velocity vector u perpendicular to the vector B i.e. u sin φ should be the same as the line velocity in the direction of the field because the field is frozen in the plasma. However, the magnetic field is attracted by the solar rotation with the rotation angle Ω The normal component magnet field velocity is (r-R_*) as according to the equation 8.

$$u \sin\varphi = \Omega\left(r - R_*\right)\cos\varphi \tag{8}$$

so it can be obtained the velocity as equation 9.

$$u = \Omega\left(r - R_*\right)\cot\varphi \tag{9}$$

IOP Conf. Series: Journal of Physics: Conf. Series **820** (2017) 012034 doi:10.1088/1742-6596/820/1/012034

The stellar that rotates very rapidly have a shape that isn't round but suffered an oblateness at the poles [21-22]. In measuring distance, it can apply a metric tensor that gives information about computing the distance between two points and characteristics of space in the coordinate systems. It is suitable for implementation to issues in which the fields are rotationally symmetric [23]. The metric tensor in solar wind for rapidly rotating stellar can be chosen oblate spheroidal coordinate systems as general coordinates (ξ, υ, ϕ) and Cartesian coordinate transformation is shown in equation 10.

$$\begin{aligned} x &= a\cosh\xi\sin\upsilon\cos\phi, \\ y &= a\cosh\xi\sin\upsilon\sin\phi, \\ z &= a\sinh\xi\cos\upsilon \end{aligned} \tag{10}$$

For each of $\vec{b}_\xi$, $\vec{b}_\upsilon$, and $\vec{b}_\phi$ is shown in equation 11, 12, and 13 respectively.

$$\begin{aligned} \vec{b}_\xi &= \frac{\partial \vec{r}}{\partial \xi} = \frac{\partial x}{\partial \xi}\hat{i} + \frac{\partial y}{\partial \xi}\hat{j} + \frac{\partial z}{\partial \xi}\hat{k} \\ &= \hat{i}\frac{\partial}{\partial \xi}\left(a\cosh\xi\sin\upsilon\cos\phi\right) + \hat{j}\frac{\partial}{\partial \xi}\left(a\cosh\xi\sin\upsilon\sin\phi\right) + \hat{k}\frac{\partial}{\partial \xi}\left(a\sinh\xi\cos\upsilon\right) \\ &= a\sinh\xi\sin\upsilon\cos\phi\hat{i} + a\sinh\xi\sin\upsilon\sin\phi\hat{j} + a\cosh\xi\cos\upsilon\hat{k} \end{aligned} \tag{11}$$

$$\begin{aligned} \vec{b}_\upsilon &= \frac{\partial \vec{r}}{\partial \upsilon} = \frac{\partial x}{\partial \upsilon}\hat{i} + \frac{\partial y}{\partial \upsilon}\hat{j} + \frac{\partial z}{\partial \upsilon}\hat{k} \\ &= \hat{i}\frac{\partial}{\partial \upsilon}\left(a\cosh\xi\sin\upsilon\cos\phi\right) + \hat{j}\frac{\partial}{\partial \upsilon}\left(a\cosh\xi\sin\upsilon\sin\phi\right) + \hat{k}\frac{\partial}{\partial \upsilon}\left(a\sinh\xi\cos\upsilon\right) \\ &= a\cosh\xi\cos\upsilon\cos\phi\hat{i} + a\cosh\xi\cos\upsilon\sin\phi\hat{j} + a\sinh\xi\sin\upsilon\left(-\hat{k}\right) \end{aligned} \tag{12}$$

$$\begin{aligned} \vec{b}_\phi &= \frac{\partial \vec{r}}{\partial \phi} = \frac{\partial x}{\partial \phi}\hat{i} + \frac{\partial y}{\partial \phi}\hat{j} + \frac{\partial z}{\partial \phi}\hat{k} \\ &= \hat{i}\frac{\partial}{\partial \phi}\left(a\cosh\xi\sin\upsilon\cos\phi\right) + \hat{j}\frac{\partial}{\partial \phi}\left(a\cosh\xi\sin\upsilon\sin\phi\right) + \hat{k}\frac{\partial}{\partial \phi}\left(a\sinh\xi\cos\upsilon\right) \\ &= a\cosh\xi\sin\upsilon\sin\phi\left(-\hat{i}\right) + a\cosh\xi\sin\upsilon\cos\phi\hat{j} + 0\hat{k} \end{aligned} \tag{13}$$

Thus from this equation could be found the squares of each equation as shown in equation 14, 15, and 16 respectively.

$$\begin{aligned} \vec{b}_\xi \bullet \vec{b}_\xi &= \left(a\sinh\xi\sin\upsilon\cos\phi\right)^2 + \left(a\sinh\xi\sin\upsilon\sin\phi\right)^2 + \left(a\cosh\xi\cos\upsilon\right)^2 \\ &= a^2\sinh^2\xi\sin^2\upsilon\cos^2\phi + a^2\sinh^2\xi\sin^2\upsilon\sin^2\phi + a^2\cosh^2\xi\cos^2\upsilon \\ &= a^2\sinh^2\xi\sin^2\upsilon\left(\cos^2\phi + \sin^2\phi\right) + a^2\cosh^2\xi\cos^2\upsilon \\ &= a^2\sinh^2\xi\sin^2\upsilon + a^2\cosh^2\xi\cos^2\upsilon \\ &= a^2\left(\sinh^2\xi\left(1-\cos^2\upsilon\right) + \left(1+\sinh^2\xi\right)\cos^2\upsilon\right) \\ &= a^2\left(\sinh^2\xi - \sinh^2\xi\cos^2\upsilon + \cos^2\phi + \sinh^2\xi\cos^2\upsilon\right) \\ &= a^2\left(\sinh^2\xi + \cos^2\upsilon\right) \end{aligned} \tag{14}$$

$$\begin{aligned}\vec{b}_{\upsilon}\bullet\vec{b}_{\upsilon} &= \left(a\cosh\xi\cos\upsilon\cos\phi\right)^2+\left(a\cosh\xi\cos\upsilon\sin\phi\right)^2+\left(-a\sinh\xi\sin\upsilon\right)^2\\ &= a^2\cosh^2\xi\cos^2\upsilon\cos^2\phi+a^2\cosh^2\xi\cos^2\upsilon\sin^2\phi+a^2\sinh^2\xi\sin^2\upsilon\\ &= a^2\cosh^2\xi\cos^2\upsilon\left(\cos^2\phi+\sin^2\phi\right)+a^2\sinh^2\xi\sin^2\upsilon\\ &= a^2\left(\cosh^2\xi\cos^2\upsilon+\sinh^2\xi\sin^2\upsilon\right)\\ &= a^2\left(\left(1+\sinh^2\xi\right)\cos^2\upsilon+\sinh^2\xi\left(1-\cos^2\upsilon\right)\right)\\ &= a^2\left(\sinh^2\xi+\cos^2\upsilon\right)\end{aligned} \tag{15}$$

$$\begin{aligned}\vec{b}_{\phi}\bullet\vec{b}_{\phi} &= \left(-a\cosh\xi\sin\upsilon\sin\phi\right)^2+\left(a\cosh\xi\sin\upsilon\cos\phi\right)^2\\ &= a^2\cosh^2\xi\sin^2\upsilon\sin^2\phi+a^2\cosh^2\xi\sin^2\upsilon\cos^2\phi\\ &= a^2\cosh^2\xi\sin^2\upsilon\left(\sin^2\phi+\cos^2\phi\right)\\ &= a^2\cosh^2\xi\sin^2\upsilon\end{aligned} \tag{16}$$

It could be proven easily that $\vec{b}_{\xi}.\vec{b}_{\upsilon}$=0, $\vec{b}_{\upsilon}.\vec{b}_{\phi}$=0, and $\vec{b}_{\upsilon}.\vec{b}_{\phi}$=0. So the metric matrix for g_{ij} as shown in equation 17.

$$g_{ij}=\begin{bmatrix} a^2\left(\sinh^2\xi+\cos^2\upsilon\right) & 0 & 0\\ 0 & a^2\left(\sinh^2\xi+\cos^2\upsilon\right) & 0\\ 0 & 0 & a^2\cosh^2\xi\sin^2\upsilon\end{bmatrix} \tag{17}$$

and for g^{ij} it will become as shown in equation 18.

$$g^{ij}=\begin{bmatrix} \dfrac{1}{a^2\left(\sinh^2\xi+\cos^2\upsilon\right)} & 0 & 0\\ 0 & \dfrac{1}{a^2\left(\sinh^2\xi+\cos^2\upsilon\right)} & 0\\ 0 & 0 & \dfrac{1}{a^2\cosh^2\xi\sin^2\upsilon}\end{bmatrix} \tag{18}$$

5. Conclusion

It is assumed that for rapidly rotating stellar the shape of the solar wind would become an oblate spheroidal. This paper has found the metric tensor for g_{ij} as shown in equation 17 and for g^{ij} as shown in equation 18. The future work for applying this tensor is able to be used in classifying of the behavior towards the transformation of space-time symmetry especially for oblate spheroidal coordinates that are believed to be more appropriate to describe the physical state of the solar wind.

6. References

[1] Carrington R C 1860 Description of a singular appearance seen on the sun on September 1, 1859 *Monthly Notices Royal Astronomical Society* **20** pp13-15
[2] Muller C 2015 The Carrington solar flares of 1859: consequences on life *Orig Life Evol Biosph.* **44** 3 pp265
[3] Meyer-Vernet, Nicole 2007 Basics of the solar winds *Cambridge University Press*
[4] Lisse C M, Dennerl K, Englhauser J, Harden M, Marshall F E, Mumma M J, Petre R, Pye J P, Ricketts M J, Schmitt J, Trümper J, West R G 1996 Discovery of x-ray and extreme ultraviolet emission from comet c/hyakutake 1996 b2 *Science* **274** 5285 pp205-209
[5] Biermann L 1951 Kometenschweife und Solare Kor-puskularstrahlung *Zeitschrift für Astrophysik* **29** pp274-286
[6] Chapman S 1916 On the kinetic theory of a gas, part II – a composite monoatomic gas: diffusion, viscosity, and thermal conduction *Phil. Trans. Royal Soc. London* **217A** pp115
[7] Parker EN 1958 Dynamics of the interplanetary gas and magnetic fields *Enrico Fermi Institute for Nuclear Studies* University of Chicago USA
[8] Hewish A 1974 Pulsars and high density physics *Nobel Lecture* pp174-175
[9] Setiawan I 2009 Geometri bintang berotasi *Physics Dept. FMIPA UGM*
[10] Weber E J 1967 The solar wind and its angular momentum *Thesis* California Institute of Technology USA
[11] Ong K K, Musielak Z E, Rosner R, Suess S T, Sulkanen M E 1997 Self consistent and time-dependent solar wind models *The Astrophysical Journal* **474** ppL143-L145
[12] Cranmer S R 2004 New views of the solar wind with the Lambert W function American *Association of Physics Teachers Am. J. Phys.* **72** 11 pp1397-1402
[13] Kim S J, Kim K S, Moon Y J, Cho K S, Park Y D 2004 Numerical calculation of two fluid solar wind model *Journal of The Korean Astronomical Society* **37** pp55-59
[14] Kavanagh A, Denton M 2007 High-speed solar wind streams and geospace interactions *A&G* **48** 6 pp6.24-6.26
[15] Parker E N 1969 Theoretical studies of the solar wind phenomenon *Space Science Reviews* **9** 3 pp325-360
[16] Lovegrove K, Dennis M 2006 Solar thermal energy systems in Australia *International Journal of Environmental Studies* **63** 6 pp791-802
[17] Wood B E 2004 Astrospheres and solar-like stellar winds *Living Rev. Solar Phys* **1** 2
[18] Parker E N 2005 The solar wind Theory of Eugene Parker
[19] Warwick 2009 Solar MHD 2016-2016: Parkers's solar wind model *Warwick PX 420*
[20] Wahyuni S 2010 Gerak pusat pemandu partikel plasma dalam magnetik heliosferik *Physics Dept. FMIPA UGM*
[21] Maedar A 2009 Physics: formation and rotating stars *Springer* New York
[22] Rozelot J P, Godier S, Lefebvre S 2000 On the theory of the oblateness of the sun *Solar Physics* **198** pp223-240 Kluwer Academic Publisher, Netherland
[23] Wall D J N 1978 Circularly symmetric green tensors for the harmonic vector wave equation in spheroidal coordinate systems *Journal of Physics A: Mathematical and General* **11** 4 pp749

Institute of Physics
Dirac House, Temple Back
Bristol BS1 6BE UK

ISSN: 1742-6588
ISBN 978-1-5108-3905-2